Praise for *Uns*

“We have too many global warming books— [illegible] one is needed. Steven Koonin has the credentials, expertise, and experience to ask the right questions and to give realistic answers.”

—Vaclav Smil, distinguished professor emeritus at the University of Manitoba

“*Unsettled* is an excellent case study on climate science, its inherent complexity and uncertainty, and a cautionary tale on how interpretive filters in the policymaking process have shaped, and sometimes misinformed, the climate policy debate. It should be on the reading list of scientists and engineers whose responsibility, as citizens, extends beyond the laboratory to communicating to a larger public often overwhelmed and confused by the media. Policymakers and politicians will find it a source of reflection for their arguments, positions, and decisions.”

—Jean-Lou Chameau, President Emeritus, Caltech

“Essential reading and a timely breath of fresh air for climate policy. The science of climate is neither settled nor sufficient to dictate policy. Rather than an existential crisis, we face a wicked problem that requires a pragmatic balancing of costs and benefits.”

—William W. Hogan, professor of global energy policy at Harvard Kennedy School

“Tough talk about climate politics from a statesman scientist—and a vision of what will actually come to pass.”

—Robert B. Laughlin, professor of physics at Stanford University

“Steve Koonin, the undersecretary for science under Obama, has written a very interesting and thoughtful book on climate. He documents how much of what you think you know about climate just ain’t so. Did you know that while the United States is now seeing many fewer cold records, absolute heat records are not increasing? *Unsettled* will definitely and rightly unsettle your climate thoughts, and all for the better. If we are to make trillion dollar investments, we deserve to be as well informed as possible.”

—Bjørn Lomborg, president of Copenhagen Consensus and visiting fellow at The Hoover Institution at Stanford University

Unsettled

Unsettled

What Climate Science Tells Us, What It Doesn't, and Why It Matters

UPDATED AND EXPANDED EDITION

Steven E. Koonin

BenBella Books, Inc.
Dallas, TX

BenBella Books, Inc.
10440 N. Central Expressway
Suite 800
Dallas, TX 75231
benbellabooks.com
Send feedback to feedback@benbellabooks.com

Printed in the United States of America
10 9 8 7 6 5 4 3 2 1

Library of Congress Control Number: 2024930879
ISBN 9781637745250 (trade paper)
ISBN 9781637745816 (ebook)

Editing by Alexa Stevenson
Copyediting by Scott Calamar
Proofreading by Sarah Vostok and Dylan Julian
Proofreading for the updated edition by Ashley Casteel and Marissa Wold Uhrina
Indexing by WordCo Indexing Services, Inc.
Text design and composition by PerfecType, Nashville, TN
Cover design by Emily Weigel
Cover image ©Shutterstock/Jemastock (globe)
Printed by Lake Book Manufacturing

For my many mentors,
who taught me the importance of scientific integrity.

CONTENTS

UNSETTLING UPDATES

Preface to the 2024 Edition

Since *Unsettled* was published in April 2021, there have been three years of advancement in climate science. Some of that progress was summarized in the IPCC's Sixth Assessment Report (AR6),* and even more is evident in the recent scientific literature.

Those three additional years of data and analysis have refined our understanding of the changing climate, but the broader conclusions remain largely the same. The globe continues to warm, and human influences continue to play a part in this warming. But it remains difficult to untangle the current effects of both natural variability and human forcings on global climate, much less what to expect in the future. Regional rather than global effects—which are perhaps most relevant—remain even more opaque.

What we do know from the science still suggests a different story than the one most have heard. *Unsettled*'s introduction mentions three possibly "surprising" climate science facts drawn directly from then recently

---

* AR6 is the sixth of the "Assessment Reports" (thus "AR6") prepared periodically by hundreds of scientists and issued by the United Nations Intergovernmental Panel on Climate Change (IPCC). These assessments are widely viewed as the most authoritative compendium of climate science. More information about these reports can be found at the beginning of this book's Part I.

published research or the then latest assessments of climate science published by the US government and the UN:

- Humans have had no detectable impact on hurricanes over the past century.
- Greenland's ice sheet isn't shrinking any more rapidly today than it was eighty years ago.
- The net economic impact of human-induced climate change will be minimal through at least the end of this century.

These statements have been largely reaffirmed by AR6 and the latest research literature. That you might still find them surprising highlights another constant—the media's often misinformed and misleading coverage of climate science.

Perhaps what has changed most over the past three years is the discussion around our *response* to the warming globe. We've seen the setting of, and then retreat from, ambitious emissions reduction goals with the growing realization of how prolonged and difficult any transition to a "low carbon future" will be. Global events, particularly following Russia's invasion of Ukraine, have emphasized the overwhelming importance of energy reliability, affordability, and security, as well as the central role of fossil fuels in today's energy mix (about 80 percent). Annual global greenhouse gas emissions resumed their upward trend after a brief pandemic-related decline in 2021. Fossil fuels are still very much with us, whatever the uncertainties and risks of future climate change. What I referred to in *Unsettled* as "the chimera of carbon-free" is increasingly recognized to be just that.

This updated edition covers the progress in both "The Science" and "The Response" since 2021, highlighting what I believe are the most important and striking developments, with a focus on topics I covered in the first edition of *Unsettled*. To avoid a lot of flipping back and forth, this new coverage chiefly comes in the form of new sections at the end of the relevant chapters.

Here's an overview of what to expect.

THE SCIENCE

The UN IPCC issued its Sixth Assessment Report (AR6) beginning with WGI (Working Group I, *The Physical Basis*)[1] in August 2021. The reports of WGII (*Impacts, Adaptation, and Vulnerability*)[2] and WGIII (*Mitigation of Climate Change*)[3] were issued in February 2022 and April 2022, respectively, and the final report of the cycle (*Synthesis*)[4] was issued in March 2023.

This update is far too short to cover even a fraction of the new science described in the WGI and WGII reports (they are 3,949 and 3,675 pages, respectively). Rather, I'll offer high-level updates on what has changed (or hasn't) and a few deeper dives into some of the topics I covered in the first edition. As usual, I cite my sources, providing links whenever possible.

The high-level message from the IPCC is unchanged from previous reports and remains at odds with the reports themselves. Here are some of the key points taken verbatim from the "headline statements"[5] in the *Synthesis* report's "Summary for Policy Makers."[6] (This "Summary" is generally what the public hears as the takeaways from the research.)

- Human activities, principally through emissions of greenhouse gases, have unequivocally caused global warming. [from headline statement A.1]
- Human-caused climate change is already affecting many weather and climate extremes in every region across the globe. This has led to widespread adverse impacts and related losses and damages to nature and people. [A.2]
- Risks and projected adverse impacts and related losses and damages from climate change escalate with every increment of global warming. [B.2]
- Some future changes are unavoidable and/or irreversible but can be limited by deep, rapid, and sustained global greenhouse gas emissions reduction. [B.3]
- There is a rapidly closing window of opportunity to secure a livable and sustainable future for all. [C.1]

Unfortunately, also unchanged from previous reports is the way the Summary misleads, for there is a significant gap between the messages put forth by these summaries and the science contained within the reports themselves.

Let's start with the first point, that human influences are unequivocally warming the climate. AR6 updated the projected global temperature rise due to human influences on the climate in several ways.

- The report endorses a narrowing of the likely range of equilibrium climate sensitivity.* The previous assessment gave a range of 1.5°–4.5°C, while in AR6 this is updated to 2.5°–4°C. That is, the low-end was raised significantly, while the high end was lowered.
- Projections of future climates rely not only on climate models but also on assumptions about future emissions made in the RCPs.† Note that the expected rise in global temperature is different from the warming influence, although it is proportional to it. For more on these topics, see Chapter 3.
- AR6 marked the first time IPCC offered judgments about the likelihood of various RCPs, stating that 8.5 and 6.0 had "become considerably less likely since AR5 but cannot be ruled out," and noting that RCP8.5 and the like "do not represent a typical 'business-as-usual projection' but are only useful as high-end, high-risk scenarios." Nevertheless—strangely—RCP8.5, which describes a high-population, emissions-heavy future, is the scenario cited most often in the AR6 report. There is a rationale for using this

* ECS, or how much the global average temperature anomaly would rise under a hypothetical doubling of carbon dioxide concentration—see Chapter 4 for more on the ECS and its significance.

† RCP is short for "Representative Concentration Pathway." Each RCP is a scenario for such future conditions as population, economics, and technology that determine emissions. The scenarios are named according to the amount of human influence of the climate at the end of the century—so RCP6.0 corresponds to 6.0 W/m^2 of warming influence by 2100, about 2.5 times what it is today.

scenario—that unrealistic extreme is very useful for researchers trying to understand how the climate responds to external influences, because forcing a model more strongly makes the response more detectable. But there is no justification to use it as a base case. Adopting extreme and unlikely emissions scenarios exaggerates projected changes in the climate and is a poor choice if your goal is to inform the public about likely future climates.

- With the reduced range of sensitivity and the dismissal of extreme emissions scenarios, the IPCC's recent estimates of the future global temperature rise above the preindustrial value are less than those a decade ago, now averaging 2.7°C for a moderate emissions scenario like RCP4.5. And the scenarios being prepared for the next IPCC assessment, AR7, suggest an even more benign future.[7]

In other words, the high-level message in the Summary that the climate is warming due in part to human CO_2 emissions remains true, as it was in AR5. As I discuss in Chapter 3, the nature of CO_2 means that even if emissions ceased tomorrow, their warming influence would continue for decades, though other forcings, human and natural alike, will also affect global temperature. However, as a summary of what this report shows, and particularly a summary of the progress in the science, that message is misleadingly incomplete. Surely the fact that the latest research suggests that the warming will be less than what was predicted in the previous report should be a prominent part of any high-level summary that is so widely cited and distributed. Yet it is mentioned only briefly in a footnote.

As for the assertion that humans are "already affecting many weather and climate extremes in every region across the globe," this edition provides updates on:

- model estimates of the climate sensitivity and the quality of regional climate projections (Chapter 4);
- the role of human influences in the recent unusually high temperatures (Chapter 5);
- trends in extreme weather events, including hurricanes (Chapter 6);

- risible predictions of rising sea levels (Chapter 8);
- losses from Greenland's ice sheet, which are largely due to internal variability, not the rising global temperature (Chapter 8);
- how a warming globe *decreases* deaths from extreme temperatures and has minimal net economic impact (Chapter 9).

Finally, there's the necessity of "deep, rapid and sustained global greenhouse gas emissions reduction . . . to secure a livable and sustainable future for all." We'll get to that shortly, but suffice it to say that the science does not at all support this part of the summary.

The US government issued its own assessment report (the fifth National Climate Assessment, or NCA5)[8] in November 2023. It contains disappointingly little climate science but does contain many of the by-now familiar sorts of misrepresentations meant to persuade rather than to inform.

The summaries of the assessment reports shape the message to decision makers, the media, and the public about climate science. So it is disappointing that their summaries, and sometimes the reports themselves, demonstrably fail to paint an accurate picture of what we know, and what we don't know, about the climate and how it might change in the future. Even more disappointing is the failure of the climate-science community and numerous scientific institutions to correct this situation—or even to acknowledge it. But, as I point out in Chapter 10, there are strong incentives not to do so.

THE RESPONSE

After a brief decline caused by the COVID-19 pandemic, global greenhouse gas emissions have continued to increase. Emissions in 2022 were equivalent to 57.4 Gt (gigatonnes) CO_2, the greatest ever, and are projected to be about the same in 2030.[9] Relatedly, the concentration of CO_2 in the atmosphere also continued to increase at historically high rates.[10] The 2022 UN Emissions Gap Report[11] that shaped discussions at the 2022 annual Conference of Parties meeting (COP27) in Sharm el-Sheikh said:

> Countries are off track to achieve even the globally highly insufficient NDCs [Nationally Determined Contributions]. Global GHG emissions in 2030 based on current policies are estimated at 58 GtCO2e. . . . To get on track for limiting global warming to 1.5°C [the aspiration of the 2016 Paris Accord], global annual GHG emissions must be reduced by 45 per cent compared with emissions projections under policies currently in place in just eight years, and they must continue to decline rapidly after 2030.

That the 2023 Emissions Gap Report[12] is titled "Broken Record—Temperatures hit new highs, yet world fails to cut emissions (again)" says it all. In short, there continue to be big disconnects between the reductions deemed necessary and those achieved, or even those planned.

Some countries *have* reduced their emissions in recent years. For example, US energy-related CO_2 emissions[13] in 2022 were only 83 percent of their 2007 peak, and comparable emissions from the EU[14] in 2022 were only 68 percent of their 1979 peak. At the political level, ambitious mitigation goals and policies continue to be emphasized, if not expanded. For example, the Biden administration would like the US electrical system to be emissions free by 2035 and put in place the Inflation Reduction Act as a step toward that goal. The UK banned the sale of new cars powered by internal combustion engines after 2030[15] (subsequently deferred to 2035), while California, New York, Washington, and Massachusetts will have a ban after 2035.[16]

At the same time, the world is increasingly realizing the enormous challenges inherent in significantly reducing *global* emissions. To play off the title of a recent film, since energy touches "everything, everywhere, all at once," changing an energy system is no simple matter. The "energy crisis" induced by the Ukraine war emphasized the primacy of reliable and affordable energy over "clean." For example, Germany, a country proudly at the forefront of greening its energy system, had to fire up its mothballed coal plants to keep the lights on and speedily authorize the importation of liquefied natural gas. For many developing countries, the crisis disrupted electricity supplies and forced factories to shut down.

And so there are already retreats from emissions policies now realized to have been overly zealous, most prominently in the EU, which, at least for the moment, has bowed to technical and economic realities in broadening its category of "green" technologies to include some natural gas and nuclear. Attempts to mandate emissions-free home heating technologies in the UK and Germany failed because of popular opposition. Popular opposition in Holland to a proposed ban on fertilizer (which is made from natural gas) induced some political changes. And there are indications that the UK "net zero by 2050" policy will be strongly moderated, if not abandoned, as energy costs rise.

Climate "solutions" through finance have largely come up empty. ESG (Environmental, Social, and Governance) investing has found disfavor for several reasons, including lack of clear metrics, lack of competitive financial return (suggesting breach of fiduciary responsibility), and lack of impact on emissions. The world spent about $4 trillion on deploying renewable energy over the past decade, yet fossil fuels accounted for 82 percent of global energy in 2022, with renewables (largely wind and solar, excluding hydropower) supplying just 7.5 percent.[17]

On the technology front, the challenges and costs of an energy system dominated by the "new" energy technologies of wind and solar generation, EVs, and heat pumps are becoming clearer to consumers, governments, and industry.

I think that those realities of energy systems, well known to those who've taken the time to study them, will become increasingly evident to policymakers, the media, and the public. Even now, the UK, Holland, and Germany are opening new oil and gas fields;[18] Japan[19] and France are renewing their commitments to nuclear energy; and automakers are far from abandoning internal combustion engine cars, especially in the US, where EV sales have stalled (at least for now) even with large government subsidies and regulatory pressures.[20]

These tensions and contradictions were on display in COP28, when some 100,000 people, including delegates from the 199 Parties, met in Abu

Dhabi in December 2023 for a "stocktake" of national and global progress in reducing greenhouse gas emissions. While most of the outcome was no different than recent past meetings, there were some notable developments. In particular, countries made non-binding pledges to triple the world's renewable energy capacity (mostly wind and solar) by 2030, and to triple its nuclear energy capacity by 2050.

But on the central subject of fossil fuels, the conference waffled, as it has in the past. It urged parties to transition away from "fossil fuels in energy systems, in a just, orderly and equitable manner . . . so as to achieve net zero by 2050 in keeping with the science."[21] Of course, there's nothing mandatory or time-specific here, there are no penalties for inaction, and "just, orderly and equitable" leaves plenty of wiggle room to do whatever a country must to secure a reliable and affordable energy supply. There is every expectation that COP28 will have as much of an impact on global emissions as its predecessors. In other words . . . not much.

In this new edition, I'll discuss:

- how much net zero would cost, how little it would get us, and the fallacy of its pursuit (Chapter 12);
- the trilemma of trying to create an electrical grid that is simultaneously reliable, affordable, and produces low emissions (Chapter 13);
- the thorny economic and ethical problems presented by the high-value materials needed for renewable technologies (also Chapter 13).

Let me now turn to a different subject—this book itself.

Unsettled's reception by the scientific community, the media, and the public has also been unfolding over the past three years. Reactions have been polarized—I've been encouraged by the many unsolicited notes from scientists and engineers thanking me for laying out what we know, simply and transparently. That more than 200,000 copies have been sold through

December 2023 suggests there was a hunger for a level-headed, fact-based overview of climate science and energy. Even so, as I predicted in the book's introduction, I've been called a shill for the fossil fuel industries and had my points misrepresented.

At the end of Chapter 11, "Fixing the Broken Science"—which discusses how to critically evaluate coverage of climate science in the media—I've included an update focusing on the reception *Unsettled* received, including responses to a few of the more prominent critical reviews. I've concentrated on the facts and refrained from name-calling, a practice one might think is a given. Unfortunately, it isn't in the climate field, even among academics.

Finally, at the end of this book, you'll find a new chapter titled "Easy on the Energy Transition."

Prior to *Unsettled*, my writing had been directed either to the scientific community or to the science-oriented policy world. Since I'd never written a book for a more general audience, I was a little naïve. I thought that *Unsettled* itself would be the primary way I'd get my analyses across and have them discussed and debated, as would normally happen in the practice of science. But I found that talking about the book and engaging with curious audiences was as important as writing it, if only to encourage people to take a closer look at what I'd written.

Unsettled's publication closed off some opportunities for me to talk about climate and energy—some of the universities that had been on my usual circuit before publication were notably silent. But even as those doors closed, many new opportunities arose. The format was often familiar—lectures to academics, national laboratories, civic groups, NGOs, and corporations let me refine the presentation and tune it to the audience. But the broadcast media was entirely unfamiliar. Since an interview was typically no more than five minutes, some of which was consumed by the interviewer's talking, my remarks had to be punchy—not at all the best way to discuss complex, nuanced topics. What's more, it's entirely unscripted, so there are ample opportunities for confusion or lack of clarity. The first few live TV appearances were a bit "unsettling." I gradually grew more comfortable with the format, but long-form podcasts were a happier medium, as they let me deliver my analyses with more fidelity.

Perhaps the most illuminating experiences I had talking about the ideas put forth in *Unsettled*—for myself as much as audiences—was a series of seven Oxford-style debates put on at the Soho Forum and at six universities.

The debates considered the proposition "*Climate science compels us to make large and rapid reductions in greenhouse gas emissions.*" If you've not thought much about climate and energy, you'll probably think "of course" based on what you've heard from the media. But I doubt most Americans realize just how much energy we consume—and take for granted. The 1.5 billion of us in the developed world enjoy abundant and affordable energy. But the globe's other 6.5 billion people are energy poor. Energy poverty isn't a matter of doing without luxuries—it is cooking with wood or dung (kitchen smoke kills some two million people a year), and lacking access to refrigeration and electricity. The further development of most of the world and some increase in population will drive a 50 percent increase in global energy demand by midcentury. For developing nations, reliable and affordable energy is the priority. We have yet to make "green" energy affordable for even the wealthiest countries. Because fossil fuels are currently the most effective—or only—way for developing nations to get their energy, total carbon dioxide emissions are not expected to decrease much in coming decades, even as the developed world's emissions slowly decline.

The argument I laid out in these debates forms the basis of this book's new Chapter 15. It considers each part of the statement—what we know about the science and about making rapid reductions in greenhouse gas emissions. It also considers who we mean by "us," and the moral dimensions of global energy policy. I hope you will find it a thought-provoking look at how the pieces of our climate and energy stories fit together.

INTRODUCTION

"The Science." We're all supposed to know what "The Science" says. "The Science," we're told, is settled. How many times have you heard it?

Humans have already broken the earth's climate. Temperatures are rising, sea level is surging, ice is disappearing, and heat waves, storms, droughts, floods, and wildfires are an ever-worsening scourge on the world. Greenhouse gas emissions are causing all of this. And unless they're eliminated promptly by radical changes to society and its energy systems, "The Science" says Earth is doomed.

Well . . . not quite. Yes, it's true that the globe is warming, and that humans are exerting a warming influence upon it. But beyond that—to paraphrase the classic movie *The Princess Bride*: "I do not think 'The Science' says what you think it says."

For example, both the research literature and government reports that summarize and assess the state of climate science say clearly that heat waves in the US are now *no more common* than they were in 1900, and that the warmest temperatures in the US have not risen in the past fifty years. When I tell people this, most are incredulous. Some gasp. And some get downright hostile.

But these are almost certainly not the only climate facts you haven't heard. Here are three more that might surprise you, drawn directly from recent published research or the latest assessments of climate science published by the US government and the UN:

- Humans have had no detectable impact on hurricanes over the past century.
- Greenland's ice sheet isn't shrinking any more rapidly today than it was eighty years ago.
- The net economic impact of human-induced climate change will be minimal through at least the end of this century.

So what gives?

If you're like most people, after the surprise wears off, you'll wonder *why* you're surprised. Why haven't you heard these facts before? Why don't they line up with the narrative—now almost a meme—that we've already broken the climate and face certain doom unless we change our ways?

Most of the disconnect comes from the long game of telephone that starts with the research literature and runs through the assessment reports to the summaries of the assessment reports and on to the media coverage. There are abundant opportunities to get things wrong—both accidentally and on purpose—as the information goes through filter after filter to be packaged for various audiences. The public gets their climate information almost exclusively from the media; very few people actually read the assessment summaries, let alone the reports and research papers themselves. That's perfectly understandable—the data and analyses are nearly impenetrable for non-experts, and the writing is not exactly gripping. As a result, most people don't get the whole story.

But don't feel bad. It's not only the public that's ill informed about what the science says about climate. Policymakers, too, have to rely on information that's been put through several different wringers by the time it gets to them. Because most government officials—and others involved in climate policy for the public and private sectors—are not themselves scientists, it's up to scientists to make sure that non-scientists making key policy decisions get an accurate, complete, and transparent picture of what's known (and unknown) about the changing climate, one undistorted by "agenda" or "narrative." Unfortunately, getting that story straight isn't as easy as it sounds.

I should know. That used to be my job.

WHERE I'M COMING FROM

I'm a scientist—I work to understand the world through measurements and observations, and then to communicate clearly both the excitement and the implications of that understanding. Early in my career, I had great fun doing this for esoteric phenomena in the realm of atoms and nuclei using high-performance computer modeling (which is also an important tool for much of climate science). But beginning in 2004, I spent about a decade turning those same methods to the subject of climate and its implications for energy technologies. I did this first as chief scientist for the oil company BP, where I focused on advancing renewable energy, and then as undersecretary for science in the Obama administration's Department of Energy, where I helped guide the government's investments in energy technologies and climate science. I found great satisfaction in these roles, helping to define and catalyze actions that would reduce carbon dioxide emissions, the agreed-upon imperative that would "save the planet."

But then the doubts began. In late 2013 I was asked by the American Physical Society—the professional organization of the country's physicists—to lead an update of its public statement on climate. As part of that effort, in January 2014 I convened a workshop with a specific objective—to "stress test" the state of climate science. In ordinary terms, that meant analyzing, critiquing, and summarizing humanity's accumulated knowledge about the past, present, and future of the earth's climate. Six leading climate experts and six leading physicists, myself included, spent a day scrutinizing exactly what we know about the climate system and how confidently we can project its future. To focus the conversation, we physicists had spent the prior two months preparing a framing document based on the UN assessment report that had just been released.[1] We posed some specific and crucial questions along the lines of: *Where is the data poor or the assumptions weakly supported—and does that matter? How reliable are the models that we use to describe the past and project the future?* Many who've read the workshop transcript were struck by how successfully—and unusually—it brought out the certainties and uncertainties of the science at that time.[2]

For my part, I came away from the APS workshop not only surprised, but shaken by the realization that climate science was far less mature than I had supposed. Here's what I discovered:

- Humans exert a growing, but physically small, warming influence on the climate. The deficiencies of climate data challenge our ability to untangle the response to human influences from poorly understood natural changes.
- The results from the multitude of climate models disagree with, or even contradict, each other and many kinds of observations. A vague "expert judgment" was sometimes applied to adjust model results and obfuscate shortcomings.
- Government and UN press releases and summaries do not accurately reflect the reports themselves. There was a consensus at the meeting on some important issues, but not at all the strong consensus the media promulgates. Distinguished climate experts (including report authors themselves) are embarrassed by some media portrayals of the science. This was somewhat shocking.
- In short, the science is insufficient to make useful projections about how the climate will change over the coming decades, much less what effect our actions will have on it.

Why were these crucial deficiencies such a revelation to me and others? As a scientist, I felt the scientific community was letting the public down by not telling the whole truth plainly. And as a citizen, I was concerned that the public and political debates were being misinformed. So I began to speak out, most publicly through a two-thousand-word "Saturday Essay" published in the *Wall Street Journal* that September.[3] In it, I outlined some of the uncertainties in climate science and argued that ignoring them could hinder our ability to understand and respond to a changing climate:

> Policy makers and the public may wish for the comfort of certainty in their climate science. But I fear that rigidly promulgating the idea that climate science is "settled" (or is a "hoax") demeans and chills the scientific enterprise, retarding its progress in these important

matters. Uncertainty is a prime mover and motivator of science and must be faced head-on.

That piece drew thousands of online comments, the great majority of them supportive. My frankness about the state of climate science was less popular in the scientific community, however. As the chair of a highly respected university earth sciences department told me privately, "I agree with pretty much everything you wrote, but I don't dare say that in public."

Many scientific colleagues, some of them my friends for decades, were outraged that I'd highlight problems with The Science and thus, as one of them said, "give ammunition to the deniers." Another said it would have been okay to publish my essay in some obscure scientific journal but reproached me for doing so in a forum with so many readers. And a prominent defender of the idea that The Science is settled enough published a response to my Op-Ed that began by calling for New York University to reconsider my employment, went on to misrepresent many of the things I had written, but then, bafflingly, acknowledged that most of the uncertainties I'd mentioned were well known and much discussed among experts.[4] It seems that by highlighting those uncertainties so plainly and publicly, I had inadvertently broken some code of silence, like the Mafia's *omerta*.

More than six years of study since the APS workshop have left me increasingly dismayed at the public discussions of climate and energy. Climate alarmism has come to dominate US politics, especially among Democrats, where I have otherwise long felt most comfortable politically. The 2020 Democratic presidential primary saw each candidate trying to outdo the other with over-the-top statements about "climate emergency" and "climate crisis" increasingly divorced from the science. The election run-up also witnessed increasingly sweeping policy proposals like the Green New Deal that would "fight climate change" with government interventions and subsidies. Not surprisingly, the Biden administration has made climate and energy a major priority, with the appointment of former secretary of state John Kerry as climate envoy and proposed spending of almost two trillion dollars to fight this "existential threat to humanity."

While I have no informed opinions on the fiscal and policy merits of proposals like the Green New Deal—I am a physicist, not an economist—I do know that any policy should be based upon what the science actually says about the changing climate. Trillion-dollar decisions about reducing human influences on the climate are, in the end, about values: risk tolerance, intergenerational and geographical equities, and a balance among economic development, environmental impact, and energy cost, availability, and reliability. But they must be informed by an accurate understanding of scientific certainties and uncertainties.

This book is an attempt to set us on the road to that understanding. And I intend to do it the only way that a scientist knows how: with documented facts, almost all drawn from the most up-to-date official assessments or quality research literature, presented in their proper context. As the late representative John Lewis, the conscience of Congress, said in his speech about the first impeachment of President Trump:[5]

> When you see something that is not right, not just, not fair, you have a moral obligation to say something, do something.

My late Caltech colleague Richard Feynman was one of the greatest physicists of the twentieth century, renowned for both the creativity and importance of his research (including Nobel Prize–winning work on quantum electrodynamics). Irreverence, showmanship, and the ability to tell a good story were also part of what made him a legend. He was a character, one with extraordinary intellectual substance.

I was one of many aspiring physicists attracted to Caltech by Feynman's presence. Before I arrived in the fall of 1968, I had already read his wonderful "red book" series of physics lectures cover to cover multiple times. My four undergraduate years at Caltech were lived pretty much as those depicted in *The Big Bang Theory*, except without the laugh track. The highlights included some one-on-one conversations with Feynman (he loved interacting with young scientists), as well as a memorable session playing bongo drums with the great man himself during my first year.

Scientific integrity is central to the Caltech ethos. Its importance is impressed upon new students from their first day on campus, and Feynman's absolute intellectual honesty demonstrated for students and faculty alike what this means for a working scientist. At the 1974 Caltech commencement, he gave a now famous address titled "Cargo Cult Science."[6] Its topic was the rigor scientists must adopt to avoid fooling not only themselves, but also others:

> In summary, the idea is to try to give all of the information to help others to judge the value of your contribution; not just the information that leads to judgment in one particular direction or another.
>
> The easiest way to explain this idea is to contrast it, for example, with advertising. Last night I heard that Wesson Oil doesn't soak through food. Well, that's true. It's not dishonest; but the thing I'm talking about is not just a matter of not being dishonest, it's a matter of scientific integrity, which is another level. The fact that should be added to that advertising statement is that no oils soak through food, if operated at a certain temperature. If operated at another temperature, they all will—including Wesson Oil. So it's the implication which has been conveyed, not the fact, which is true, and the difference is what we have to deal with.

Much of the public portrayal of climate science suffers from Feynman's Wesson Oil problem—in an effort to persuade rather than inform, the information presented withholds either essential context or what doesn't "fit." (And coincidentally, as with cooking oil, it's mostly a matter of temperature.)

Most of the climate researchers I've met pursue their work with the objectivity and rigor that are the norm in every field of science. But because the potential impact of a changing climate strikes at human existence itself, the issue understandably engenders passion and emotion. Some people argue that there's no harm in a bit of misinformation if it helps "save the planet," and indeed, when phrases like this (however unwarranted or inaccurate) are being used to describe the stakes, perhaps it isn't surprising that some climate scientists are less than objective when talking to the public. The late Stephen Schneider, a prominent climate researcher, said it explicitly as early as 1989[7]:

> On the one hand, as scientists we are ethically bound to the scientific method, in effect promising to tell the truth, the whole truth, and nothing but—which means that we must include all the doubts, the caveats, the ifs, ands, and buts. On the other hand, we are not just scientists but human beings as well. And like most people we'd like to see the world a better place, which in this context translates into our working to reduce the risk of potentially disastrous climatic change. To do that we need to get some broad based support, to capture the public's imagination. That, of course, entails getting loads of media coverage. So we have to offer up scary scenarios, make simplified, dramatic statements, and make little mention of any doubts we might have. This "double ethical bind" we frequently find ourselves in cannot be solved by any formula. Each of us has to decide what the right balance is between being effective and being honest. I hope that means being both.

Many others have made similar points, or commented on the dark side of Schneider's supposed "double bind." For example:

- "It doesn't matter what is true, it only matters what people believe is true."

  —PAUL WATSON, COFOUNDER OF GREENPEACE[8]

- "We've got to ride this global warming issue. Even if the theory of global warming is wrong, we will be doing the right thing in terms of economic and environmental policy."

  —TIMOTHY WIRTH, PRESIDENT OF THE UN FOUNDATION[9]

- "Some colleagues who share some of my doubts argue that the only way to get our society to change is to frighten people with the possibility of a catastrophe, and that therefore it is all right and even necessary for scientists to exaggerate. They tell me that my belief in open and honest assessment is naïve."

  —DANIEL BOTKIN, FORMER CHAIR OF ENVIRONMENTAL STUDIES AT THE UNIVERSITY OF CALIFORNIA AT SANTA BARBARA[10]

And so the media is filled with scary climate predictions. Here are a few old enough to have been proven wrong:

- "[Inaction will cause] . . . by the turn of the century [2000], an ecological catastrophe which will witness devastation as complete, as irreversible as any nuclear holocaust."

  —Mostafa Tolba, former executive director of the United Nations Environment Program, 1982[11]

- "[Within a few years] winter snowfall [in the UK] will become a very rare and exciting event. Children just aren't going to know what snow is."

  —David Viner, senior research scientist, 2000[12]

- "European cities will be plunged beneath rising seas as Britain is plunged into a Siberian climate by 2020."

  —Mark Townsend and Paul Harris, quoting a Pentagon report in the *Guardian*, 2004[13]

Although Schneider later spent many words trying to explain his statement about the "double ethical bind," I believe the underlying premise is dangerously wrong. There should be no question about "what the right balance is between being effective and being honest." It is the height of hubris for a scientist even to consider deliberately misinforming policy discussions in service of what they believe to be ethical. This would seem obvious in other contexts: imagine the outcry if it were discovered that scientists were misrepresenting data on birth control because of their religious beliefs, for instance.

Philip Handler, a former president of the National Academy of Sciences, identified the problem in a 1980 editorial that resonates eerily four decades later:

> Difficulty arises in the scientific community from confusion of the role of scientist *qua* scientist with that of scientist as citizen, confusion of the ethical code of the scientist with the obligation of the citizen, blurring the distinction between intrinsically scientific and intrinsically political questions. When scientists fail to recognize

these boundaries, their own ideological beliefs, usually unspoken, easily becloud seemingly scientific debate.[14]

With scientists' unique role comes a special responsibility. We're the only people who can bring objective science to the discussion, and *that* is our overriding ethical obligation. Like judges, we're obligated to put personal feelings aside as we do our job. When we fail to do this, we usurp the public's right to make informed choices and undermine their confidence in the entire scientific enterprise. There's nothing at all wrong with scientists as activists, but activism masquerading as The Science is pernicious.

We scientists shouldn't be selling cooking oil.

ABOUT THIS BOOK

Unsettled tells two related stories. The first (Part I) is about the science of the changing climate, while the second (Part II) is about the response that society could make to those changes.

Part I begins by clarifying the important questions society asks of climate science—how the climate has changed, how it will change in the future, and what the impact of those changes will be. It also offers some basics about the official assessment reports that we look to for answers to those questions.

To understand why it's changing today and how it might change in the future, we need to know how the climate has changed in the past, and Chapter 1 starts in on the science itself by exploring this. The chapter explains both the importance and challenges of obtaining quality observations of the earth's climate (which is not the same as its weather) over many decades; it also reviews some of the indications of a warming globe and puts them in geological context.

Chapter 2 then turns to how the earth's temperature arises in the first place—from a delicate balance between warming sunlight and cooling heat radiation. We'll see that this balance is disturbed by both human and natural influences, with greenhouse gases playing an important role. Because the climate is very sensitive, we need an accurate and precise understanding of those influences and how they've changed over time.

The most important human influence on the climate is the growing concentration of carbon dioxide (CO_2) in the atmosphere, largely due to the burning of fossil fuels. This is the focus of Chapter 3—particularly, how the connection between CO_2 emissions and concentration diminishes the prospect of even stabilizing growing human influences.

Computer models of how the climate responds to human and natural influences are the subject of Chapter 4. Drawing upon my half-century involvement with scientific computing and the authorship of a pioneering text on that subject, we'll look at how they work, what they tell us, and some of their deficiencies. These dozens of sophisticated models are what scientists use to make their projections and what the media cites in their coverage—alas, they give results that differ significantly not only from each other but from observations (that is, they're right in a few ways, but wrong in many others). In fact, the results have become more divergent with each generation of models. In other words, as our models have become more elaborate, their descriptions of the future have become less certain.

Chapter 5 is the first of five chapters dealing with contradictions between the science and the prevailing notion that "humans have already broken the climate," exploring areas where the facts and popular perception are at odds (and probing the source of those discrepancies). This chapter focuses on record high temperatures in the US—they're no more common today than they were in 1900, yet you wouldn't know that from the misrepresentations of an allegedly authoritative assessment report. Chapter 6 likewise explains why experts conclude that human influences haven't caused any observable changes in hurricanes, and how assessment reports obscure or distort that finding. In Chapter 7, I describe the modest changes seen in precipitation and related phenomena over the past century, discussing their significance and highlighting some points likely to surprise anyone who follows the news—for instance, that the global area burned by fires each year has declined by 25 percent since observations began in 1998.

Chapter 8 offers a levelheaded look at sea levels, which have been rising over the past many millennia. We'll untangle what we really know about human influences on the current rate of rise (about one foot per century) and explain why it's very hard to believe that surging seas will drown the coasts anytime soon. Chapter 9 covers a trio of oft-cited climate-change impacts (fatalities, famine, and economic ruin), predictions of which are belied by the historical record and assessment report projections, even if it's hard to discern this when reading the reports themselves.

Having demonstrated that the science doesn't support what's portrayed in most popular discussions, in Chapter 10 I take up the question of "Who broke it?"—why the science has been communicated so poorly to decision makers and the public. We'll see how overwrought portrayals of a "climate crisis" serve the interests of diverse players, including environmental activists, the media, politicians, scientists, and scientific institutions. Chapter 11 closes out Part I by describing how we might improve communication and understanding of climate science, including adversarial ("Red Team") reviews of the assessment reports, best practices for media coverage, and what non-experts can do to be better informed and more critical consumers of all science media—but especially about the climate.

Part II begins its discussion of the response story by drawing a distinction between what society *could* do, what it *should* do, and what it *will* do in response to a changing climate—three very different issues often conflated, even by experts. Chapter 12 illuminates the *will* issue by discussing the formidable challenges in meaningfully reducing human influences on the climate, including the lack of progress toward the goals of the Paris Agreement. Chapter 13 sheds some light on the *could* issue by discussing the tremendous changes it would take to create a "zero-carbon" energy system in the US. The response story wraps up in Chapter 14, with a discussion of "Plan B" strategies that allow the world to respond to a climate changing from either human or natural causes—adaptation, which *will* happen, and geoengineering, which *could* be deployed *in extremis*.

The book concludes with some closing thoughts on climate and energy, including what I believe to be prudent steps society *should* take, both to improve climate science and the way it's presented to non-experts, as well as to prepare for future climate changes, whether natural or caused by humans.

Some practical points about the book:

Scientists work in the metric system—temperatures in degrees Celsius, distances in meters or kilometers, etc. However, the "imperial" system is more common in the US—temperatures in degrees Fahrenheit, distances in feet or miles, and so on. To make the material accessible to the broadest audience, I'll usually quote quantities in both sets of units.

It's important to know when to be precise and when approximate is good enough. Let's say you're hoping your pond will ice over for some skating. Water freezes at 0°C (32°F), so if I told you the temperature was about 10°C (50°F), that's much too warm for ice to form, and it wouldn't matter if the actual temperature were 9°C or 11°C. However, if I told you the temperature was *about* 1°C (34°F), whether it was −1°C or +3°C would make a big difference, and it would be important for me to tell you that it's actually, say, −0.3°C. So the precision with which I'll quote numbers will depend upon the context. For example, I might use a phrase like "the US population is about 330 million," even if the official count was actually 329.135 million on January 1, 2020, because the difference doesn't affect the point I'm considering.[15] In other cases, such as the discussion of sea level rise in Chapter 8, a difference between 2.5 mm/year and 3.0 mm/year (0.10 inches/year and 0.12 inches/year) does indeed matter, and I'll be appropriately precise.

One of the advantages of writing a book as opposed to an Op-Ed is that it allows not only a deeper discussion, but also the more liberal use of graphs. Please take them in stride. Graphs are the language of data and data is central to both the science and how it is communicated. Virtually all of the graphs I've chosen are taken from (or are directly adapted from) the assessment reports, the underlying scientific literature, or other official data sources. I've sometimes used versions of official graphs to emphasize that they're what the science says, not what I say. And, of course, I'll always provide information about the source of the graphs or their data.

Being a critical reader of scientific graphs is a skill well worth honing—I've included a few figures from popular media to illustrate just how misleading they can be.

My grade school class visited the United Nations headquarters building on a field trip some sixty years ago. I remember being impressed by an enormous Iranian carpet hanging in the lobby and being told that the weavers had deliberately introduced a hard-to-notice imperfection in the elaborate design to signify that it was a product of humans. While there are surely imperfections in this book, they are not deliberate. I've done my best to accurately represent the state of the science going into 2021.

Even if this book is error free, alas, I'll be attacked for writing it. Some will question my credentials, saying I'm not "a climate scientist." In other words, that I am not formally trained in the earth sciences, even though I've published several papers in the field. In truth, climate science involves many different scientific fields, encompassing the quantum physics of molecules and the classical physics of moving air, water, and ice; the chemical processes in the atmosphere and ocean; the geology of the solid earth; and the biology of ecosystems. It also includes the technologies used to "do" the science, including computer modeling on the world's fastest machines, remote sensing from satellites, paleoclimate analysis, and advanced statistical methods. Then there are the related areas of policy, economics, and the energy technologies aimed at reducing greenhouse gas emissions.

This enormous swath of knowledge and methods makes the study of climate and energy the ultimate multidisciplinary activity. No single researcher can be an expert in more than two or three of its aspects, so the challenge in assessing and communicating the state of the science is to read widely and critically enough to put together—and convey, which requires a skill set all its own—a coherent, fact-based picture of the whole. Like many other climate researchers with a background in physics, including James Hansen and Michael Mann, I've found satisfaction in applying a physicist's tools and sensibilities to create that kind of picture, with the additional benefits of my experience in energy technology and advising government and private-sector decision makers on not only climate policy, but also other important national matters, including the quality standards for the human genome project[16] and testimony to then senator Joe Biden's Foreign Relations Committee on the post-9/11 dangers of dirty bombs.[17]

Even if they accept my credentials, some of this book's critics will say that I've ignored the bigger picture, that this book is too focused on aspects of the science that don't support the alleged consensus. Given the great breadth of climate science, however, it has to focus somewhere—after all, each of the assessment reports alone runs to more than one thousand pages. My focus is on significant points where the popular perception about climate and energy is very different from what the science says. In that way, this book is about more than what's scientifically correct and what isn't; it's also about how the science, with all of its certainties and uncertainties, becomes The Science—how it gets summarized and communicated, and what's lost in the process. Not everything you've heard about climate science is wrong, and I've done my best to provide a balanced presentation for each subject I treat within the limits of length and technical level; the references I cite can be consulted for even more information.

Yet another criticism will be that my points are inconsequential. But that can't be, since the media, politicians, and even some scientists are constantly highlighting their opposites to support the prevailing narrative: "Record high temperatures are becoming more common," "Hurricanes are strengthening under human influences," "Climate change will be an economic disaster." Imagine instead headlines like "Record high temperatures are becoming rarer," "Hurricanes show no sign of human influence," or "Global warming won't have much impact on the economy." I think you're unlikely ever to see those headlines, even though they're a lot closer to what the science actually says, as I'll show in the following chapters.

Less serious critics will attack me personally. Some will call me a shill for the fossil fuel industries, even as my résumé shows otherwise. Others will say I'm a "climate denier." An actual "climate denier" would be, say, an antiscience politician who refuses to accept the evidence of the data—quite the opposite of my position. How can I be denying the science if what I'm saying is straight out of the official data and reports? I find it particularly abhorrent to have a call for open scientific discussion equated with Holocaust denial, especially since the Nazis killed more than two hundred of my relatives in Eastern Europe.

Leaving name-calling aside, I also expect this book to be criticized by (perhaps former) friends in the scientific community, who will question—as they did with that first Op-Ed of mine—why I'm saying such things to a broad audience, even if my points are well known to experts. The reason is one I've already discussed: for a scientist, I believe it is a responsibility, almost an act of conscience, to portray without bias just how settled—or unsettled—the science truly is.

I hope that readers approach this book with an open mind. There has been far too little serious public discussion about the knowns and unknowns of climate science—perhaps unsurprisingly, given the pitch of the rhetoric. In a speech in which he compared human-caused climate change to weapons of mass destruction, then secretary of state John Kerry (now the Biden administration's climate envoy) said, "The science is unequivocal . . . President Obama and I believe very deeply that we do not have time for a meeting anywhere of the Flat Earth Society."[18] But the science is *not* settled. Open debate is at the heart of the scientific process; it is absurd that scientists should fear being labeled *antiscience* for engaging in it. In that light, this book issues a challenge and solicits, indeed welcomes, informed argument and disagreement. This would be an important step toward wiser societal decisions on climate and energy—and less heat in the debate over the science of our warming planet.

PART I

THE SCIENCE

My wife and I have three children. Like most parents, we tried to guide their childhood development, both by example and by rewarding good behavior and admonishing bad, in the hopes that they would grow into happy, productive adults. Of course, human nature being what it is, each of our three kids responded to our influences differently, depending upon the mix of genes they inherited and other experiences they had growing up. Those responses have had impacts on their lives—while there have been bumps in the road for each of them, we're very proud of the unique adults our children have become.

Those same three issues—influence, response, and impact—form the core questions of climate science:

- How have humans *influenced* the climate—and how will those influences change in the future?
- How does the climate *respond* to human (and natural) influences?
- How will the climate's response *impact* ecosystems and societies?

The past few decades have seen an enormous international effort to answer these questions. Of course, science being what it is, none of these answers is, or ever will be, entirely certain. And since the answer to each question depends upon the answer to the one before it, we can expect that answers to the final—and perhaps most significant—question will be the most uncertain.

UNDERSTANDING UNCERTAINTIES

The science we learn in grade school is a collection of certainties about the natural world—the earth revolves around the sun, DNA carries the blueprint of an organism, and so on. Only when you start to learn the practice of science do you realize that each of these "facts" was hard won through a succession of logical inferences based upon many observations or experiments. The process of science is less about collecting pieces of knowledge than it is about reducing the uncertainties in what we know. Our uncertainties can be greater or lesser for any given piece of knowledge depending upon where we are in that process—today we are quite certain of how an apple will fall from a tree, but our understanding of turbulent fluid flow (such as convection in the atmosphere) remains a work in progress after more than a century of effort.

Every measurement of the physical world has an associated uncertainty interval (usually denoted by the Greek letter sigma: σ). We can't say what the measurement's true value is precisely, only that it is likely to be within some range specified by σ. Thus, we might say the global mean surface temperature in 2016 was 14.85°C with a σ of 0.07°C. That is, there is a two-thirds chance that the true value is between 14.78 and 14.92°C.

For a scientist, knowing the uncertainty in a measurement is as important as knowing the measurement itself, because it allows you to judge the significance of differences between measurements. If the temperature in 2016 were 14.85 ± 0.07°C (the first number is the value and the second its σ) and it was measured at 14.54 ± 0.07°C in 2005, a scientist would declare the difference of 0.31°C significant, since it's more than four times the uncertainties in the measurements themselves. On the other hand, the measured annual increase of 0.04°C between 2015 (14.81 ± 0.07°C) and 2016 is insignificant since it is smaller than the uncertainties—about half as large, in fact. The media might well still scream "Temperatures Continue to Rise," either out of ignorance or to capture readers' attention, but this is like political commentators getting into a tizzy about a one percentage point change in a poll when the poll's margin of error (its σ) is three points.

Uncertainties and significance in measurements are a common language among all scientists. But talking about the uncertainties in our higher-level understanding of climate, particularly to non-scientists, is trickier. To more precisely convey the extent of the unknowns involved, the assessment reports have developed a formal language, as shown in this table[1]:

IPCC LIKELIHOOD SCALE	
Term	**Likelihood of the Outcome**
Virtually certain	99–100% probability
Very likely	90–100% probability
Likely	66–100% probability
About as likely as not	33–66% probability
Unlikely	0–33% probability
Very unlikely	0–10% probability
Exceptionally unlikely	0–1% probability

In this language, a statement that is *virtually certain* has at most a 1 percent likelihood of being incorrect, while a statement that's *likely* has about a two-thirds chance of being true, and a *very unlikely* statement has at most a 10 percent chance of being true.

Because climate science is complex, uncertainties aren't always easy to quantify in terms of probabilities. The United Nations' Intergovernmental Panel on Climate Change (IPCC) has therefore established a second set of calibrated terms to indicate *confidence* in a given finding. Confidence is a qualitative judgment that depends upon the number, quality, and agreement of different lines of evidence. The five levels of confidence are *Very high, High, Medium, Low, and Very low*, as illustrated in the IPCC chart below.[2]

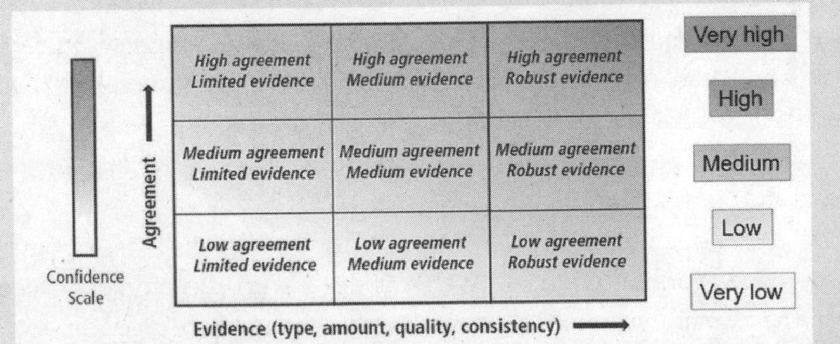

The IPCC reports make many explicit confidence assessments.

Climate science is a lively field. Thousands of researchers supported by billions of dollars work to observe the climate, understand it, and project its future. They report their results in scientific journal articles, publishing more than ten thousand each year. In most other fields of science, that would be the end of the story.

However, climate science isn't just any other field. Because the answers to its core questions are so important, with huge potential impact on human society, the United Nations and the US government regularly convene large groups of researchers to prepare formal "assessment" reports meant to provide "best answers" for non-experts, including scientists in other fields, decision makers in government and the private sector, and the public. Those reports, which run to many hundreds of pages each, review and summarize recent research and interpret its findings for non-scientists. The most recent are the Fifth Assessment Report (AR5) released by the IPCC[3] in 2013, and the Fourth National Climate Assessment (NCA2018) released by the United States Global Change Research Program (USGCRP)[4] in two installments in 2017 and 2018. Great fanfare and intense media coverage accompany the release of each of these reports.

ABOUT ASSESSMENT REPORTS

The most prominent series of assessment reports is produced under the auspices of the United Nations' Intergovernmental Panel on Climate Change (IPCC), which was established in 1988. The IPCC issued its first assessment in 1990; the Fourth Assessment Report (termed AR4) was issued in 2007,[5] the Fifth (AR5) in 2013,[6] and the Sixth (AR6) is expected in the summer of 2021.

The foundational section of each of the AR reports is that from the so-called Working Group I (WGI). It deals with the physical aspects of the climate system, in particular the changes observed in recent decades, and how the climate responds to human and natural influences. Other working groups build upon the WGI assessment to describe the impacts of a changing climate and society's response to them. Each working group also prepares a distillation of their section called a Summary for Policymakers (SPM); a synthesis of all sections is published as well. Along with its comprehensive AR series of assessments, the IPCC also publishes more focused special reports, such as those on Extreme Events,[7] the Ocean and Cryosphere,[8] or Climate Change and Land.[9]

The US government also issues its own independent series of assessment reports. The Global Change Research Act of 1990 requires a National Climate Assessment (NCA) every four years.[10] Those reports are produced by the US Global Change Research Program (USGCRP). The NCA reports have much the same purpose as the IPCC's ARs, although with more of a US focus. The content of the NCAs generally aligns with the substance of the ARs, but there can be differences in emphasis and language.

The first three NCAs were issued in 2000, 2009, and 2014. (The George W. Bush administration was less than diligent about it.) The fourth, NCA2018, comprises two volumes. Volume I, focused on physical climate science, was released in November 2017 as the Climate Science Special Report (CSSR).[11] Volume II, released in November 2018, focused on the impacts and risks of a changing climate, as well as on how we might adapt.[12] The analysis of future climate impacts in Volume II naturally builds upon projections of future climate change in the CSSR; its credibility therefore depends crucially upon the extent to which the CSSR faithfully portrays the certainties and uncertainties in climate science. The fifth NCA was released in 2023.

The AR and NCA assessments are drafted and reviewed by similar processes. The sponsoring organization (the IPCC or USGCRP) identifies teams of expert authors for each chapter. Those teams produce successive drafts that are refined in response to comments from still other experts; the NCA also undergoes a formal review by a panel convened by the National Academies. The entire process takes years. For example, the first meeting of the lead authors for AR6 took place in June 2018, about three years before the report's planned release. Production of the CSSR ran more quickly, but it still took about twenty months to draft and review.

The assessment reports literally define The Science for non-experts. Given the intensive authoring and review processes, any reader would naturally expect that their assessments and summaries of the research literature are complete, objective, and transparent—the "gold standard." In my experience, the reports largely do meet that expectation, and so much of the detail in the first part of this book, the science story, is drawn from them. But a careful reading of the most recent assessment reports also reveals some elementary failures that mislead or misinform readers on important points. What those failures are, how they came about, how the media promulgates them, and what can be done to correct them is another dimension of the science story.

Organizations and individuals further along in the "game of telephone" rely on the assessment reports when talking about climate. For example, a 2019 report by the American Association for the Advancement of Science (AAAS) entitled "How We Respond," references NCA2018 when it opens with this high-level summary of the science:

> Our nation, our states, our cities and our towns face an urgent problem: climate change. Americans are already feeling its effects and will continue to do so in the coming decades. Rising temperatures will impact farmers in their fields and transit riders in cities. Across the country, extreme weather events such as hurricanes, floods, wildfires and drought are occurring with greater frequency and intensity. While these problems pose numerous risks to society and the planet, undoubtedly the biggest risk would be to do nothing. Science tells us that the sooner we respond to climate change, the lower the risks and the costs will be in the future.[13]

I have been a member of the AAAS for almost five decades and was named a Fellow of the organization many years ago. So I can tell you that the statement above was never submitted for comment, let alone endorsement, by the organization's 120,000 members. Had I been asked to comment, I would have offered a somewhat different statement, based upon my familiarity with the assessment reports and literature:

> *The earth has warmed during the past century, partly because of natural phenomena and partly in response to growing human influences. These human influences (most importantly the accumulation of CO_2 from burning fossil fuels) exert a physically small effect on the complex climate system. Unfortunately, our limited observations and understanding are insufficient to usefully quantify either how the climate will respond to human influences or how it varies naturally. However, even as human influences have increased almost fivefold since 1950 and the globe has warmed modestly, most severe weather phenomena remain within past variability. Projections of future climate and weather events rely on models demonstrably unfit for the purpose.*

Later, I'll further explore some of the reasons why individuals and organizations (the AAAS among them) tend toward unsupported hyperbole when communicating on the subject of climate, and I'll outline some steps that would move the discussion away from an unseemly posture of persuasion toward a more professional stance of imparting information impartially, completely, and with context. The following chapters will support the more factual, cautious, and less alarming tone of my take—after all, I'm not selling cooking oil.

WHAT WE KNOW ABOUT WARMING

I've had an urge to measure the world since I was a child. Temperature was one of my early fascinations, and I became intrigued with the small alcohol thermometer in my kindergarten classroom. *How did it work? Why did it change?* Eventually, my five-year-old mind became curious about what the thermometer would show if I took it into the school hallway and then out of the building. So one winter day as the class was about to be dismissed, I pocketed the thermometer and walked out of the classroom. I was delighted to see the reading drop as I walked home from school and then rise again as I entered my house. Unfortunately, my mother discovered my unauthorized borrowing of experimental apparatus when I put the thermometer into our freezer. The next morning I was made to return it, apologize to my teacher, and promise not to take it again—all in all, a good life lesson. And a few days later, my parents presented me with my own thermometer.

As I knew instinctively in childhood, understanding the natural world starts with measurements—the data. But gathering useful global climate data is much more involved than simply carrying around a pocket thermometer. The earth is large and not easy to cover (particularly the 70 percent that's oceans), and since we're looking for small changes over decades, we need records that are accurate, precise (having small uncertainties), and span a long period. And even when we've got good data, the story it tells is seldom simple. We'll see in this chapter that there is much more to the story of global temperature change than "humans are warming the earth."

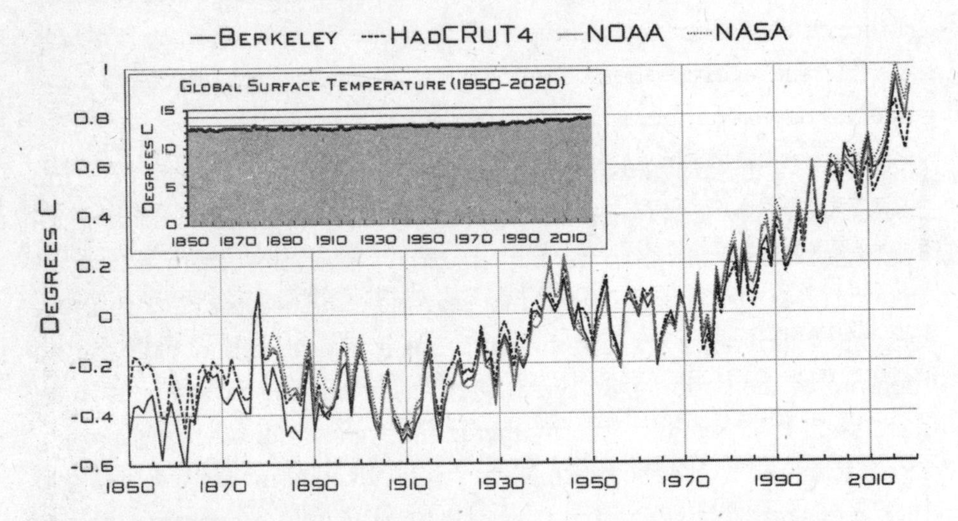

FIGURE 1.1 **Annual global surface temperature anomalies as determined by four independent analyses.** Anomalies are the deviation of temperatures from a baseline (average) value. Though there are minor differences among them, all four analyses show similar trends and fluctuate in sync. Typical uncertainties in the data points are ± 0.1°C.[1] The inset shows global average temperatures, rather than anomalies. Differences among the four data sets are too small to display there.

Most everyone has seen some version of the iconic graph in Figure 1.1, showing the earth's "temperature" rising by about 1°C (1.8°F) since 1850, with what appears to be a sharp uptick in slope beginning around 1980. It sure looks like something is changing. But exactly what is that graph? (After all, you'll notice it reads "Temperature Anomalies" rather than "Temperature.") Moreover, the annual average temperature in, say, New York City (about 13°C or 55°F) can vary from year to year by more than 2°C (3.6°F), greater than the entire range of the graph. So should we be concerned by these long-term changes, which the inset shows are quite small in terms of the globe's actual temperature? What is this graph really telling us?

Even as a five-year-old experimenting with a purloined thermometer, I was able to see that temperature varies from place to place and changes over time. Today, thousands of observing stations around the world and dozens of satellites overhead are continuously documenting those changes (and much else about the weather) all over the planet. Weather bureaus assemble and analyze these observations to produce the forecasts that guide our day-to-day plans.

However useful they are in helping us decide whether to take a sweater when we leave the house in the morning, using weather observations to learn something about the climate is altogether more complicated, because climate is not the weather—a distinction often missing in popular discussion. The weather anywhere varies constantly in ways both predictable and unexpected—through the day (it's usually warmer at 4 PM than it is at 4 AM), across days (as when a front passes through), with the seasons, and from year to year. On the other hand, *a location's climate is the average of its weather over decades.* In fact, the UN's World Meteorological Organization defines climate as a thirty-year average, although climate researchers will sometimes discuss averages over a period as short as ten years. So changes in the weather from one year to another do not constitute changes in climate.

Non-experts often confuse climate and weather (experts sometimes do as well, occasionally deliberately). Here's an example that clarifies the difference between the two: If you are moving from Wisconsin to southern Arizona, knowledge about your new home's climate tells you to invest in air conditioning for the summers, that it's probably safe to leave your heavy winter coats behind, and that the water-loving plants that flourish in Madison are unlikely to fare as well in Tucson. But knowledge about the weather tells you that, according to Tuesday's forecast, you'll need an umbrella when you arrive on Thursday. An aphorism traceable to 1901 captures it well: *Climate is what you expect, weather is what you get.*

Because climate is an average over many years, it changes slowly. It takes at least a decade of observations to define a climate, and so two or more decades to identify a change in it. Those long times verge on the limit of human memory, particularly when changes are small, so we need records to keep from being fooled. What garners most popular attention are extreme weather events like storms and heat waves—their numbers and intensities also vary from year to year but, again, it's their average properties over decades that define the climate.

Not only can we be fooled by failing to take a "big picture" view of climate over time, but we can also be fooled by failing to take a big picture view of the planet. Climate varies from place to place depending upon things like latitude (it's colder toward the poles), elevation (it's colder on mountains), and proximity to water (a moderating influence). The average daily high temperature in Singapore is about 33°C (91°F) all year, while it's –4°C (25°F) in Moscow during January and 24°C (75°F) in July. To assess the effects of human influences, it's best to consider the temperature over large regions of the globe, both because the influences of greatest concern—such as greenhouse gases—act worldwide and because averaging over large areas makes small changes in climate more evident by "bringing them out of the noise."

Alas, it isn't easy to measure the surface temperature over the whole earth, particularly when you're looking for changes of a fraction of a degree over decades. You have to worry about variations in the thermometers themselves, how they're housed, and exactly where they're located. And even if a station hasn't been moved over the years, urbanization around a site is a concern, since buildings, roads, and concentrated human activity make cities a few degrees warmer than their rural surroundings. Most importantly, how can we possibly measure the temperature of the globe if we don't have a thermometer everywhere? Fortunately, a landmark paper in 1987 by Hansen and Lebedeff showed that locations less than 1,200 km (750 miles) apart have, on average, similar temperature changes.[2] In other words, if the average temperature rises by a degree in one location, it's likely (but not guaranteed) to do the same in other locations nearby. This lets us judge temperature changes over a large area using only a sparse network of stations, and the gaps in coverage can be filled in probabilistically. Figure 1.1 shows that four independent groups using different analysis methods come to very similar global surface temperature records based on this general idea.

Speaking of Figure 1.1, let's return to our earlier question: Why "Temperature Anomalies" instead of "Temperature"? "Anomalies" measure how much the observed condition—in this case, temperature—at a location deviates from the baseline (average) value at that location. Using values relative to a baseline puts changes in the Arctic on the same footing as changes in the Tropics, allowing us to compare changes in climate between locations as well as over time, and so tease out large-scale trends. The baseline is usually chosen as the average over a thirty-year period in the middle of the time span being considered (chosen as 1951–1980 in Figure 1.1). This choice is somewhat arbitrary, but the arbitrariness is unimportant since choosing a different baseline would simply shift the curve up or down, but it wouldn't change its shape or size (remember, we're interested in *changes* in climate).

So that's the global temperature anomaly graph of Figure 1.1—the deviation of the average daily temperature from its expected value, averaged over each day of the year and over the whole globe. Looking at it, we can see that the global average anomaly has gone up over the past century. Of course, that doesn't mean the temperature everywhere has gone up all of the time or by the same amount. But its change over time does give us some important and interesting information.

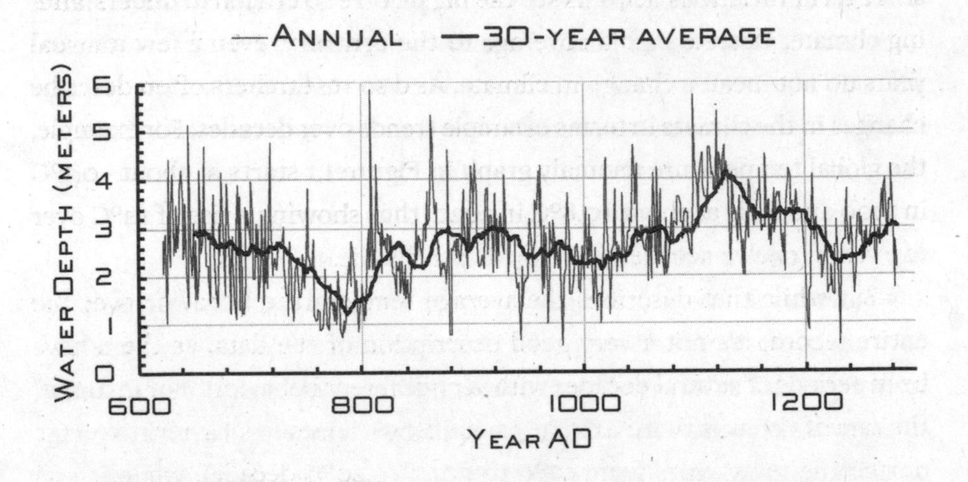

FIGURE 1.2 **A climate record with fluctuations and trends.** The annual minimum depth of the Nile River near Cairo over more than 650 years from 622 to 1284 AD. The data, measured in meters, shows a characteristic pattern of year-to-year fluctuations around longer-term trends.[3]

One striking feature of the global average surface temperature record shown in Figure 1.1 is that, even though the data fluctuates from year to year, there are clear trends over decades, with ups and downs superimposed on an overall warming trend. That is, the history is not just a jumbled sequence of random values.

Most climate-related time series show a pattern like this: that is, of long-term trends visible despite year-to-year variation. The phenomenon was discovered by the British hydrologist H. E. Hurst, who was studying the water level of the Nile River, which is sensitive to rainfall over about 10 percent of the African continent. Figure 1.2 shows the minimum Nile water depth observed each year by the Roda Nilometer near Cairo over the more than 650 years from 622 to 1284 AD. As you can see, even as the year-to-year values fluctuate, sometimes dramatically, the thirty-year averages indicated by the solid line show clear trends over decades, just as in the global temperature data from Figure 1.1.

Looking at the signal of longer-term trends rather than the noise of short-term variations helps us see the big picture so crucial to understanding climate. Despite media coverage to the contrary, even a few unusual years do not mean a change in climate. And so researchers often describe changes in the climate in terms of simple trends over decades. For example, the global temperature anomaly graph in Figure 1.1 starts at about −0.3°C in 1900 and ends at about +0.8°C in 2020, thus showing a rise of 1.1°C over 120 years (twelve decades), or 0.09°C per decade.

But while that describes the average temperature behavior over the entire record, it's not a very good description of the data, as there have been periods of several decades with very different behaviors. For instance, the rate of rise was twice as large as our 0.09°C/decade long-term average during the forty years from 1980 to 2020 (0.20°C/decade), while it was *negative* during the forty years from 1940 to 1980 (−0.05°C/decade). And for the thirty years before that, from 1910 to 1940, the rate of rise was again nearly twice the 0.09°C/decade average (about 0.17°C/decade).

So trends are often highly dependent on the time span being considered; here we can get almost any trend we want depending upon which interval we choose. Such "cherry picking" of data is unfortunately quite common in the media (and occasionally in the assessment reports) when the goal is to persuade. But if the goal is to inform, it is crucial to present and discuss the entire data set, with all the ups and downs that are significant on whatever scale you're talking about.

Scale—that big picture we keep referencing—is also important because it helps us untangle global and local climate changes. August is high season for hot-weather coverage in the Northern Hemisphere media, and on August 13, 2019, the *Washington Post* published a front-page story under the headline "Extreme climate change has arrived in America." To make its point, the paper prominently displayed the maps you see reproduced in Figure 1.3, which depicted (in flaming reds and oranges) the county-by-county change in temperature across the contiguous United States between 1895 and 2018.

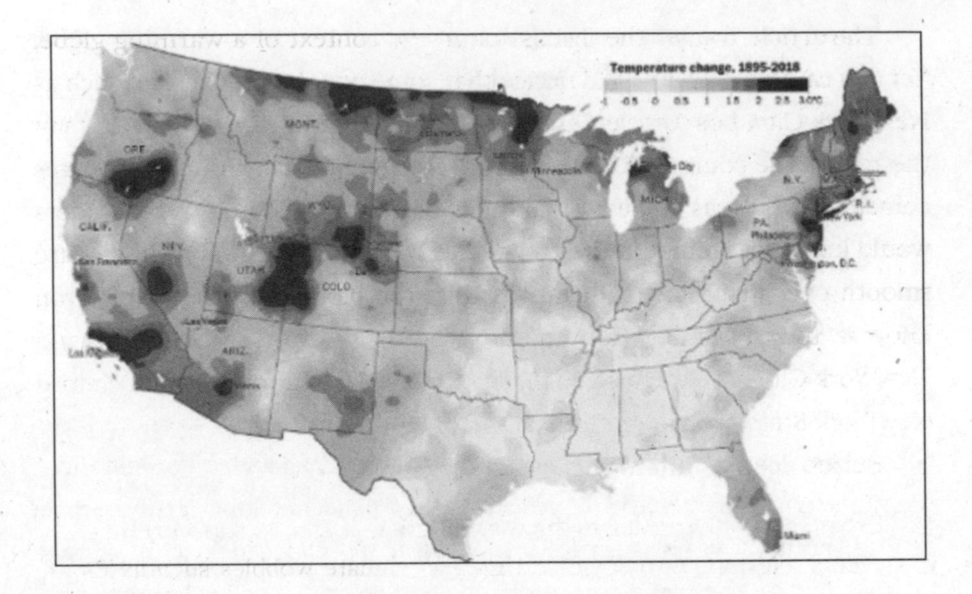

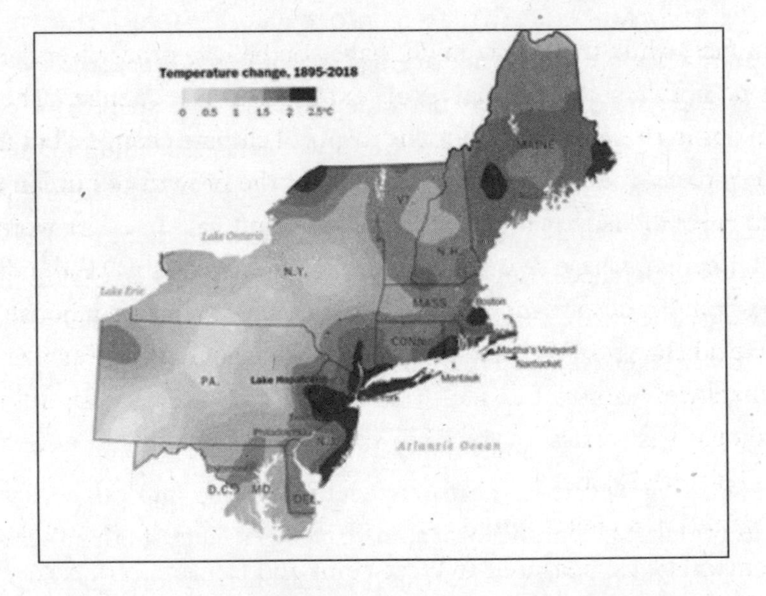

FIGURE 1.3 **County-by-county changes in US surface temperatures between 1895 and 2018.** Created from NOAA data and published in the *Washington Post* on August 13, 2019, under the headline "Extreme climate change has arrived in America."[4]

The article frames the discussion in the context of a warming globe. Yet any careful reader would notice that some population centers (such as New York City, Los Angeles, and Phoenix) have warmed much faster than the rest of the country. And some blotches in the Rocky Mountain states coincide with areas of new oil and gas production. More expert readers would know that temperature trends due to a changing global climate are smooth over much larger distances than the blobs on these maps—you know it, too: remember Hansen and Lebedeff's 1,200 km? So how could New York City have warmed so much more rapidly than a region in central New York State 250 km (150 miles) away?

Buried deep within the article one finds:

> Urban heat effects, changing air pollution levels, ocean currents, events like the Dust Bowl, and natural climate wobbles such as El Niño could all be playing some role, experts say.

In fact, while the article might have you believe otherwise, the *Post*'s maps do not illustrate the arrival of "extreme climate change." The localized blobs in these maps are not due to global climate changes, but instead are very likely the result of urbanization or the growth of human activities in rural areas that began producing oil and gas. In other words, the local climates of these areas may indeed have changed since the Industrial Revolution. Yet despite the article's frequent mention of greenhouse gases, these local changes have very little to do with global influences like these. For instance, carbon dioxide—the most important human-influenced greenhouse gas—exists in the atmosphere at roughly the same concentration all over the globe.

Figure 1.4, a chart I often use in my lectures, shows the annual average temperatures measured at West Point and Central Park NYC, 70 km (42 miles) apart.[5] The (not quite perfectly) correlated fluctuations are clear, as is the warming influence of urbanization in the NYC record in the forty years after 1920 (West Point was, and remains, relatively rural). As you can see, while temperatures are indeed different in the two locations, their

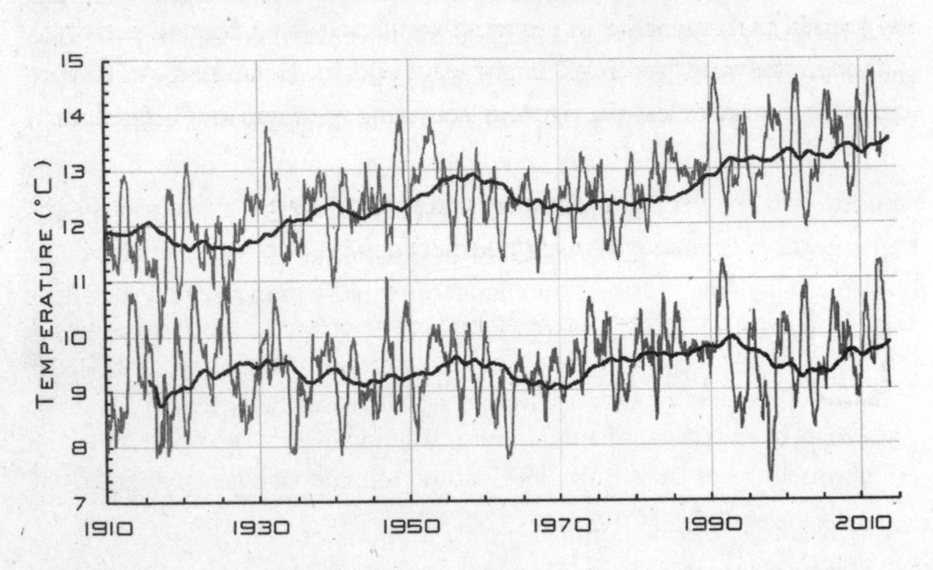

FIGURE 1.4 **Annual average temperatures from 1910 to 2013 for New York City (top) and West Point, NY (bottom), 42 miles to the north.** The gray lines are the annual values while the black lines are the ten-year trailing averages. The West Point data has been shifted downward by 1.5°C for clarity.

fluctuations pretty well match in both direction and size, just as Hansen and Lebedeff would predict. Of course, if we took two stations very far apart (say NYC and Beijing), the fluctuations wouldn't match up at all.

Avoiding confusion of local and global effects is why professionals routinely smooth maps of temperature changes over distances of 1,200 km (750 miles).[6] One of my students would not fare well if they produced a map like the *Post*'s without explaining which warming is due to a changing global climate and which isn't.

The varying trends of the global temperature anomaly in Figure 1.1 are thought to have several different causes. One is that the climate system shows internal variability—ebbs and flows over decades, largely associated with slow ocean currents. Then there are natural phenomena, like changes in the sun's brightness, that "force" (influence) the climate system to change. And finally, and of greatest interest to us, there are those changes and trends that might be in response to forcing from human activities. (Although the word "forcings" might seem odd, in the language of climate science, it is more or less the rigorous-sounding synonym for "influences.")

"CLIMATE CHANGE" VS. A CHANGING CLIMATE

Using the term "climate change" promotes (and sometimes perhaps deliberately exploits) a confusion of meaning. The UN Framework Convention on Climate Change defines "climate change" as:

> . . . a change of climate which is attributed directly or indirectly to human activity that alters the composition of the global atmosphere and which is in addition to natural climate variability observed over comparable time periods . . .[7]

That definition explicitly *excludes* changes due to natural causes, which differs from the plain-language meaning of the term. So when the average person hears "climate change" (as in the commonly shouted credo *Climate change is real!*), they are likely to assume it means change we are responsible for. The media is not particularly precise or consistent in their usage of the term, sometimes employing it one way and other times another, often without clarification; articles about changes due to multiple or unknown causes get "climate change" headlines, and calls to *fight climate change!* make it sound as if reducing human influences would keep the climate from changing.

To avoid confusion, this book will be specific and precise and avoid ambiguous language. If we are talking about changes in response to human influences, we will use a phrase like "human-caused climate change." Along those same lines, "a changing climate" will mean just what it sounds like, referring to changes from any source. Precise terminology is one of the most powerful tools a scientist has for reasoning and communicating clearly, while imprecise terminology is just as powerful a tool for those seeking to persuade.

As we'll see in the next chapter, human influences on the climate were negligible prior to 1900. There weren't many people around in 1900 (only one-fifth of today's count), and they were mostly farming; industrialization was just getting underway for most of the globe. Human influences remained quite small as late as 1950, when they were less than one-quarter of what they are today. Variations in the climate before 1950, then, show that other phenomena must have been at play, if not dominant, since the earth actually cooled a bit between 1940 and 1980 even as warming human influences grew. And since those natural variations (both internal variability and natural forcings) are presumably still present, it is vitally important to understand them if we're to have confidence in attributing even part of the recent warming to human influences, much less project how the climate will change in the future.

Another point worth noting as we consider the recent temperature rise is that, despite the misleading Figure 1.3, the warming of the past forty years on large scales hasn't been uniform over the globe. That's evident in Figure 1.5, reproduced from the US government's 2017 CSSR (Climate Science Special Report, described earlier). As you can see, the land is warming more rapidly than the ocean surface, and the high latitudes near the poles are warming faster than the lower latitudes near the equator. More generally, the coldest temperatures (at night, during the winter, and so on) are rising more rapidly than the warmest temperatures—the climate is getting milder as the globe is getting warmer.

So what? you might wonder. *The earth IS getting warmer; why does it matter if that warming hasn't been steady or uniform over the globe? Or that low temperatures are rising more rapidly than the high ones?* But details like these are important. They help us isolate and quantify the relative roles of human-caused and natural changes, both in the past and in the coming decades. They also help us to understand the impacts of a changing climate. How have ecosystems changed already as the globe has warmed? How have societies adapted to the climate changes we've already seen, and how adaptable might they be to future changes? As in all science, the

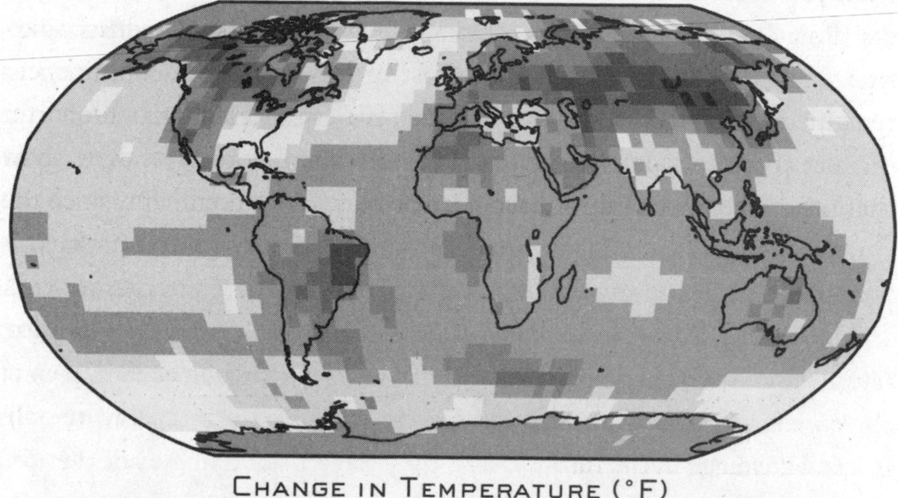

FIGURE 1.5 **Surface temperature change (in °F) for the period 1986–2015 relative to 1901–1960.** Changes are generally significant over most land and ocean areas. Changes are not significant in parts of the North Atlantic Ocean, the South Pacific Ocean, and the southeastern United States. There is insufficient data in the Arctic Ocean and Antarctica to compute long-term changes there. (Adapted from CSSR Figure 1.3.)[8]

details deepen our understanding of what's happening, why it's happening, and what might happen in the future.

Of course, there's much more to climate than changes in the surface temperature, or even changes throughout the atmosphere. In fact, the atmosphere is a relatively minor part of a much larger and more complex system that includes the water (oceans, lakes, and so on), snow and ice on land and sea, the solid earth, and living things (microbes, plants, animals, and humans).

The world's oceans are both the most important and the most problematic piece of the earth's climate system. They hold more than 90 percent of the climate's heat and are its long-term memory. Conditions in the atmosphere swing wildly from day to day and year to year in response to any number of influences—that's part of what makes untangling weather and climate so difficult. The oceans, on the other hand, change—and respond to changes—over decades to centuries.

Yet, as mentioned, gathering ocean data with enough precision and coverage to detect climate changes is even more of a challenge than it is on land. The oceans are vast and largely unpeopled, and while the near surface is accessible, the depths are far less so (the average ocean depth is 3,700 meters or 12,000 feet). Satellites can measure the temperature only at and above the ocean surface, and they have been doing even that for less than half a century. Before satellites, there were only surface measurements or soundings from passing ships (which don't go everywhere) and buoys (which are at few fixed locations).

In 2000, an international program called Argo began setting a fleet of robotic floats adrift on the oceans to record their properties.[9] The Argo float system first achieved worldwide coverage in 2005, and there are now more than 3,900 of those drifters covering the world's oceans. The floats usually drift at a depth of 1 km (3,300 feet), but every ten days, they descend to more than 2 km (6,600 feet) and measure the temperature and salinity profile through the water column during a six-hour ascent to the surface, where they transmit their results via satellite before returning to 1 km depth.

Argo has vastly improved our knowledge of ocean conditions. Before the year 2000, no more than 40 percent of the ocean had been sampled to 400 meters (1,300 feet), and less than *10 percent* had been sampled below 900 meters (3,000 feet). (Remember, the average ocean depth is 3,700 meters.) Argo improved coverage dramatically during the past two decades, and some 60 percent of the ocean is now being sampled at least yearly to a depth of 2 km.[10]

While Argo data will be essential to understanding changing ocean conditions in the coming decades, data on the ocean's past is limited in coverage and quality. Even so, we're confident that the oceans have been

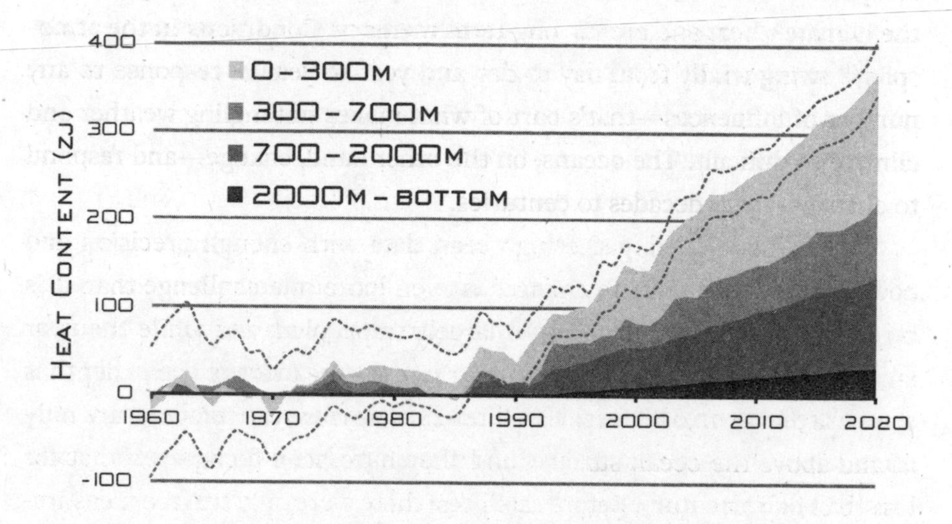

FIGURE 1.6 **Ocean heat content from 1960 to 2019.** The anomalies are related to a 1958–1962 baseline, and the time series are smoothed over twenty-four months. The gray dashed lines are the 95 percent confidence interval of the total ocean heat budget.[11]

warming over decades, if not centuries. Figure 1.6 is another anomaly graph—this time showing how the heat content of various layers of the ocean has increased over the past sixty years. The heat energy is measured in zetajoules (one ZJ is equal to 10^{21} joules), and the anomalies are relative to a 1958–1962 baseline. The dashed lines give the 95 percent confidence interval of the oceans' total heat—in other words, the graph says that there is a 95 percent chance that the true value falls within those dashed lines. (The lines are farther apart further back in time because we have less reliable information then, making the uncertainties greater.) There is a clear upward trend in the oceans' heat energy, with the upper 300 meters (1,000 feet) warming more rapidly than the deeper layers, as would be expected if heat is being absorbed from the warming surface. In contrast, the slowly changing deeper layers better reflect past conditions.

Hundreds of ZJ sounds like a lot of energy (and it is, at least on a human scale—all of the world's energy derived annually from fossil, nuclear, and renewable sources amounts to only about 0.6 ZJ). However, when that heat is spread out over all the water in the oceans, the temperature rise it represents is very modest: a few hundredths of a degree Celsius per decade. Nevertheless, the growing ocean heat content is the surest indication that the planet has indeed been warming in recent decades.

But lest you think the matter is settled, consider that a different peer-reviewed analysis found that the heat content increased between 1990 and 2015 at only half the rate shown in Figure 1.6, with the same smaller rate of increase also between 1921 and 1946, when human influences on the climate were much smaller.[12] And yet another paper found that the upper 2 km of ocean warmed from 1750 to 1950 at a rate about one-third of that shown in the figure.[13] So as with the surface temperature record, poor historical data and large natural variability complicate efforts to understand the role of human influences.

Rising temperatures at the surface and in the ocean are not the only indicators of recent warming. The ice on the Arctic Ocean and in mountain glaciers has been in decline, and growing seasons have been lengthening slightly. Satellite observations show that the lower atmosphere is warming as well. Later chapters will discuss some of these indicators in more detail.

But to what extent are human influences driving this warming? One clue involves looking at the climate centuries or more in the past, when human influences were truly negligible. We've already seen that there were significant trends in ocean heat well before we came on the scene. What did the global temperature anomaly graph look like before 1850, the earliest date shown in Figure 1.1? Were there long-term rising trends before humans influenced the climate? And how much did the temperature anomaly "wiggle" over the decades due to natural forcings and/or internal variability? Answers to questions like those are essential to knowing how the climate has already responded to human influences and how it might respond in the future.

Although crude thermometers were around in the mid-seventeenth century, Daniel Fahrenheit created the first reliable instrument in just 1714, and these weren't widespread until the mid-1800s. So past climates have to be inferred in other ways. We have historical records such as weather diaries and crop yields going back a few millennia, though obviously the extent of these varies both from time to time and place to place. But paleoclimatologists can infer climate even further back by using proxies—that is, measuring a temperature-sensitive property of material preserved from the past.[14] One example is the analysis of the thickness and composition of tree rings, which are laid down annually as a tree grows. Another is by temperature measurements of the water in boreholes—holes drilled deep into the ground—at various locations around the globe. Just as in the oceans, the deeper water carries information about surface temperatures further in the past.

Figure 1.7 shows several different proxy reconstructions of the global temperature anomaly over the past 1,500 years, alongside our modern instrumental record, which begins in the late 1800s. This time our baseline is the average temperature between 1881 and 1980, represented by the dashed line.

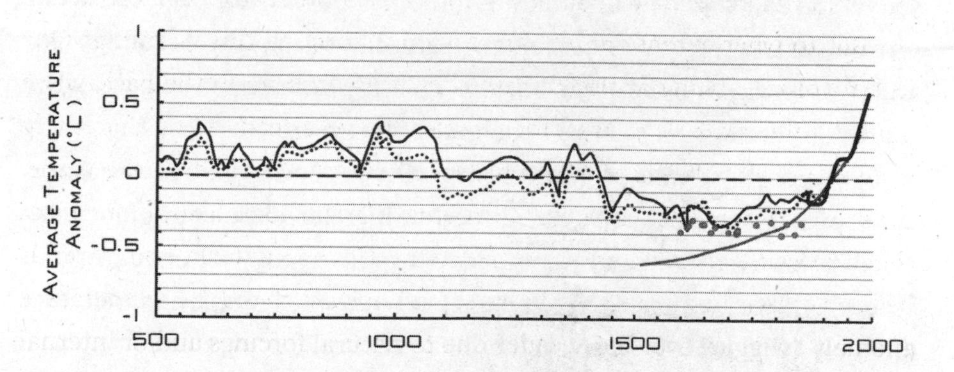

FIGURE 1.7 **Global average surface temperature anomalies during the last fifteen hundred years as reconstructed by different proxy methods, together with the modern instrumental record (black line).** Anomalies are relative to an 1881–1980 baseline and have been smoothed to reduce variations on timescales shorter than about fifty years.[15]

As you can see, centuries of warmer temperatures gave way gradually around the year 1000, leading up to the Little Ice Age, an unusually cold period that ran from about 1450 to 1850. This was followed by more rapid warming that has continued until today.

Because of the limited evidence, the most recent assessment report (AR5, from which the data in Figure 1.7 is taken) has only *low confidence* that the global warming of the past thirty years has exceeded the range of reconstructed temperatures. But proxy data are better and more plentiful for the Northern Hemisphere (where there's more land), so AR5 has medium confidence that the last thirty years were likely (a two-out-of-three chance) the warmest thirty-year period of the last fourteen hundred years for the Northern Hemisphere.

What happens if we zoom out to look at even longer timescales? Tree rings go back some fifteen thousand years into the past, but cores drilled into the ice sheets of the Antarctic or Greenland can take us much further. Those ice sheets build up in layers, year by year, and the properties of each layer (including trapped gases, isotopic composition, and dust) carry information about the climate conditions when it was formed. Analyses of layers deep in these cores can tell us about the distant past—the oldest cores now go back almost three million years.[16] And cores into ocean-bottom sediments can take us one hundred million years into the past: the shells of tiny marine organisms continually rain down from the ocean surface and are deposited on the ocean bottom, creating a continuous record of the conditions under which they were created and grew.

None of these proxies for temperature is as good as a direct measurement with a thermometer. They directly reflect conditions at only a single location, and they can be complicated to interpret—for example, tree growth is sensitive not just to temperature but to precipitation as well. And uncertainties grow as you go further back in time. But proxies do give us a sense of how the climate changed before there were humans systematically observing and recording the weather.

Our accumulated knowledge of the earth's surface temperature over the past five hundred million years is summarized in Figure 1.8, which

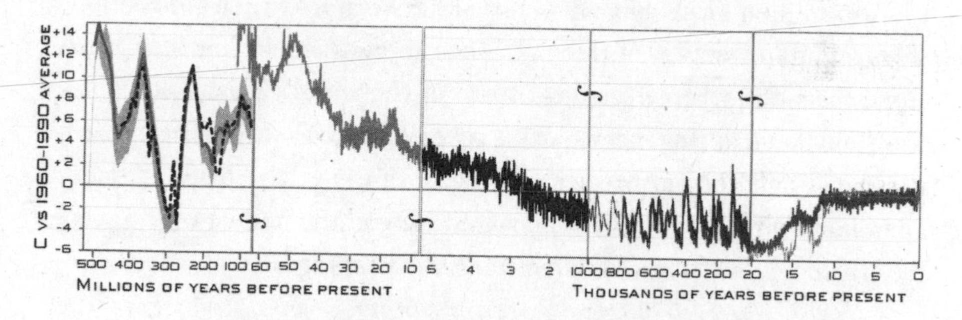

FIGURE 1.8 **Global average surface temperature anomaly as determined by various geological proxies for five periods extending to five hundred million years ago.**

shows global average surface temperature anomalies in degrees Celsius relative to the 1960–1990 baseline.[17] There are five different panels, each spanning an interval between geologically significant events. The timescale of each successive panel is about ten times shorter than the one before it.

Starting at the right-hand (most recent) panel, we see that the globe warmed about 5°C (9°F) starting some 20,000 years ago, when ice sheets last covered a large fraction of the earth. The relatively warm and stable temperature over the past 10,000 years supported the rapid development of civilization.

Over the past million years, beginning in the second panel from the right, periods of rapid warming alternated with periods of slower cooling, every 40,000 years or so early on and then every 100,000 years starting about 500,000 years ago. Those variations were driven by slight changes in the earth's orbit around the sun and the tilt of its axis; the most recent warm period before the present began about 127,000 years ago and lasted for some 20,000 years. During that time, the global surface temperature anomaly was up to 2°C (3.6°F) warmer, and the upper ocean was 2–3°C (3.6–5.4°F) warmer than today.

Going further back in time, as shown in the three leftmost panels, there are even stronger swings, some of which have had impacts on the modern world. For example, the Carboniferous period extended from ~360 to ~300 million years ago, the interval between evolution's invention of trees and its invention of tree rot. Since there was nothing then to consume the wood when trees died, some of the world's great coal deposits were laid down during that time. It's sobering to realize that the entire figure shows only the most recent 10 percent of the earth's history, and that anatomically modern humans appeared only in the middle of the second most recent panel (a few hundred thousand years ago).

I hope this chapter has given you a new appreciation for some of the many moving parts involved in climate science—and in drawing conclusions from it. We've explored these foundational concepts and problems in the context of global warming. Past variations of surface temperature and ocean heat content do not at all disprove that the ~1°C (1.8°F) rise in the global average surface temperature anomaly since 1880 is due to humans, but they do show that there are powerful natural forces driving the climate as well, and they illuminate the scientific challenge of understanding those natural influences well enough to confidently identify the climate's response to human ones. In other words (as we've made clear throughout this chapter), the real question is not whether the globe has warmed recently, but rather to what extent this warming is being caused by humans.

To answer this, we need to know more about how (and how much) humans influence the climate, which is the subject of the next chapter.

HUMBLE HUMAN INFLUENCES

Like many people, I keep an eye on my weight. Eating more and exercising less makes me heavier, while eating less and staying active makes pounds disappear. How much I eat and move aren't the only variables—there are factors like health, hormones, and genetics that influence how fast I burn or store calories—but in short, my weight is determined by the balance between calories consumed and calories burned, and any imbalance quickly shows up on the scale. In the same way, the earth's temperature results from a crucial balance between warming by sunlight and cooling by heat radiated back out into space.

On the warming side of this balance is the sunlight energy that the planet absorbs. As the planet heats up, it emits infrared radiation back into space, which makes up the cooling side of the balance. A fundamental physical law—the Stefan-Boltzmann law—discovered around 1880 by two physicists working in Austria tells us that the amount of infrared radiation an object emits increases with its temperature in a very predictable way. So as a planet's temperature rises due to solar warming, the cooling by infrared radiation also increases until the infrared cooling is equal to the solar warming. The technical term for this Goldilocks condition—where the planet is neither gaining nor losing energy and its temperature is steady—is "radiative equilibrium." The temperature at which that balance is achieved, the "equilibrium temperature," depends upon several things, most obviously the planet's distance from the sun.

Let's look more closely at the warming side of the balance—the sunlight energy absorbed. Because the earth is not completely black, it absorbs only 70 percent of the sunlight that reaches it; the other 30 percent is reflected back into space and doesn't contribute to the planet's warming. That 30 percent number, corresponding to the earth's reflectivity, is called the "albedo" (from the Latin word *albus,* meaning "white"). When the albedo is higher, the earth reflects more sunlight and so is a bit cooler, and conversely when the albedo is lower, the earth absorbs more sunlight and is warmer. While the planet's average albedo is 0.30, its value at any given moment depends upon which part of the earth is facing the sun (oceans are darker, land is brighter, clouds are brighter still, and snow or ice is very bright), and the monthly average varies by about ± 0.01 with the seasons (it's larger in March and smaller in June/July).

EARTHSHINE AND THE ALBEDO

Precision measurements of the global albedo are important to understanding the climate system. If the average albedo were to increase from 0.30 to 0.31, say because of a 5 percent increase in cloudiness, that additional reflectivity would largely compensate for the warming influence of *doubling* the atmosphere's CO_2.

I developed a personal interest in albedo measurements in the summer of 1991 when I participated (along with physicist Freeman Dyson and astronaut Sally Ride) in a study conducted by JASON, an independent organization of scientists that advises the US government on sensitive and pressing science and technology issues. This study focused on the possible uses of small satellites in observing the climate.[1] One thing such satellites could measure is the fraction of sunlight that a patch of the earth's surface reflects back into space. With enough satellites covering enough of the globe, we could average those fractions to determine the global albedo. In fact, that's how the albedo has been measured for more than forty years, except by a few large, expensive satellites instead of many small ones.[2]

The JASON study thirty years ago also motivated me to revive an older, simpler way to study the earth's albedo. The French astronomer André Danjon first measured the earth's albedo in the early 1930s. His very clever method was to observe "earthshine," the faint glow of the "dark" part of the lunar disk most visible when the moon is less than half-full, as shown in Figure 2.1. As that light stems from sunlight reflected by the earth and then reflected again by the lunar surface, its brightness depends upon the earth's reflectivity, and so is a measure of the global albedo.

FIGURE 2.1 **Earthshine and sunshine visible on a crescent moon.** The upper-right portion of the image shows the thin lunar crescent illuminated by sunshine, easily visible to the naked eye. A strong filter placed over that region (causing the line through the lower part of the image), makes the earthshine visible on the rest of the lunar disk.[3]

Because Danjon got an implausibly high value for the albedo, 0.39, earthshine studies became just a curiosity after about 1950. But in the summer of 1991, we JASONs realized that when Danjon's data were reanalyzed taking into account that the porosity of the lunar surface enhances its reflection of earthshine, the global albedo came out about right.[4] So I enlisted some astronomer colleagues in a program of modern earthshine observations beginning in 1995. Among the advantages of measuring albedo via earthshine is that it requires only a small telescope and standard camera, and that it's self-calibrating, since the brightness of the earthshine is compared with the sunshine in the same image. That would allow such measurements to be reproduced by future researchers decades or even centuries hence using whatever instrumentation is available then.

Precision climate observations, whether by satellite or other means, often go through a process of successive refinement until researchers get it right. Our earthshine work was no different. But eventually we were able to determine annual average albedos accurate to ± 0.003 from 1999 to 2014 that showed no significant trend, in agreement with the satellite values.[5] That uncertainty is about twice that of the satellite-derived values, but at one-thousandth the cost. Understanding changes in the earth's reflected light visible on the moon also turns out to be a good test of our ability to study other planets that are orbiting other stars, which are visible through the starlight they reflect.[6] Such are the unexpected connections of science.

Knowing the earth's albedo (as averaged over the globe and daily and seasonal cycles), we can determine its equilibrium temperature by balancing the sunlight absorbed against the infrared cooling. As we discussed, that cooling becomes stronger as the temperature increases—if the earth gets hotter, it emits more heat—making it a kind of thermostat. Determining the earth's equilibrium temperature by calculating that balance is a basic problem assigned at the start of every serious climate course. It gives an average surface temperature of . . . –18°C (0°F).

But –18°C (0°F) is wrong, far below the earth's actual global average temperature of 15°C (59°F). What's missing is the insulation provided by the greenhouse gases in the atmosphere, which raises our planet's surface temperature to its observed value. How that insulation works is best illustrated by a story.

In January 2010, when I was serving as the Department of Energy's undersecretary for science, I had the privilege of traveling to the South Pole.[7] The Department of Energy had helped install three wind turbines on a ridge between the US McMurdo Station and New Zealand's Scott Base on the periphery of Antarctica. The electricity generated by these turbines would reduce the amount of diesel fuel that had to be brought in by tanker ship. There was to be a ceremonial dedication, and I was part of the delegation doing the honors.

I flew from Washington, DC, to Los Angeles, and then on to Christchurch, New Zealand, through Auckland. There our delegation was fitted with Extreme Cold Weather (ECW) gear—insulated overalls, parka, fleece, heavy boots, woolen cap, gloves, and goggles. The following morning we boarded a military cargo jet operated by the New York National Guard. The next five hours were uncomfortable ones: we had to wear our ECW gear just in case the plane had a problem and we needed to exit in a hurry. That long flight south brought us to the Pegasus runway, about twenty miles across the ice shelf from McMurdo Station.

We dedicated the wind turbines that evening, and the next morning we flew further south on a cargo prop plane to visit the Pole. The temperature was −33°C (−27°F); we stayed for eight hours. I was able to enjoy this amazing experience thanks to the insulation of the cold weather gear, which intercepted my body heat, impeding its flow away from me and into the surrounding air.

Just like that cold weather gear, greenhouse gases in the atmosphere intercept and impede the flow of infrared heat from the earth's surface into space. Some of that heat finds its way back down to the surface, where it causes additional warming (the greenhouse effect), as shown in Figure 2.2. It's often said that greenhouse gases "trap" heat, which gives the impression that the heat never escapes. But all of the heat must eventually be radiated to space to keep the planet in energy balance, as discussed earlier. The radiated heat must balance the absorbed solar energy very precisely, to within less than one-half a percent. If it didn't, we'd see the earth warming or cooling much more rapidly than we do. So when discussing the effect of greenhouse gases on the heat flowing from the earth's surface, a more appropriate metaphor is "catch and release." For this reason, I'll use the terms "intercept" and "impede" rather than "trap."

The most common of the gases making up the earth's atmosphere are nitrogen (78 percent) and oxygen (21 percent). Combined, then, these two account for 99 percent of the dry atmosphere, and because of the peculiarities of molecular structure, heat passes through them easily. The largest part of the remaining 1 percent is the inert gas argon. But while even less abundant, some of the other gases—most significantly water vapor, carbon dioxide, methane, nitrous oxide, and ozone—intercept, on average, about 83 percent of the heat emitted by the earth's surface.[8] So the earth does indeed emit energy equivalent to what it absorbs from the sun, but instead of directly flowing off into space, cooling our planet to a chilly average of 0°F, much of that energy is intercepted by the atmosphere blanketing us.

Water vapor is the most important of the greenhouse gases. Of course, the amount in the atmosphere at any given place and time varies greatly (the humidity changes a lot with the weather). But on average, water vapor

FIGURE 2.2 **Flows of sunlight and heat through the earth's climate system.** About 30 percent of the incoming solar radiation is reflected, while the atmosphere intercepts more than 80 percent of the infrared radiation emitted from the surface.

amounts to only about 0.4 percent of the molecules in the atmosphere. Even so, it accounts for more than 90 percent of the atmosphere's ability to intercept heat. John Tyndall, the Irish physicist who was the first to study the infrared properties of gases, eloquently expressed its importance in an 1863 public lecture:

> Aqueous [water] vapor is a blanket, more necessary to the vegetable life of England than clothing is to man. Remove for a single summer-night the aqueous vapor from the air which overspreads this country, and you would assuredly destroy every plant capable of being destroyed by a freezing temperature. The warmth of our fields and gardens would pour itself unrequited into space, and the sun would rise upon an island held fast in the iron grip of frost.[9]

The next most significant greenhouse gas, carbon dioxide (CO_2), is different from water vapor in that its concentration in the atmosphere is much the same all over the globe. CO_2 currently accounts for about 7 percent of the atmosphere's ability to intercept heat. It's also different in that human activities have affected its concentration (that is, the fraction of air molecules that are CO_2). Since 1750, the concentration has increased from 0.000280 (280 parts per million or ppm) to 0.000410 (410 ppm) in 2019, and it continues to go up 2.3 ppm every year. Although most of today's CO_2 is natural, there is no doubt that this rise is, and has been, due to human activities, primarily the burning of fossil fuels.

The CO_2 that humans have added to the atmosphere over the past 250 years increases the atmosphere's ability to impede heat (it's like making the insulation thicker), and is exerting a growing warming influence on the climate. The exact increase in insulation at any place and time depends upon temperature, humidity, cloudiness, and so on. Taking average clear-sky (no clouds) conditions as an example, the CO_2 added from 1750 until today increases the fraction of heat intercepted from 82.1 percent to 82.7 percent. And as the amount of CO_2 continues to increase, the atmosphere's heat-intercepting ability (and hence its warming influence) will also increase; doubling the CO_2 concentration from the 1750 value of 280 ppm to 560 ppm would increase it to 83.2 percent under clear-sky conditions. Such an increase in concentration would amount to an increase of just 2.8 molecules per 10,000—in other words, an increase of fewer than three molecules of CO_2 out of every 10,000 molecules of air would increase the amount of heat intercepted from 82.1 percent to 83.2 percent, or by about 1 percent.

If you've followed this far, you might be puzzled by two things. First, how could changing fewer than three molecules out of 10,000, a 0.03 percent change, increase the atmosphere's heat-intercepting ability by about thirty times that amount (1 percent)? And second, how could a mere 1 percent increase in heat-intercepting ability be such a big deal?

The answer to the first question depends upon the details of the infrared (heat) radiation the planet emits to keep cool. While we've talked about how the overall amount of that radiation has to balance the warming sunlight, the radiation is actually spread over a spectrum of different wavelengths. Think of those like "colors," although not visible to our eyes. Water vapor, the most significant greenhouse gas, intercepts only some colors, but because it blocks almost 100 percent of those it does, adding more water vapor to the atmosphere won't make the insulation much thicker—it would be like putting another layer of black paint on an already black window. But that's not true for carbon dioxide. That molecule intercepts some colors that water vapor misses, meaning a few molecules of CO_2 can have a much bigger effect (like the first layer of black paint on a clear window). So the greater potency of a CO_2 molecule depends upon relatively obscure aspects of how it, and water vapor, intercept heat radiation—another example of why the details are important when attempting to understand human influences on the climate.

Figure 2.3 illustrates some of those details. It shows how the amount of heat radiation leaving the top of the atmosphere varies with the radiation's color (that is, the spectrum of infrared radiation). Were there no atmosphere, the spectrum would correspond to the smooth gray line in the graph, a curve which is described by the basic physics of the Stefan-Boltzmann law. The area under that curve corresponds to the cooling power of the radiation. The lighter, jagged gray line shows what the spectrum would look like with all major greenhouse gases present *except* CO_2

(so with CO_2 at 0 ppm). Combined, these gases reduce the radiation's cooling power by about 12.1 percent. All of the line's ups and downs arise from the detailed properties of the various greenhouse gas molecules, most importantly water vapor, but also methane and ozone. The solid black line shows a further 7.6 percent reduction in cooling power (increase in insulation) when CO_2 is present at 400 ppm (about today's concentration). Finally, the dashed black line shows an additional 0.8 percent loss of cooling power when the CO_2 concentration is raised to 800 ppm, roughly twice what it is today; this change is barely visible in the sides of the large dip.

There are two takeaways from this graph. One is the complexity of the spectra—hundreds of thousands of molecular properties, many measured in the laboratory, go into creating these simulated spectra, which agree

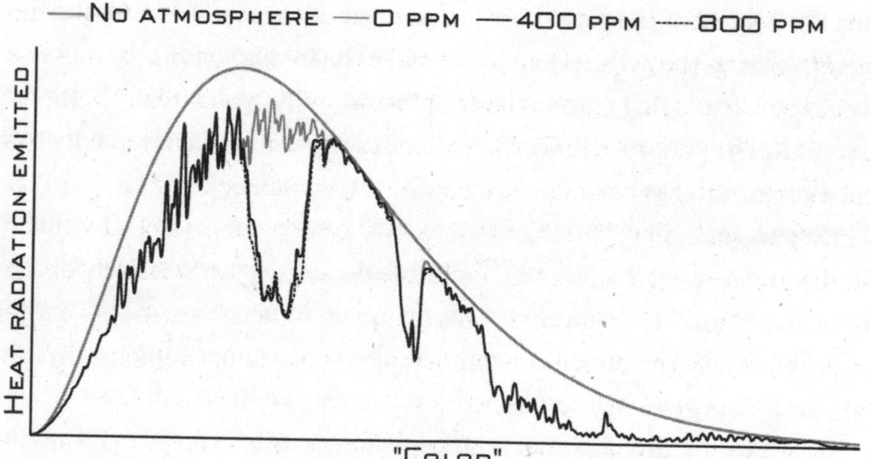

FIGURE 2.3 **The spectrum of heat leaving the top of the atmosphere.** The smooth gray curve corresponds to having no atmosphere, while the spiky gray curve (0 ppm) corresponds to having all of the major greenhouse gases except CO_2 (water vapor, methane, ozone, and nitrous oxide). The solid black and dotted black lines show how the spectum changes when CO_2 is included at concentrations of 400 and 800 ppm, respectively. Where only one curve is visible, all curves coincide.[10]

very well with satellite observations. Second, although the effect of CO_2 at today's concentration is significant (7.6 percent), doubling it doesn't change things much (an additional 0.8 percent) due to the "painting a black window" effect we've already discussed.

But now let's return to that second question from a bit ago: How could a 1 percent change in the heat intercepted possibly matter?

The IPCC's climate models predict that doubling the CO_2 concentration from preindustrial levels—causing the 1 percent change in heat intercepting we've discussed—would increase the average surface temperature by about 3°C (5.5°F). Since we've said the globe's average surface temperature is 15°C (59°F), a rise of 3°C would represent a 20 percent increase in temperature (3°C out of 15°C). But on the Fahrenheit scale, this same temperature change is 5.5°F from the average of 59°F: a 10 percent rise. Why should the rise depend upon which temperature scale we use? And, in any event, either of these numbers, whether 20 percent or 10 percent, seems too large. How could a 1 percent change in the atmosphere's heat interception produce such an outsized effect?

Physicists generally expect changes to be commensurate—a 1 percent change in heat interception should produce something like a 1 percent change in temperature—and when they're not, it's a sign that we're missing a piece of the puzzle.

In this case, that missing piece is the temperature scale. The Stefan-Boltzmann law, which, as you'll recall from the opening of this chapter, describes the relationship between the heat radiated by the earth and its temperature, is framed in terms of absolute temperatures, which are measured on the Kelvin scale. Both the Celsius and Fahrenheit scales are anchored to the properties of water—freezing at 0°C (32°F) and boiling at 100°C (212°F). The Kelvin scale is anchored to absolute zero, the temperature at which matter is so cold it doesn't emit any heat at all (0 K = −273.15°C or −459.67°F). Kelvin degrees are just as large as Celsius degrees (each is 1.8 Fahrenheit degrees), so the earth's average surface temperature of 15°C (59°F) corresponds to about 288 K. A rise of 3°C (or 3 K or 5.5°F) in that surface temperature then corresponds to a warming of 3 K out of

288 K, or about 1 percent, in accord with the 1 percent increase in the atmosphere's ability to intercept heat when the CO_2 concentration is doubled.

So on the scales that matter to us, the climate system is quite sensitive—the few-degree changes in average surface temperature we've seen over the past few centuries (and could see in this one) correspond to physically small (about 1 percent) influences. That sensitivity greatly complicates the task of figuring out how the earth will respond to rising greenhouse gas levels, especially since they're not the only influence at work.

Understanding how the climate system responds to human influences is, unfortunately, a lot like trying to understand the connection between human nutrition and weight loss, a subject famously unsettled to this day. Imagine an experiment where we fed someone an extra half cucumber each day. That would be about an extra twenty calories, a 1 percent increase to the average 2,000-calorie daily adult diet. We'd let that go on for a year and see how much weight they gained. Of course, we would need to know many other things to draw any meaningful conclusions from the results: What else did they eat? How much did they exercise? Were there any changes in health or hormones that affect the rate at which they burn calories? Many things would have to be measured precisely to understand the effect of the additional cucumbers, although we would expect that, all else being equal, the added calories would add some weight.

The problem with human-caused carbon dioxide and the climate is that, as in the cucumber experiment, all else *isn't* necessarily equal, as there are other influences (forcings) on the climate, both human and natural, that can confuse the picture. Among the other human influences on the climate are methane emissions into the atmosphere (from fossil fuels, but more importantly from agriculture) and other minor gases that together exert a warming influence almost as great as that of human-caused CO_2.

Not all human influences are warming. Aerosols are fine particles in the atmosphere such as those produced by the burning of low-quality coal. They cause severe health problems, contributing to millions of deaths per year. But they also make the globe more reflective both by directly reflecting sunlight and by inducing the formation of reflective clouds.

Human-caused aerosols, together with changes in land use like deforestation (pasture is more reflective than forest), increase the albedo and so exert a net cooling influence that cancels about half of the warming influence of human-caused greenhouse gases.

Then there are natural forcings: erupting volcanoes loft aerosols high into the stratosphere, where they remain for several years reflecting a bit more sunlight than usual and so exerting a cooling influence. Such eruptions are unpredictable, but they're sometimes significant enough to negate human influences completely for a few months and therefore have to be taken into account. (For example, the earth was about 0.6°C cooler during the fifteen months that followed the eruption of Mt. Pinatubo in June 1991.[11]) And changes in the sun's intensity of even a fraction of a percent over decades (due to its own internal variability) can change the amount of sunlight reaching the earth, further complicating our attempts to account for all the human and natural forcings affecting the planet's delicate energy balance. But if we're to understand the climate's response to growing CO_2 levels, it is important to know what those other influences are, how big they are, and how and when they come into play.

The energy that flows in and out of the climate system is measured in watts per square meter (W/m^2). The sunlight energy absorbed by the earth (and hence the heat energy radiated by the earth) amounts to an average of 239 W/m^2. Since a 100-watt incandescent light bulb gives off, well, one hundred watts (almost all as heat), this means the planet radiates heat as if there were a bit more than two light bulbs in every square meter (eleven square feet) of its surface. Human influences today amount to just over 2 W/m^2, or slightly less than 1 percent of that natural flow (about the same influence as half a cucumber on the daily human diet).

People are often curious about heat put into the climate system by two other non-sunlight sources. One is the geothermal heat flowing out of the earth's surface. While it can be quite large at localized sources (such as volcanoes, hot springs, and vents on the seafloor), the global average of only

0.09 W/m^2 is too small to have a meaningful direct effect on the climate's energy balance. However, there can be indirect effects, such as the melting of ice by subglacial Antarctic volcanoes.[12]

Another source of heat input into the climate system is the energy that humans derive from fossil fuels and nuclear material. After that energy is used for heating, mobility, and generating electricity, the Second Law of Thermodynamics guarantees that virtually all of it ends up as heat in the climate system, ultimately to be radiated into space along with the earth's natural heat emissions. (A very small fraction winds up as visible light that escapes directly to space through the transparent atmosphere, but even that ultimately winds up as heat somewhere "out there.") That human heat can indeed affect the local climate where this energy use is concentrated (for example in cities and near power plants). But averaged over the globe, it currently amounts to only 0.03 W/m^2, some ten thousand times smaller than the natural heat flows of the climate system, and about one hundred times smaller than the other human influences.

The totality of human and natural influences on the climate is shown in Figure 2.4. It illustrates much of what we've discussed already. We can see the growth of greenhouse gas warming (predominantly from rising CO_2 and methane concentrations, but also other human-emitted greenhouse gases), and that this has been partially offset by growing aerosol cooling. The episodic cooling by large volcanic eruptions is also evident. We can also see that, before 1950, total human influences (the sum of "CO_2," "Other GHG," and "Human cooling") were less than one-fifth of what they are today.

Also shown in Figure 2.4 is how uncertain we are about these various forcings. While the warming effects of CO_2 and other greenhouse gases are known to within 20 percent, the uncertainty in the cooling influence of human-caused aerosols is much larger, making the total human-caused forcing uncertain by 50 percent—that is, the best we can say is that the net human influence today is very likely to be between 1.1 and 3.3 W/m^2.

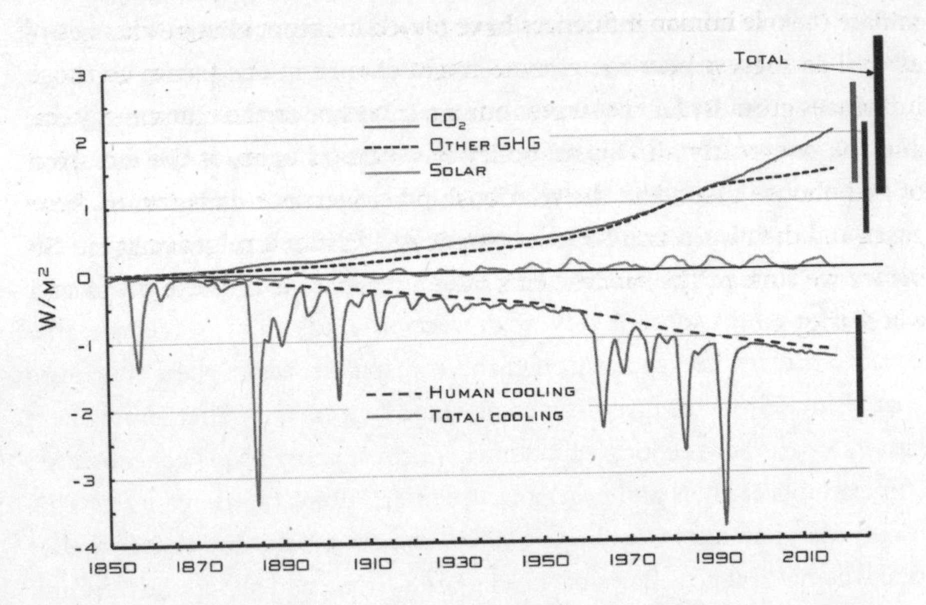

FIGURE 2.4 **Changes in human and natural influences on the climate, 1850–2018.** Human-caused CO_2 and other greenhouse gases (including methane, halocarbons, ozone, and oxides of nitrogen) exert a warming influence, while human-caused aerosols and changes in land albedo exert a cooling one. Episodic natural cooling by large volcanic eruptions and small variations in the sun's intensity complete the picture. The bars at right show 2σ uncertainties in each forcing today, and also for the total of all forcings.[13]

The fact that human influences currently amount to only 1 percent of the energy that flows through the climate system has important implications, and means there's a lot to understand. To usefully measure them and their effects, we have to observe and understand the larger parts of the climate system (the other 99 percent) with a precision better than 1 percent. Small natural influences must also be understood to that same precision, and we've got to be sure they're all accounted for. This is an enormous challenge in a system for which we have limited observations for a limited time, and about which our uncertainties are still large.

Comparing climate models with and without these forcings can illuminate the role human influences have played in recent climate changes—as well as suggest how the climate might change in the future as those influences grow. By far the largest human influence on the climate system, and the one nearly all climate policy has focused upon, is the emission of greenhouse gases. But the relationship between our emission of these gases and their influence is more complicated than you might imagine. So before we turn to the models, let's take a closer look at these gases and where they go.

EMISSIONS EXPLAINED AND EXTRAPOLATED

In 2004, I became BP's chief scientist, working to accelerate renewable energy technologies. I was soon invited to attend a small dinner hosted by Prince Philip at Buckingham Palace. I arrived in black tie at the palace courtyard via London cab; a quick security check and I was ushered into a reception room along with other guests. After predinner drinks and small talk, we, a group of about fourteen that included Prince Philip, Princess Anne, BP CEO John Browne, and other notables from UK academia, business, and government, moved into a grander room and settled around a large dining table.

The chitchat among tablemates quieted as Prince Philip made his welcome and reminded us that the topic of the evening was climate and energy. He then opened the conversation by asking the group a question about the relationship between carbon dioxide emissions and rising global temperatures. The prince's framing was sufficiently technical that there was awkward silence around the table—until yours truly, the cheeky American scientist, spoke up in a Brooklyn accent to deliver a mini-lecture on infrared-active molecules, the "black window" effect, and the connection between atmospheric concentration and emissions. I earned an appreciative nod from the Duke of Edinburgh, whom I found to be quite knowledgeable.

I suspect that the Duke already knew the answer to his conversation-starting question when he asked it. In any event, the lively discussion that followed over a fine dinner mirrored many others that I've been involved in—finding non-experts eager to understand the complex and nuanced subjects of climate and energy, as well as a confusion about the nature and scale of the problems facing us.

The most significant human-caused greenhouse gases influencing the climate are carbon dioxide (CO_2) and methane (CH_4). Their concentrations in the atmosphere are increasing because we're emitting them; that's why efforts to reduce human influences on the climate focus on reducing emissions. But here's the key point: the connection between concentrations and emissions isn't a simple one and, particularly for CO_2, the complications of this relationship profoundly increase the challenge of reducing the concentration.

This chapter is about movement—primarily the movement of carbon. Human-emitted CO_2 is a relatively small add-on to a vast natural cycle of carbon moving among the earth's crust, oceans, plants, and atmosphere. As you will see, our addition to that cycle will increase for decades under any scenario. But despite the precision claimed by climate models, the impact of this on the climate is highly uncertain.

Charles David Keeling, a geochemist at the Scripps Institution of Oceanography in La Jolla, California, began precision measurements of CO_2 concentration in the 1950s while he was a postdoctoral researcher at Caltech. One of his early surprises was finding a 1 percent increase in the concentration between 1957 and 1959. That finding motivated a more concerted, longer-term program to monitor CO_2 concentrations, which later expanded to include other gases in the atmosphere. I had the pleasure of talking with Dr. Keeling ("Dave," as he was known) some forty years later during one of my JASON summers in La Jolla. I found him to be a quiet, thoughtful, precise man, focused on doing what he was doing very well, knowing it was important to the world.

Figure 3.1 shows what is now called the "Keeling curve," the concentration of carbon dioxide in the atmosphere as measured monthly at Mauna Loa, Hawaii. Data from that remote island location, far from significant localized sources that might skew the observations, is a good measure of the globe's "background" concentration. The measurements show a steady rise from 310 ppm in 1960 to 410 ppm in 2019; during the past decade, the concentration has risen about 2.3 ppm each year. But superimposed on that decades-long trend is an annual cycle—the concentration moves up and down seasonally by 2.4 ppm, as shown in the inset. Both the trend and the

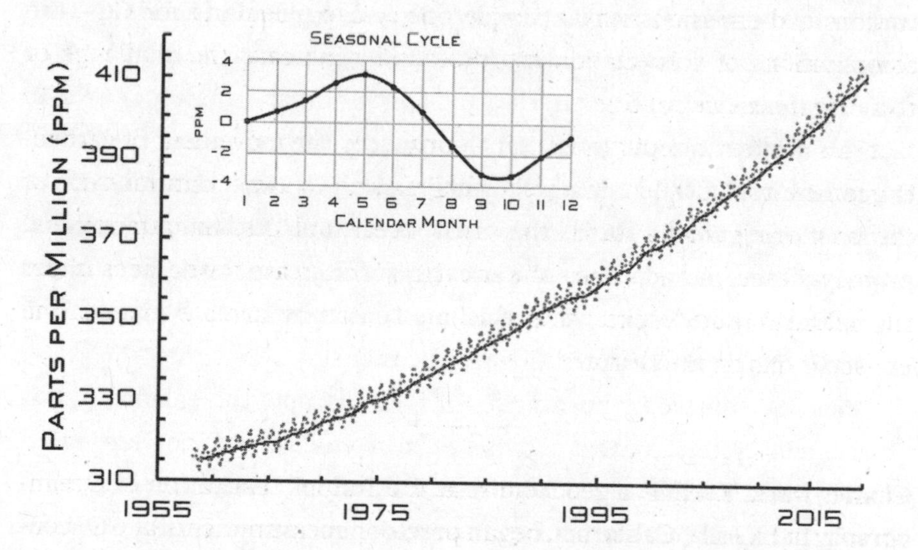

FIGURE 3.1 **Monthly average concentration of carbon dioxide as measured in Mauna Loa, Hawaii, from 1958 through 2020.** The inset shows the average seasonal variation.[1]

cycle, which are measured at a number of locations around the world, help tell a story.

Let's start at the beginning. The earth formed 4.5 billion years ago with a fixed endowment of carbon. Today, that carbon is found in several different circumstances around the planet—what are called "reservoirs." The largest reservoir by far is the earth's crust, which contains almost all of the planet's carbon, about 1.9 billion gigatons (1 gigaton, abbreviated Gt, is one billion tons).[2] The next largest amount, about 40,000 Gt, is in the oceans, almost all of that far below the surface. There are about 2,100 Gt more stored on land in soils and living things, and 5,000–10,000 Gt in fossil fuels underground. The roughly 850 Gt of carbon in the atmosphere, almost all in the form of carbon dioxide, is equal to about 25 percent of the carbon at or near the earth's surface (in the soils, plants, and shallow ocean) but is only 2 percent of the total carbon in the oceans.[3]

Powerful natural processes move the earth's carbon among those reservoirs, often by changing its chemical form. The most important of these processes is the seasonal flow of about one-quarter of the atmosphere's carbon to the surface as plants grow—they use photosynthesis to turn atmospheric CO_2 into organic matter—then return that carbon back to the atmosphere through respiration and as their organic matter decays. In fact, plant growth in the Northern Hemisphere is what causes the February to July rise in the Mauna Loa background CO_2 concentration seen in the inset of Figure 3.1; this is the earth "breathing." Other, much slower, processes move carbon from the oceans' surfaces into their depths and then ultimately into rock, such as the limestone and marble that form from the shells of marine creatures.

The CO_2 emitted by burning fossil fuels disrupts the balance of this great annual cycle, since that carbon has been pulled out of the deep underground, where it was isolated from these natural processes. The amount of carbon that fossil fuel use adds to the cycle is currently about 4.5 percent of what flows each year. About half of that increase is taken up annually by the surface (the rising CO_2 has increased vegetation over much of the planet), and the remainder stays in the atmosphere, increasing its CO_2 concentration. The situation is not unlike what we've already seen in the planet's energy flow—small but steady human influences gradually adding to a much, much larger natural process.

Global emissions of all greenhouse gases are rising rapidly, as can be seen in Figure 3.2. For the past fifty years, they've gone up 1.3 percent per year, though a bit more slowly (1.1 percent per year) during the decade ending 2018. If the longer-term trend were to continue, emissions in 2075 would be twice as large as they are today. Nearly all of the increase in emissions has been due to CO_2 from fossil fuel use (changes in land use such as deforestation emit far lesser amounts of carbon stored in plants and soils). The next largest contributor after CO_2 is methane (CH_4), while nitrous oxide (N_2O) and the fluorinated gases (F-gases such as HFCs) make up much smaller portions of the total.

I don't know of any expert who disputes that the rise in CO_2 concentration over the past 150 years is almost entirely due to human activities, since there are five independent lines of evidence supporting that conclusion.

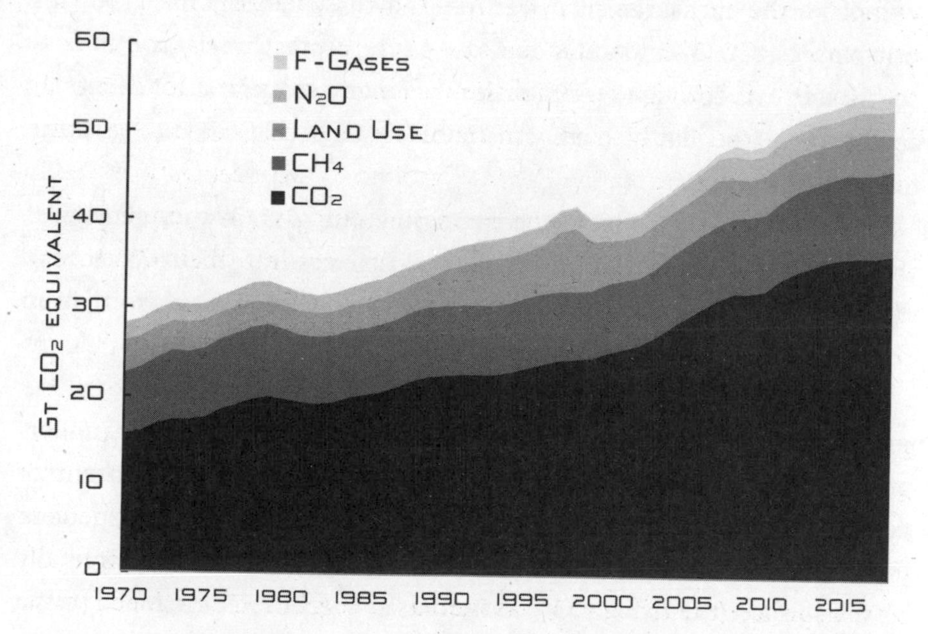

Figure 3.2 **Annual global greenhouse gas emissions from 1970 to 2018.** Emissions of non-CO_2 gases are expressed as CO_2-equivalent amounts.[4]

One is the timing of the rise—concentrations in air samples over the past ten thousand years varied between 260 and 280 ppm before a sharp uptick began in the mid-nineteenth century. The second is that the size of the rise is in the ballpark of what we'd expect from the CO_2 emitted by burning fossil fuels. A third is that the rise in the Northern Hemisphere leads that in the South by about two years—most fossil fuels are burned in the north, which has more land and people—and that lead is increasing as emissions grow.

A fourth, more subtle confirmation comes from carbon isotopes—the relatively rare carbon atoms that are about 8 percent heavier than ordinary carbon atoms. About 1.1 percent of the earth's carbon is the isotope ^{13}C; the rest is the lighter isotope ^{12}C. But the proportion of ^{12}C and ^{13}C isn't the same in all forms of carbon. In particular, the chemical reactions of life have a very slight preference for ^{12}C, so that the carbon in living things (as opposed to mineral carbon in the earth's crust) is "light"; that is, it has a slightly lesser proportion of ^{13}C. Since the carbon in the atmosphere's CO_2 has become progressively "lighter" over the decades, we can infer that it arises from the burning of fossil fuels, which, after all, were once living things. Finally, measurements over the past three decades have shown a tiny but detectable and steady decrease in the atmosphere's oxygen concentration. The decrease is too small to raise any concern at all about our ability to breathe, but it roughly matches what's been needed to turn the fossil carbon into CO_2.

As is usual in climate science, zooming out to look over geological times gives us a quite different perspective. The natural processes moving the earth's carbon around were different in the past, so much so that by geological standards, today's Earth is starved for atmospheric CO_2. Figure 3.3 shows estimates of past CO_2 concentrations. The horizontal scale is geological time extending back to the Cambrian period, about 550 million years ago. The vertical axis is the ratio of past atmospheric CO_2 concentrations to the average concentration over the past few million years (about 300 ppm). This particular proxy record comes from analyzing the fraction of ^{13}C relative to ^{12}C in carbonate sediments and paleosols (fossilized soils). Other proxies give qualitatively similar results.

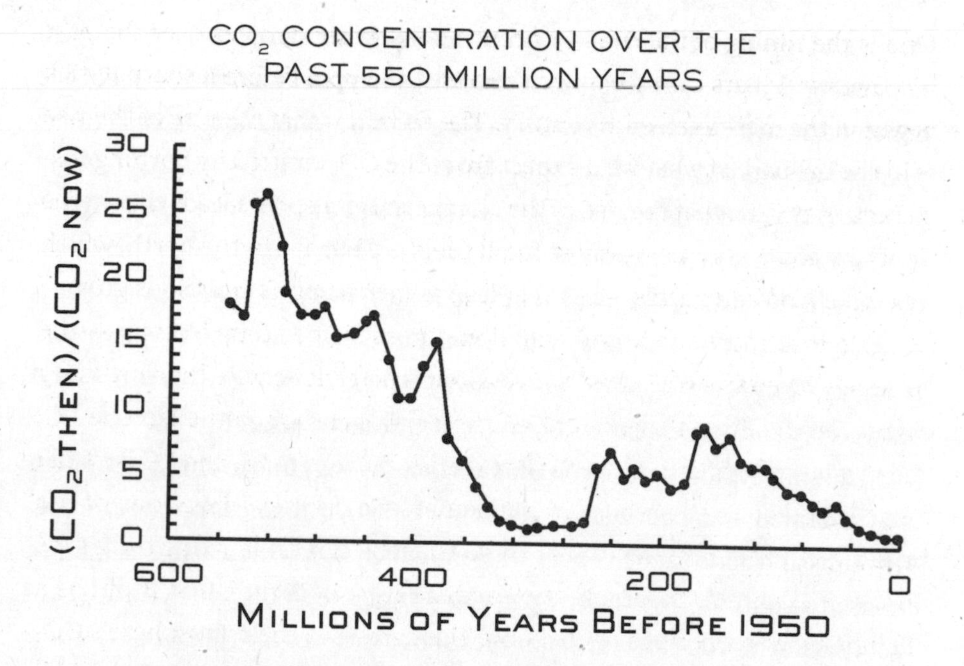

FIGURE 3.3 **Atmospheric concentration of carbon dioxide beginning at 550 million years ago.** Values determined from the isotopic ratio in carbonate sediments and fossilized soils are relative to the average over the past few million years; today's concentration would be about 1.3 on this scale, down in the lower right-hand corner.[5]

Only once in the geological past—the Permian period, 300 million years ago—have atmospheric CO_2 levels been as low as they are today. Plant and animal life flourished abundantly during times when CO_2 levels were five or ten times higher than today's. But those were different plants and animals. So while carbon dioxide, in and of itself, is not particularly a concern for the planet, what *is* a concern is that, because life today has evolved to be well-suited to a low level of CO_2 (anatomically modern humans appeared only some 200,000 years ago, at the extreme right of this chart), the rapid increases of the past century might prove disruptive. Concentrations up to 1,000 ppm (2.5 times that in open air today) are common in classrooms or auditoriums. Humans start to feel drowsy above that level, so when students start to nod off in my classroom, I like to believe it's that 1,000 ppm, not the quality of my lecture. More serious physiological effects begin above 2,000 ppm. However, if the trends of the past decade continue, it will be some 250 years before the concentration reaches 1,000 ppm, which would be at 3.3 on this chart.[6]

Carbon dioxide is the single human-caused greenhouse gas with the largest influence on the climate. But it is of greatest concern also because it persists in the atmosphere/surface cycle for a very long time. About 60 percent of any CO_2 emitted today will remain in the atmosphere twenty years from now, between 30 and 55 percent will still be there after a century, and between 15 and 30 percent will remain after one thousand years.[7]

The simple fact that carbon dioxide lasts a long time in the atmosphere is a fundamental impediment to reducing human influences on the climate. Any emission adds to the concentration, which keeps increasing as long as emissions continue. In other words, CO_2 is not like smog, which disappears a few days after you stop emissions; it takes centuries for the excess carbon dioxide to vanish from the atmosphere. So modest reductions in CO_2 emissions would only slow the increase in concentration but not prevent it. Just to stabilize the CO_2 concentration, and hence its warming influence, *global emissions would have to vanish.*

Methane, the second most important human-caused greenhouse gas, has also been increasing over the past century and so also exerts a growing warming influence on the climate. Like CO_2, methane concentrations (seen in Figure 3.4) show a long-term rising trend and an annual cycle. The cause of the plateau between 1998 and 2008 is yet another uncertainty in climate science. And as with CO_2, today's methane concentrations are dramatically higher than those of the past few million years, beginning a sharp rise about four thousand years ago.[8]

But there are several important differences between methane and carbon dioxide. One is that methane concentrations are much lower (2,000 parts per *billion*, which is about 1/200th that of CO_2's 400 parts per million). Another difference is that a methane molecule lasts in the atmosphere for only about twelve years—though after that, chemical reactions convert it to CO_2. And a third difference is that, because of the peculiarities

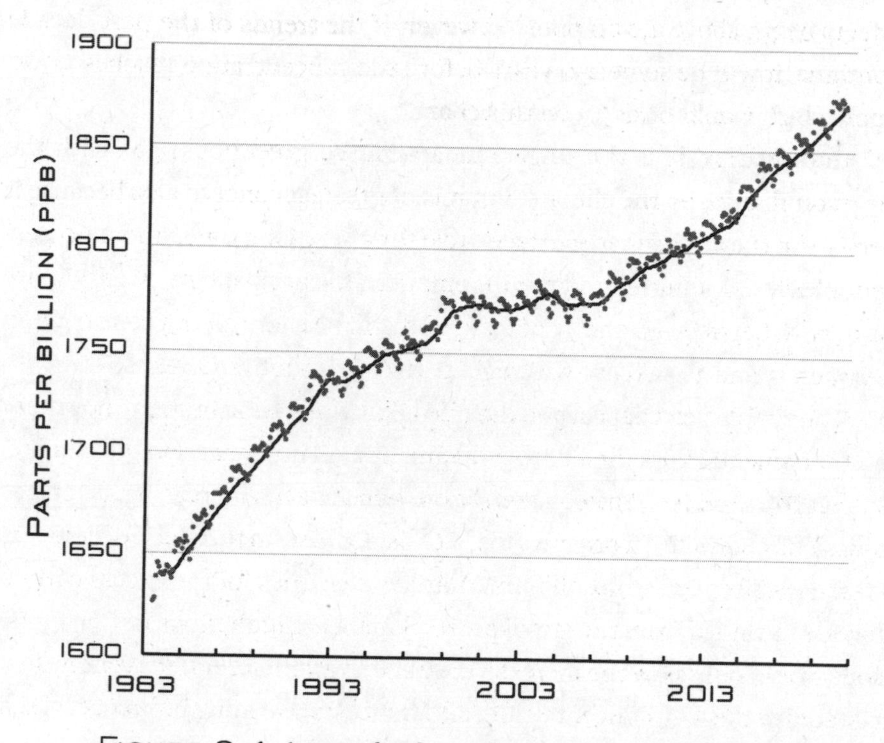

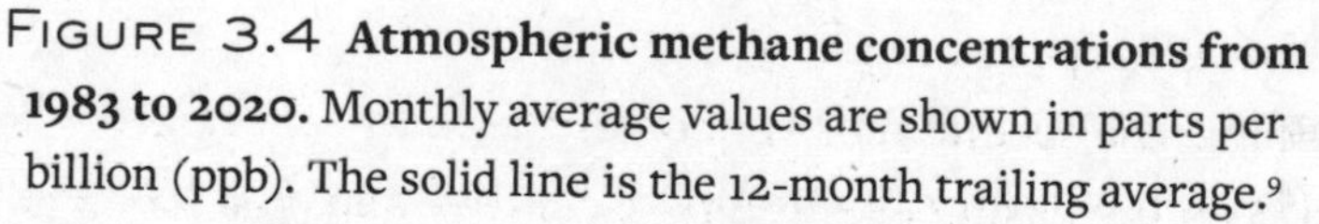
FIGURE 3.4 **Atmospheric methane concentrations from 1983 to 2020.** Monthly average values are shown in parts per billion (ppb). The solid line is the 12-month trailing average.[9]

of how molecules interact with the different colors of infrared radiation, every additional methane molecule in the atmosphere is thirty times more potent in warming than a molecule of carbon dioxide. These differences—lower concentration and shorter lifetime, but greater warming potency—must be taken into account when comparing CH_4 and CO_2 emissions. For instance, the 300 million tons of methane humans emit each year is only 0.8 percent of the 36 gigatons of CO_2 emitted by burning fossil fuels. But as shown in Figure 3.2, that methane has a disproportionate warming influence, equivalent to ten gigatons of CO_2.

One additional point about methane that surprises many people is that fossil fuels account for only about one quarter of global human-caused

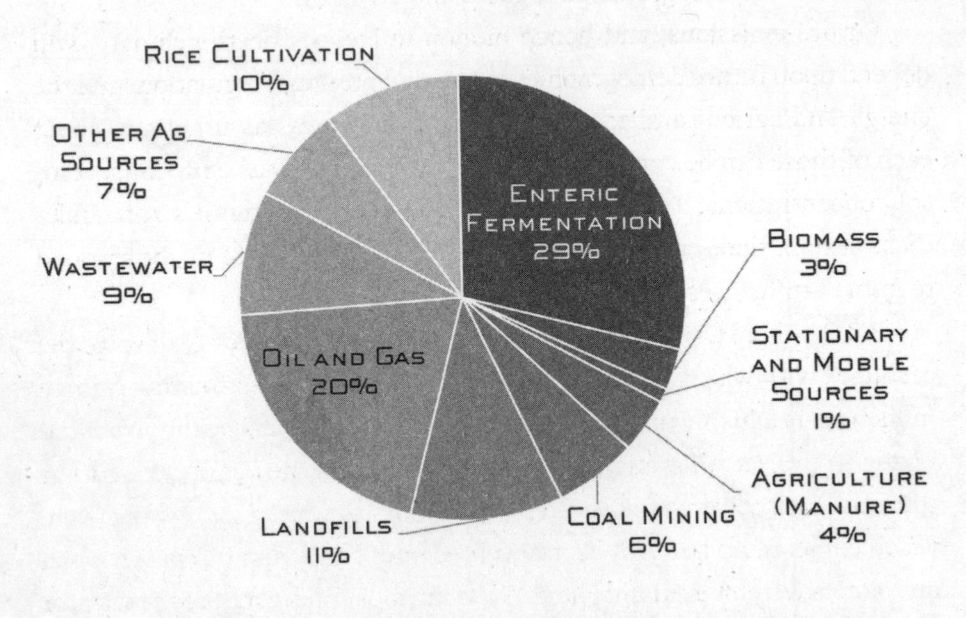

FIGURE 3.5 **Sources of global methane emissions due to human activities in 2010.**[10]

methane emissions, as shown in Figure 3.5. Rather, most methane emissions arise from enteric fermentation (digestion in cattle—mostly emitted from the front of the animal, not the back) and other agricultural activities, particularly rice cultivation; the decay of material in landfills is also significant. So any effort to drastically reduce emissions must also address those sources.

Future climates will be determined by the climate's response to both human and natural influences, as well as by its internal variability—as we've seen, the climate is quite capable of varying without any help from us. While we've little detailed knowledge, much less control, over internal variability or natural influences (volcanoes, the sun, and deep ocean currents have minds of their own, just like the climate), we can make plausible assumptions about the range of what humans will do, particularly in regard to their emissions of greenhouse gases and aerosols.

Future emissions, and hence human influences on the climate, will depend upon future demographics, economic progress, regulation, and the energy and agricultural technologies in use. Various assumptions about each of those can be combined to project greenhouse gas emissions, aerosol concentrations, and changes in land use. Climate models run under these assumptions can give some sense of how the climate might respond to human influences in the coming decades.

But it is here that the warning light should start blinking. Despite the certainty with which projections are reported as facts, estimating human influences is a highly uncertain business. Imagine being back in 1900 and trying to project what civilization would be like in the year 2000. At the time, the first powered flight and the first mass-produced automobile were yet to come, radio had just been invented and X-rays just discovered, and antibiotics weren't even imagined. Even the most prescient prognosticator back then would have missed most of what transpired in the subsequent century as the global population quadrupled and the global economy grew by a factor of forty! They'd be amazed at the scale and rapidity with which people, goods, and information now move around the globe, at how we manufacture, and at our advances in agriculture and medicine.

Because of the great uncertainties about the decades to come, instead of making precise predictions of future concentrations, the IPCC created a set of scenarios. They have the rather complicated name of "Representative Concentration Pathways," or RCPs. These are meant to span a plausible range of possibilities for population, economy, technology, and so on.[11] Each RCP has a number indicating the amount of warming human influence expected in 2100 under that scenario, so that RCP6 corresponds to 6 W/m^2 of human-induced radiative forcing (warming) at the end of the century. (Remember, net human influences are currently about 2.2 W/m^2 of warming, as shown in Figure 2.4.) These scenarios are not meant to be predictions, but rather are schematic descriptions of distinct, but plausible, future worlds. The RCPs have since been elaborated into "Shared

Socioeconomic Pathways" that also describe society's ability to reduce emissions and adapt to a changing climate, but the important takeaways can be understood by discussing the simpler RCPs.[12]

Historically, two of the most important drivers of emissions have been growing population and growing economic activity. Figure 3.6 shows the assumptions made about these factors in four RCP scenarios. In the low-emissions RCP2.6 scenario, which has 2.6 W/m^2 radiative forcing at century's end, global population grows from today's 7.8 billion to peak at 9 billion in 2070 and then declines by a few hundred million by 2100. At the other extreme, RCP8.5 assumes that population grows steadily to more than 12 billion in 2100. Global real GDP is assumed to grow strongly through the twenty-first century in all scenarios, by a factor of six in the higher-emitting scenarios, but by a factor of ten in the lower-emitting scenarios, presumably because a more prosperous world is able to place a higher priority on environmental matters. Since GDP grows by a larger multiple than population in all scenarios, the world in 2100 is projected to be more prosperous on a per capita (that is, per person) basis in *any* future (a detail often omitted from discussions of model results).

The corresponding RCP assumptions about CO_2 emissions, CO_2 concentration, and total human-caused radiative forcing are shown in Figure 3.7. (The last is the net effect of all human-caused greenhouse gases and aerosols.) The diversity of emissions assumptions among the scenarios leads to a similar diversity of human influences on the climate at century's end. As expected, lower emissions lead to lower concentrations and hence weaker human influence on the climate (forcing). The populous coal-heavy world of RCP8.5 has annual CO_2 emissions more than tripling by century's end, the concentration soaring to more than 900 ppm, and the radiative forcing more than triple what it is today. In contrast, the prosperous low-population world of RCP2.6 has CO_2 emissions vanishing by 2080, so that both the concentration and the forcing stabilize at today's values, declining very slowly afterward.

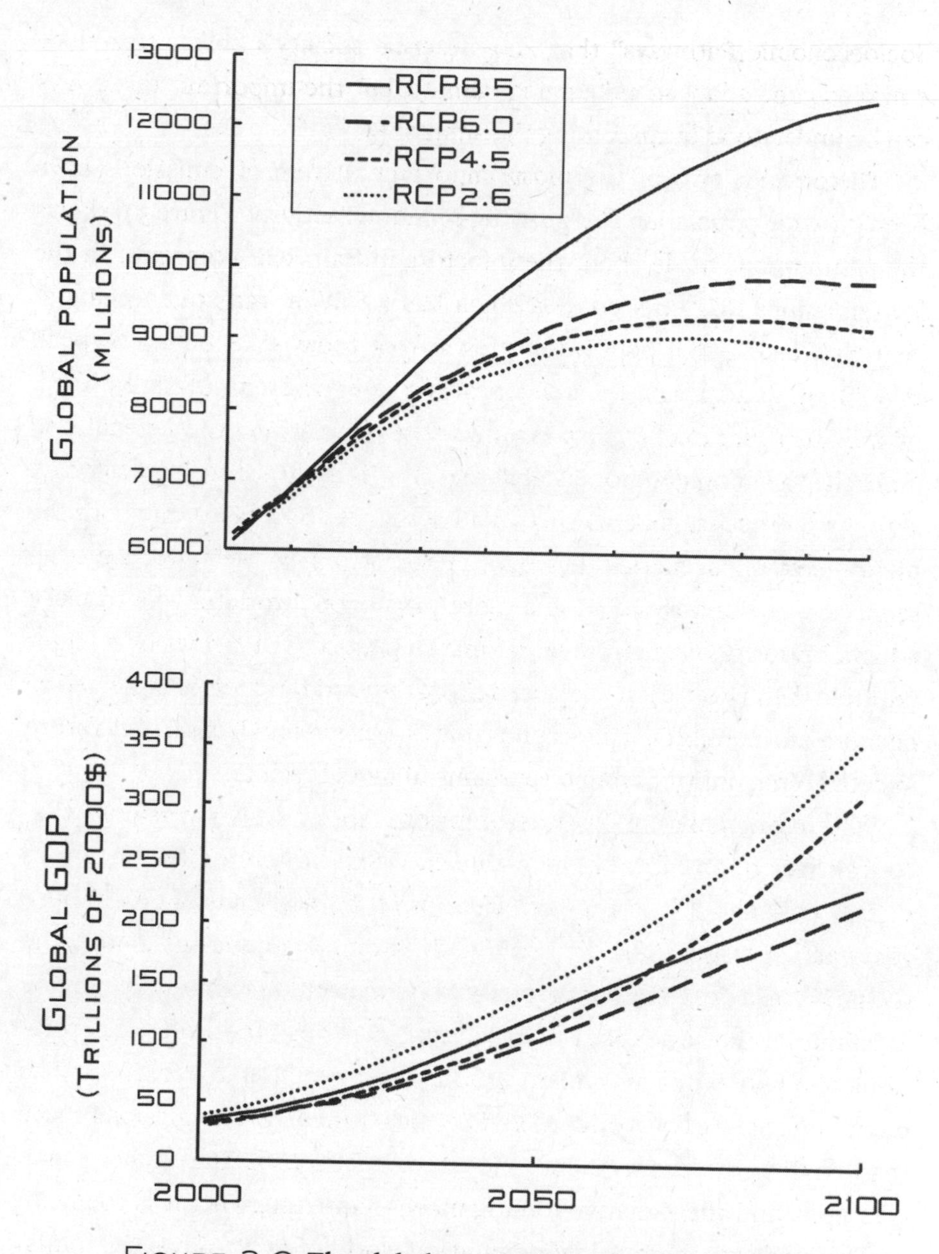

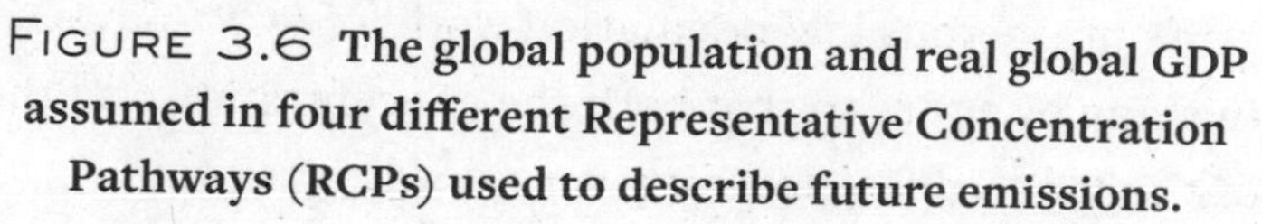
FIGURE 3.6 **The global population and real global GDP assumed in four different Representative Concentration Pathways (RCPs) used to describe future emissions.**

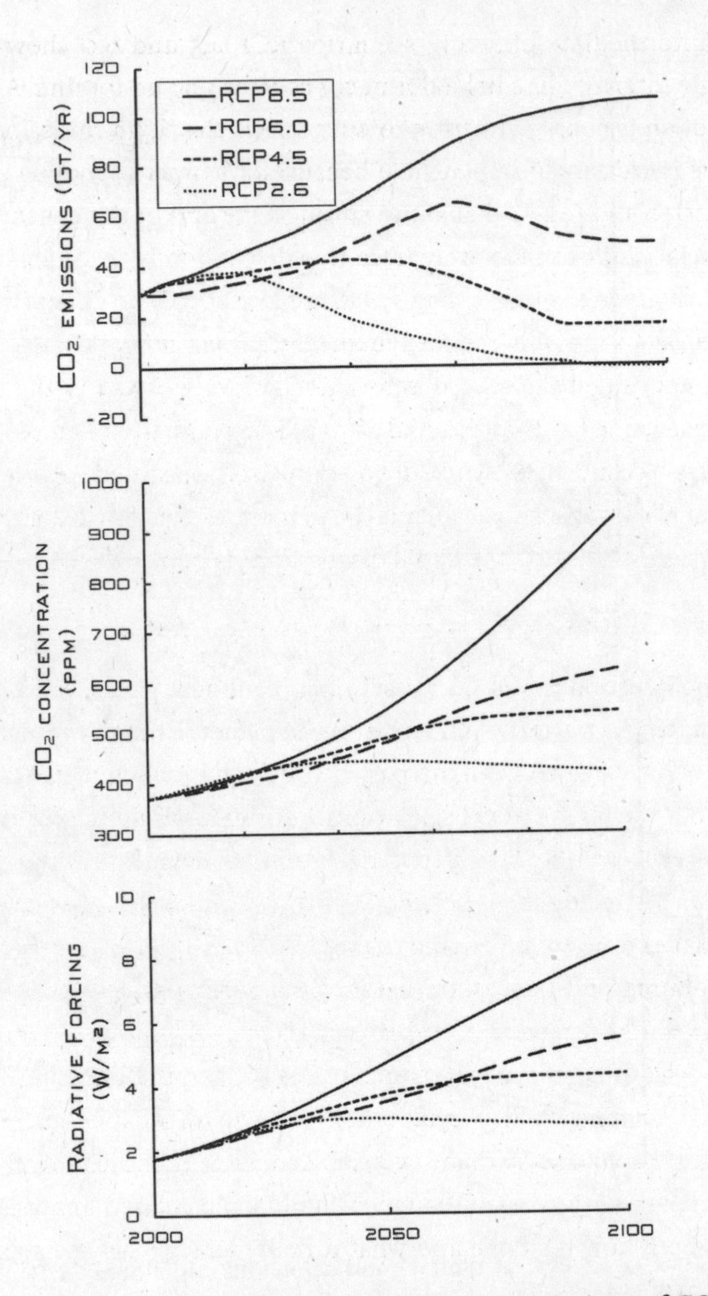

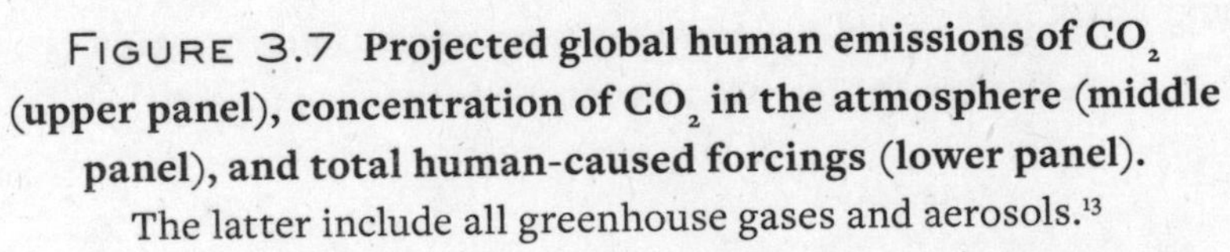
FIGURE 3.7 **Projected global human emissions of CO_2 (upper panel), concentration of CO_2 in the atmosphere (middle panel), and total human-caused forcings (lower panel).** The latter include all greenhouse gases and aerosols.[13]

The intermediate-emitting scenarios RCP 4.5 and 6.0 show correspondingly intermediate behavior in concentration and forcing. A recent analysis of emissions from 2005 to 2017 shows that high-emissions scenarios are increasingly implausible because of slower economic growth through 2040 and reduced coal use through the end of the century.[14]

A crucial point to take away from this discussion is that *human influences will continue to grow in any scenario short of ceasing all emissions.* If future emissions are only reduced modestly, human influences on the climate will continue to grow. Fifteen years ago, when I was in the private sector, I learned to say that the goal of stabilizing human influences on the climate was "a challenge," while in government it was talked about as "an opportunity." Now back in academia, I can more forthrightly call it what it is: "a practical impossibility," as I'll discuss in this book's Part II.

There is no question that our emission of greenhouse gases, in particular CO_2, is exerting a warming influence on the planet. Human influences on the climate have grown over the past decades and will continue to grow under all but the most radical scenarios for future emissions. Not only are human influences difficult to disentangle from other aspects of the climate system, but the relationship between emissions and atmospheric concentrations makes it very challenging to moderate our influences.

When human influences are fed into a model to project a future climate, the results are, at some level, unremarkable—higher greenhouse gas emissions lead to higher global temperatures sooner. But knowing exactly how much warming would occur, when and where, what other changes there might be in the climate system, and how those changes might actually impact society requires much more sophisticated analysis. This analysis—both how it's done and what it really tells us—is the subject of the next chapter.

MANY MUDDLED MODELS

You've likely heard about "climate models" that predict this or that—and no doubt that such models prove this or that as well. But exactly what *are* climate models? The short answer is that they are computer programs that perform mathematical simulations of the climate system. As University of Wisconsin statistician George Box said famously in 1978: "All models are wrong, but some are useful."

I have been involved with scientific computing for my entire career. I learned to code more than fifty years ago on an IBM 1620 as a high school sophomore thanks to a Columbia University science enrichment program. One of my first published papers, in 1974, involved computer modeling of the nuclear reaction in stars that produces the oxygen in the universe.[1] In 1981, an IBM representative unexpectedly showed up in my office. He had a gift for me—one of the first personal computers—and he had only one request: that I "do something interesting" with it. I eventually developed a course at Caltech on computational physics—that is, computer modeling—and wrote one of the first textbooks on the subject.[2] Even now, almost four decades later, I'm pleased when a researcher tells me how much they learned from the book about how to translate pencil-and-paper physics into useful simulations.

Some of my most widely cited research involved developing and making use of novel algorithms for simulating quantum mechanical systems like the electrons in an atom or the protons and neutrons in a nucleus. And for the past three decades I have helped guide simulations that give the US confidence in its stockpile of nuclear weapons, even in the absence of exploding nuclear weapons to test them, which is now prohibited by international treaties.

This long and varied experience has given me an appreciation for the power of computer models—but also a respect for their limitations. Professor Box's aphorism hits the nail on the head.

Computer modeling is central to climate science. The models help us understand how the climate system works, why it has changed in the past, and—most importantly—how it might change in the future. The most recent United Nations assessment report on climate science, AR5 WGI ("WGI," you will remember, standing for "Working Group I"), devotes four of its fourteen chapters entirely to models and their results; those results underpin the reports of the other UN working groups that assess the impacts of a changing climate on ecosystems and society.

I first learned about the details of climate modeling almost thirty years ago in the course of a JASON study of how the then-new massively parallel computers, which coordinate thousands of processors working on a single problem, might improve the predictive capabilities of climate models.[3] Much of that promise has indeed been realized in the three decades since the study. But there's a big caveat—usefully describing the earth's climate remains one of the most challenging scientific simulation problems there is.

So . . . how good are our climate models? And how much confidence should we have in what they say about future climates? To answer those questions, we have to dig a bit deeper into the details.

COMPUTING CLIMATE

Scientific computers are machines for doing arithmetic—they can store many, many numbers (today's largest machines are approaching 10^{17}, or 100 million billion) and can manipulate them at blinding speed (today some 10^{18}, or a billion billion, operations per second). Since we have a very solid understanding of the physical laws that govern matter and energy, it's easy to be seduced by the notion that we can just feed the present state of the atmosphere and oceans into a computer, make some assumptions about future human and natural influences, and so accurately predict the climate decades into the future.

Unfortunately, that's just a fantasy, as you might infer from weather forecasts, which can be accurate only out to two weeks or so. That *is* better than they were thirty years ago, largely due to more computer power, as well as to improved observations of the atmosphere that provide a more accurate starting point for the models.[4] But the scant two-week-long accuracy of weather forecasts reflects a fundamental problem described by Ed Lorenz at MIT in 1961. The weather is chaotic—small changes in how we start the model can lead to very different predictions after a few weeks. So no matter how precisely we might specify current conditions, the uncertainty in our predictions grows exponentially as they extend into the future. More computer power cannot overcome this basic uncertainty.

But remember: *Climate is not weather.* Rather, it's the average of weather over decades, and that is what climate models try to describe. There's reason to believe that's possible. After all, while we can't predict with much detail how individual bubbles will arise in a boiling pot of water, we *can* confidently predict how the average level of the water will decline as a result of all that boiling. Of course the climate system is a lot more complicated than a boiling pot of water, and a host of vexing practical problems means that climate model results require at least a pinch, if not a pound, of salt.

So let's talk about what climate models and climate modelers actually do.

All but the simplest computer models of the climate begin by covering the earth's atmosphere with a three-dimensional grid, typically ten to twenty layers of grid boxes stacked above a surface grid of squares that are typically 100 km × 100 km (60 miles × 60 miles), as shown in Figure 4.1. But because the height of the atmosphere that needs to be modeled is comparable to the size of one surface grid square, the layered boxes atop the grid are much more like pancakes than the cubes shown in the figure (more about that shortly). The grid covering the oceans is similar, but with smaller surface grid squares, typically 10 km × 10 km (6 miles × 6 miles), and more vertical layers (up to thirty). With the entire Earth covered this way, there are about one million grid boxes for the atmosphere and one hundred million grid boxes for the ocean.

Grid in place, the computer models use the fundamental laws of physics to calculate how the air, water, and energy in each box at a given time move to neighboring grid boxes at a slightly later time; this time step can be as small as ten minutes. Repeating this process millions of times over simulates the climate for a century (just over five million times if the time step

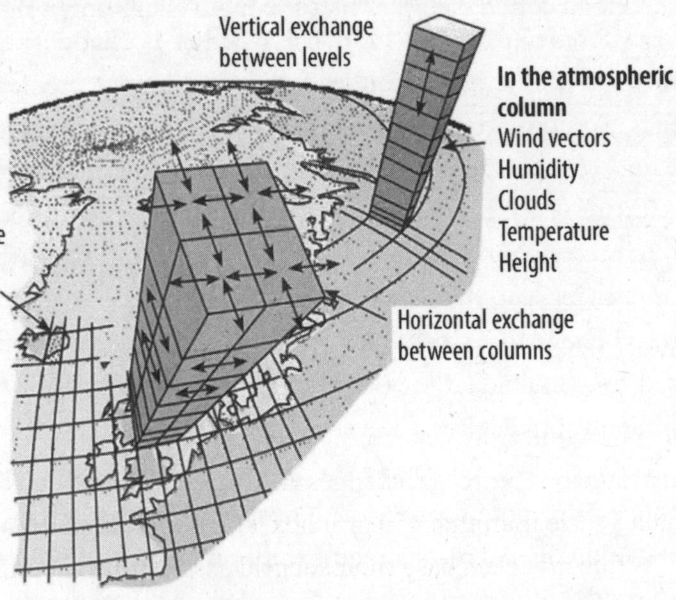

FIGURE 4.1 **Schematic of the grid used in computer models of the atmosphere.**[5]

is ten minutes). These many time steps in a simulation can take months of computer time on even the world's most powerful supercomputers; the amount depends upon the number of grid boxes and how many time steps are taken, as well as the sophistication of the model's description of what goes on in the grid boxes (its "physics"). Researchers can make trade-offs among these different factors depending upon the purpose of the model. For the same amount of computer time, a less sophisticated model can be run with a finer grid and/or to simulate a longer period. Comparing the results of computer runs with what we know about past climates (both the average and year-by-year variations) gives some sense of how good a model is. Once the model is in place, a series of computer runs into the future under assumed human and natural influences then attempt to describe the climate decades hence.

This all sounds straightforward, but it's not at all easy. In fact, it's excruciatingly difficult, and anyone who says that climate models are "just physics" either doesn't understand them or is being deliberately misleading. One major challenge is that the models use only single values of temperature, humidity, and so on to describe conditions within a grid box. Yet many important phenomena occur on scales smaller than the 100 km (60 mile) grid size (such as mountains, clouds, and thunderstorms) and so researchers must make "subgrid" assumptions to build a complete model. For example, flows of sunlight and heat through the atmosphere are influenced by clouds. They play a key role—depending upon their type and formation, clouds will reflect sunlight or intercept heat in varying amounts. Physics tells us that the numbers and types of clouds present in each of the layers of atmosphere above a grid square (the stacked boxes) will generally depend upon conditions there (humidity, temperature, and so on). Yet as illustrated by Figure 4.2, changes and differences in clouds occur on a much smaller scale than that of a grid box, and so assumptions are necessary.

While modelers base their subgrid assumptions upon both fundamental physical laws and observations of weather phenomena, there is still considerable judgment involved. And since different modelers will make different assumptions, results can vary widely among models. This is not

FIGURE 4.2 **Clouds are much smaller than model grid boxes and so require that modelers make subgrid assumptions.** Note that this figure is misleading in that the actual grid boxes are much thinner than what's shown.

at all an unimportant detail, since ordinary fluctuations in the height and coverage of clouds can have as much of an impact on flows of sunlight and heat as do human influences. In fact, the greatest uncertainty in climate modeling stems from the treatment of clouds.[6]

So why not use a finer grid to make the subgrid assumptions less ambiguous? Unfortunately, that would dramatically increase the size of the computation, most obviously because there would be more grid boxes to deal with. But aside from the number of boxes, a finer grid introduces another complication: Any computation is only accurate if things don't change too much over one time step (that is, don't move more than one grid box). So if the grid is finer, the time step has to be smaller as well, meaning even more computer time will be required. As an illustration, a simulation that takes two months to run with 100 km grid squares would take more than a century if it instead used 10 km squares. The run time would remain at two months if we had a supercomputer one thousand times faster than today's—a capability probably two or three decades in the future.

Another issue arises because of the difference in the way our grid divides the earth into meaningful chunks horizontally versus vertically. The atmosphere and ocean are both thin shells covering the earth's surface—the average ocean depth (4 km or 2.5 miles) is very small compared to the earth's radius (6,400 km or 4,000 miles), as is the height of the relevant atmosphere (about 100 km or 60 miles). To accurately describe vertical variations, the several dozen grid boxes stacked up into the atmosphere above each square (or below it into the ocean) have to be more like very flat pancakes than like cubes, typically some one hundred times wider than they are high. For comparison, a dime is only thirteen times wider than it is thick.

Pancake boxes generally make for a more accurate simulation where the atmosphere is flowing in layers, for example in the upper parts of the stack (there's a reason the high-altitude atmosphere is called the stratosphere). But these flat boxes become a problem in the atmosphere below 10 km (6 miles), where turbulent weather happens. Upward flows of energy and water vapor (think thunderhead clouds) occur over areas much smaller than the 100 km (60 miles) of our grid. This is particularly troublesome in the tropics, where upward flows are important in lofting energy and water vapor from the ocean surface into the atmosphere. In fact, the flow of energy carried into the atmosphere by evaporation of the ocean waters is more than thirty times larger than the human influences shown back in Figure 2.4. So subgrid assumptions about this "moist convection"—how air and water vapor move vertically through the flat grid boxes—are crucial to building accurate models.

Any simulation also needs to be "initialized"—that is, we need to somehow specify the state of the ocean and atmosphere at the start of the time stepping: the temperature, humidity, winds, etc., in every one of the grid boxes covering the atmosphere, as well as the temperature, salinity, current, etc., in every ocean grid box. Unfortunately, even with our sophisticated observation systems, that kind of detail isn't available today, let alone for decades in the past. And even if it were, the level of chaos in the simulation (remember our discussion of weather prediction) would render most of the details irrelevant after two weeks or so. So the initialization needs to correctly capture only the gross features of the climate system (such as the atmosphere's jet stream or the major ocean currents).

Even with the grid, the basic physics, the subgrid assumptions, and initialization in hand, we're still not ready to generate a useful climate simulation. The last remaining step is to "tune" the model. Each subgrid assumption has numerical parameters that have to be set—somehow. Cloud cover and convection are only two examples out of dozens. *How much water evaporates from the land surface depending upon the soil properties, plant cover, and atmospheric conditions? How much snow or ice is on the surface? How do ocean waters mix?*

Subgrid assumptions are inexact by nature because they are, well, subgrid—there isn't a "number" modelers can take from reality. So modelers set subgrid parameters based on what they know about the physics and then run their model. Since the results generally don't much look like the climate system we observe, modelers then adjust ("tune") these parameters to get a better match with some features of the real climate system. Most important are the near exact balance between solar heating and infrared cooling we discussed in Chapter 2, and what the surface temperature is, determined by how sunlight and heat flow through the atmosphere.

Although "tuning" sounds like a minor detail, as in "fine-tuning," there isn't anything "fine" or minor about it. It's the process of adjusting the model to deal with troublesome inconsistencies or paper over irksome uncertainties. And sometimes modelers are tuning subgrid parameters in ways that aren't based on their "knowledge" of the parameter, but rather are aimed at producing a desired result. For example, UK researchers tuned their latest model in part by adjusting how partial snow cover changes the albedo of northern forests (snow reflects more than treetops). They also chose to adjust how much dimethyl sulfide is produced by microorganisms on the ocean surface—that chemical produces aerosols and so increases the albedo over the oceans.[7] Who would have thought those details are important to the climate?

In any event, it is impossible—for both practical and fundamental reasons—to tune the dozens of parameters so that the model matches the far more numerous observed properties of the climate system. Not only does this cast doubt on whether the conclusions of the model can be trusted, it makes it clear that we don't understand features of the climate to anywhere near the level of specificity required given the smallness of human influences.

Among the most important things that a model has to get right are "feedbacks." The growing greenhouse gas concentrations that raise the global temperature can also cause other changes in the climate system that either amplify or diminish their direct warming influence. For example, as the globe gets warmer, there'll be less snow and ice on the surface, which decreases the planet's albedo. The less reflective earth will then absorb more sunlight, causing even more warming. Another example of a feedback is that as the atmosphere warms, it will hold more water vapor, which further enhances its heat-intercepting ability. But more water vapor will also change the cloud cover, enhancing both heat interception (high clouds) and reflectivity (low clouds). On balance, the reflectivity wins, and the net cloud feedback somewhat diminishes the direct warming. The size, or in some cases even the sign, of these feedback effects—that is, whether they enhance or diminish the direct influence—cannot be understood precisely enough from first principles but must emerge from a model as it's tuned, and each model will give somewhat different answers. The average of results over many different models suggests that the net effect of all feedbacks is to double or triple CO_2's direct warming influence.

So tuning is a necessary, but perilous, part of modeling the climate, as it is in modeling of any complex system. An ill-tuned model will be a poor description of the real world, while overtuning risks cooking the books—that is, predetermining the answer. A paper co-authored by fifteen of the world's leading climate modelers put it this way:

> Choices and compromises made during the tuning exercise may significantly affect model results . . . In theory, tuning should be taken into account in any evaluation, intercomparison, or interpretation of the model results . . . Why such a lack of transparency? This may be because tuning is often seen as an unavoidable but dirty part of climate modeling, more engineering than science, an act of tinkering

> that does not merit recording in the scientific literature. There may also be some concern that explaining that models are tuned may strengthen the arguments of those claiming to question the validity of climate change projections. Tuning may be seen indeed as an unspeakable way to compensate for model errors.[8]

Indeed. A paper laying out the details of one of the most esteemed models, that of Germany's Max Planck Institute, tells of tuning a subgrid parameter (related to convection in the atmosphere) by a factor of ten because the originally chosen value resulted in twice as much warming as had been observed.[9] Changing a subgrid parameter by a factor of ten from what you thought it was—that's really dialing the knob.

A RANGE OF RESULTS

By now you should have a pretty clear picture of how the models come to be, and of why any glimpse they offer of the future might be less clear than we'd like. But let's have a look at the results. Because no one model will get everything right, the assessment reports average results from an "ensemble" made up of a few dozen different models from research groups around the world. The Coupled Model Intercomparison Project—otherwise known as CMIP—compiles the ensembles.[10] Its CMIP3 ensemble informed the IPCC's AR4 report, while CMIP5 underpinned the 2013 AR5 report, and CMIP6 will be the basis for the upcoming AR6 assessment.

But here we need to pause. The implication is that the models generally agree. But that isn't at all the case. Comparisons among models within any of these ensembles show that, on the scales required to measure the climate's response to human influences, model results differ dramatically both from each other and from observations. But you wouldn't know that unless you read deep into the IPCC report. Only then would you discover that the results being presented are "averaging" models that disagree wildly with each other. (By the way, the discordance among the individual ensemble members is further evidence that climate models are more than "just physics." If they weren't, multiple models wouldn't be necessary, as they'd all come to virtually the same conclusions.)

One particularly jarring failure is that the simulated global average surface temperature (not the anomaly) varies among models by about 3°C (5.6°F), three times greater than the observed value of the twentieth-century warming they're purporting to describe and explain. And two models whose average surface temperatures differ by that much will vary considerably in their details. For example, since you're not allowed to tune the freezing temperature of water (as it's determined by nature), the amounts of snow and ice cover, and hence the albedos, might be very different.

The assessment reports downplay this embarrassment of unphysical average temperatures by focusing on the rise in the average temperature and displaying the temperature changes calculated by each model, rather than the temperatures themselves. This makes differences among the ensemble members less apparent; the result is Figure 4.3 (adapted from the AR5 report). This graph looks at the global average surface temperature anomaly, comparing the averages and spreads of the ensembles used in AR4 and AR5 with the observed values (the information we encountered back in Figure 1.1). The agreement of the ensemble averages with the observations looks impressive, but these results need to be taken with some of that salt I mentioned at the beginning of the chapter. One of the world's most accomplished climate modelers has said that "it's reasonable to assume that there has been some tuning, implicit if not explicit, in models that fit the [temperature history] well."[11] And the details of Figure 4.3 reveal a few other problems.

One stunning problem is that the spread of the CMIP5 ensemble in the years after 1960 is larger than that of the models in CMIP3—in other words, the later generation of models is actually *more* uncertain than the earlier one. So here is a real surprise: even as the models became more sophisticated—including finer grids, fancier subgrid parametrizations, and so on—the uncertainty increased rather than decreased. Having better tools and information to work with *should* make the models more accurate and more in line with each other. That this has not happened is

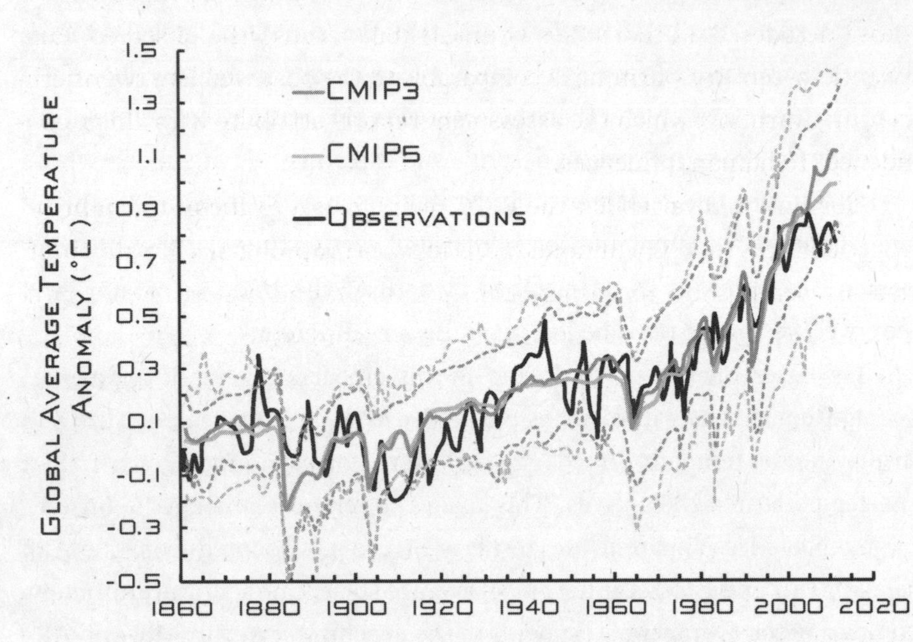

FIGURE 4.3 **The global average surface temperature anomaly as simulated in the CMIP3 and CMIP5 model ensembles.** The solid gray lines show the ensemble averages, while the corresponding dotted lines show the ensemble spreads. The black line shows the observed anomalies.[12]

something to keep in mind whenever you read "Models predict . . ." The fact that the spread in their results is increasing is as good evidence as any that the science is far from settled.

But another equally serious issue is also illustrated here: Figure 4.3 shows that the ensembles fail to reproduce the strong warming observed from 1910 to 1940. On average, the models give a warming rate over that period of about half what was actually observed. As the IPCC noted in measured and somewhat antiseptic language:

> It remains difficult to quantify the contribution to this warming from internal variability, natural forcing and anthropogenic forcing, due to forcing and response uncertainties and incomplete observational coverage.[13]

More bluntly, they're saying that we've no idea what causes this failure of the models. They cannot tell us why the climate changed during those decades. And that's deeply unsettling, because the observed early twentieth-century warming is comparable to the observed late twentieth-century warming, which the assessment reports attribute with "high confidence" to human influences.

That internal variability the IPCC refers to as a "difficult to quantify" contributor, as though a minor issue, is in fact a big problem. Climate observations clearly show repeating behaviors over decades—and even centuries. At least some of these are due to slow changes in ocean currents and the interaction between the ocean and the atmosphere. The best-known example is that of El Niño events (technically, the El Niño-Southern Oscillation), a shift in heat across the equatorial Pacific Ocean that occurs irregularly every two to seven years and influences global weather patterns. A slower behavior that's less well known is the Atlantic Multidecadal Oscillation (AMO), which involves cyclic temperature changes in the North Atlantic.[14] Figure 4.4 shows that the strength of the AMO as inferred from sea surface temperatures repeats over sixty- to eighty-year cycles.

The Pacific Ocean shows a similar, though unrelated, cyclic behavior known as the Pacific Decadal Oscillation (PDO), which has a cycle length of about sixty years. Because we have only about 150 years of good observations, systematic behaviors that occur over longer timescales are less well known—there could be (and almost certainly are) other natural cyclic variations occurring over even longer periods.

Cycles like these influence global and regional climates and are superimposed upon any trends due to human or natural forcings like greenhouse gas emissions or volcanic aerosols. They make it difficult to determine which observed changes in the climate are due to human influences and which are natural. For example, the spikes in the global temperature anomaly during 1998 and 2016 (as shown in Figure 1.1) are due to especially large El Niño events.

While today's models can reproduce some aspects of El Niño events, they aren't very good at reproducing the strength, duration, pattern on the globe, or timing of slower cycles. The AR5 reported that, while a number of

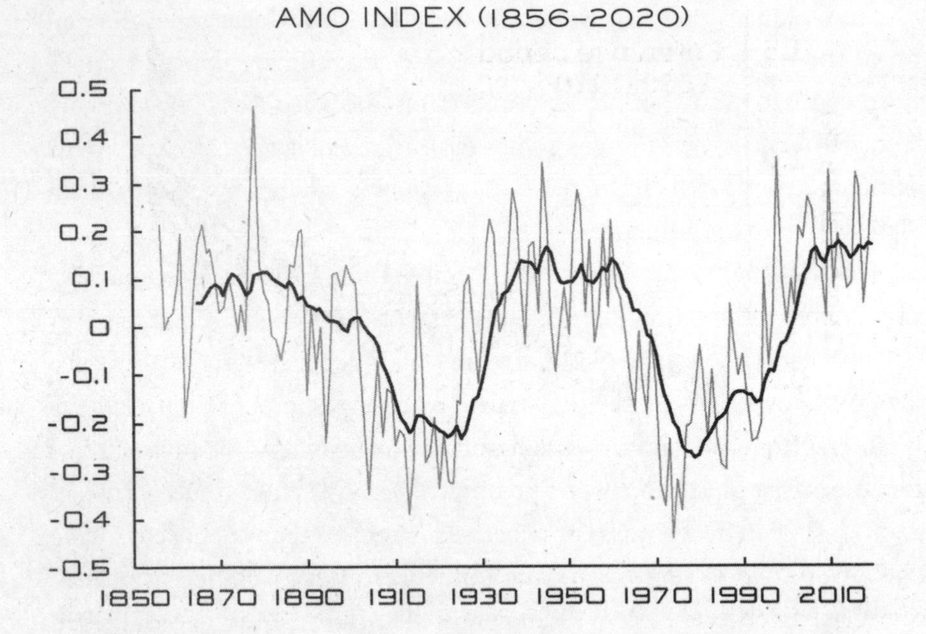

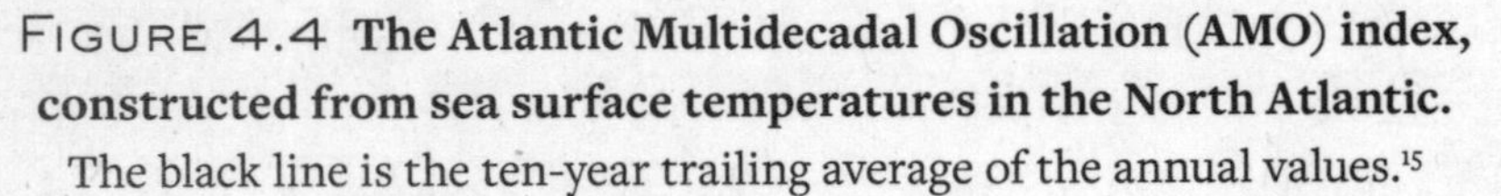
FIGURE 4.4 **The Atlantic Multidecadal Oscillation (AMO) index, constructed from sea surface temperatures in the North Atlantic.** The black line is the ten-year trailing average of the annual values.[15]

models do produce something like the AMO, their versions of this phenomenon differed from model to model and from observations in many respects. Most notably, the timescales for the AMO-like cycle produced by the models ranged from forty years to a century or more.[16] The models were not any better at reproducing the multidecadal variability in the Pacific basin.[17]

And the CMIP6 models that inform the IPCC's upcoming AR6 don't perform any better than those of CMIP5, at least by these measures. Figure 4.5 compares temperature anomaly results for the CMIP6 models and observations, much as Figure 4.3 did for the CMIP3 and CMIP5 ensembles. An analysis of 267 simulations run by twenty-nine different CMIP6 models created by nineteen modeling groups around the world shows that they do a very poor job of describing warming since 1950 and continue to underestimate the rate of warming in the early twentieth century.[18]

The failure of even the latest models to warm rapidly enough in the early twentieth century suggests that it's possible, even likely, that internal

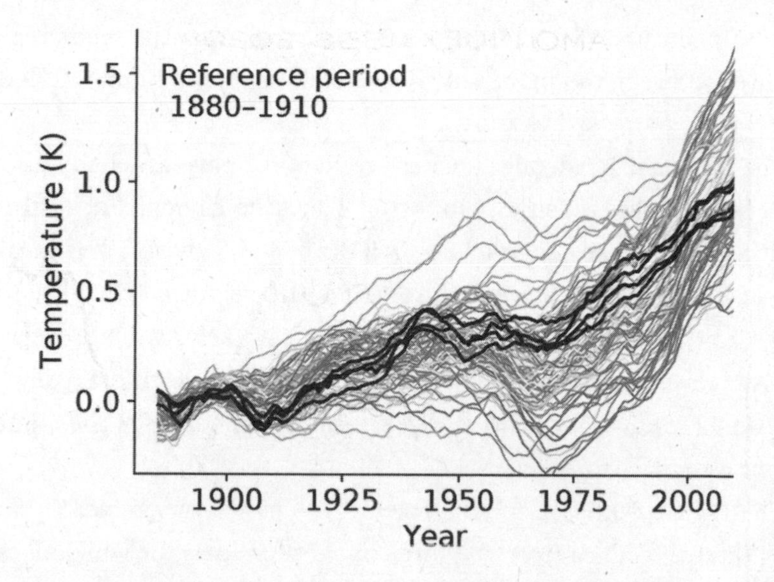

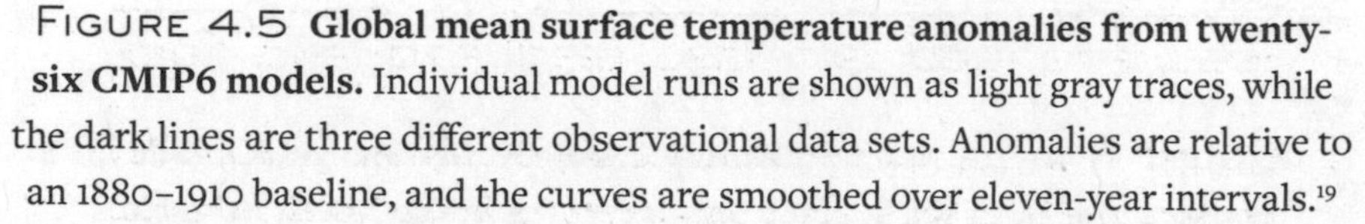
FIGURE 4.5 **Global mean surface temperature anomalies from twenty-six CMIP6 models.** Individual model runs are shown as light gray traces, while the dark lines are three different observational data sets. Anomalies are relative to an 1880–1910 baseline, and the curves are smoothed over eleven-year intervals.[19]

variability—the natural ebbs and flows of the climate system—has contributed significantly to the warming of recent decades.[20] That the models can't reproduce the past is a big red flag—it erodes confidence in their projections of future climates. In particular, it greatly complicates sorting out the relative roles of natural variability and human influences in the warming that has occurred since 1980.

A common measure of how the climate system responds to human influences, and an important piece of information we hope to learn from models, is the equilibrium climate sensitivity, or ECS. That's how much the average surface temperature anomaly (recall that the anomaly is the deviation from the expected average) would increase if the CO_2 concentration were hypothetically doubled from its preindustrial value of 280 ppm. If emissions continue at their current pace and the carbon cycle doesn't change much, that doubling would happen in the real world toward the end of this century. The higher the ECS (i.e., the larger the predicted temperature increase), the more sensitive the climate is to human influences (or at least to increased CO_2).

The National Academies' "Charney Report" gave a benchmark estimate of the climate sensitivity in 1979.[21] That blue-ribbon panel estimated the ECS to be in the range of 1.5–4.5°C (2.7–8.1°F), with a "most likely" value of 3°C (5.5°F).[22] In 2007, the IPCC's AR4 narrowed the likely range (2–4.5°C, or 3.6–8.1°F) but gave the same "most likely" value. Seven years later, AR5 reverted to the 1.5–4.5°C range, and it gave no "most likely" estimate. So in 2014, we were no more certain of how sensitive the climate is than we were in 1979.

The IPCC's AR6 will rely on the CMIP6 model ensemble; Figure 4.6 shows the ECS values from the forty of those models for which data was available as of May 2020. About one-third of the models (those shown in black) result in simulated climates that are *more* sensitive than the likely upper limit of 4.5°C (8.1°F) given previously by the IPCC. The more

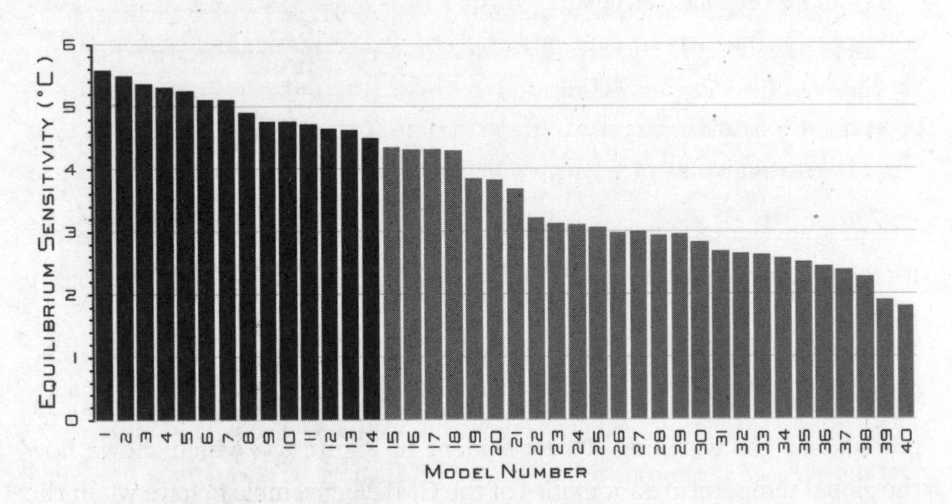

FIGURE 4.6 **Equilibrium climate sensitivities of forty models from the CMIP6 ensemble.** Models are arranged in order of decreasing sensitivity. Those shown in black are more sensitive than the likely upper limit given in AR5.[23]

sensitive models also warm more rapidly[24, 25] than the observations in recent decades and are inconsistent with paleoclimate data.[26]

These higher sensitivities seem[27] to arise from these models' subgrid representations of clouds and their interaction with aerosols.[28] As one of the lead researchers said:

> Cloud-aerosol interactions are on the bleeding edge of our comprehension of how the climate system works, and it's a challenge to model what we don't understand. These modelers are pushing the boundaries of human understanding, and I am hopeful that this uncertainty will motivate new science.[29]

In other words, we don't really understand an influence on the climate system that's about the same size as the human-caused warming influence. It will be interesting to see how—or whether—the upcoming AR6 report addresses this.

We shouldn't be too comforted by those CMIP6 models whose sensitivities are more in line with the older ones, either. Consider this from the Max Planck Institute modelers:

> We have documented how we tuned the MPI-ESM1.2 global climate model to match the instrumental record warming; an endeavor which has clearly been successful. Due to the historical order of events, the choice was to do this practically by targeting an ECS of about 3 K [3°C] using cloud feedbacks, as opposed to tuning the aerosol forcing.[30]

In other words, the researchers tuned their model to make its sensitivity to greenhouse gases what they thought it should be. Talk about cooking the books.

One of the reasons the climate sensitivity is so uncertain is that aerosols currently exert a cooling influence that partially offsets (or masks) the greenhouse gas warming. That's evident in Figure 4.7, which shows how the global temperature anomalies of the CMIP6 ensemble behave when the simulations are run with different forcings in place. Greenhouse gases alone caused a warming of 1.5°C (2.7°F) from 1900, which is partially offset by a cooling of about 0.6°C (1.1°F) from human-caused aerosols (solar variation

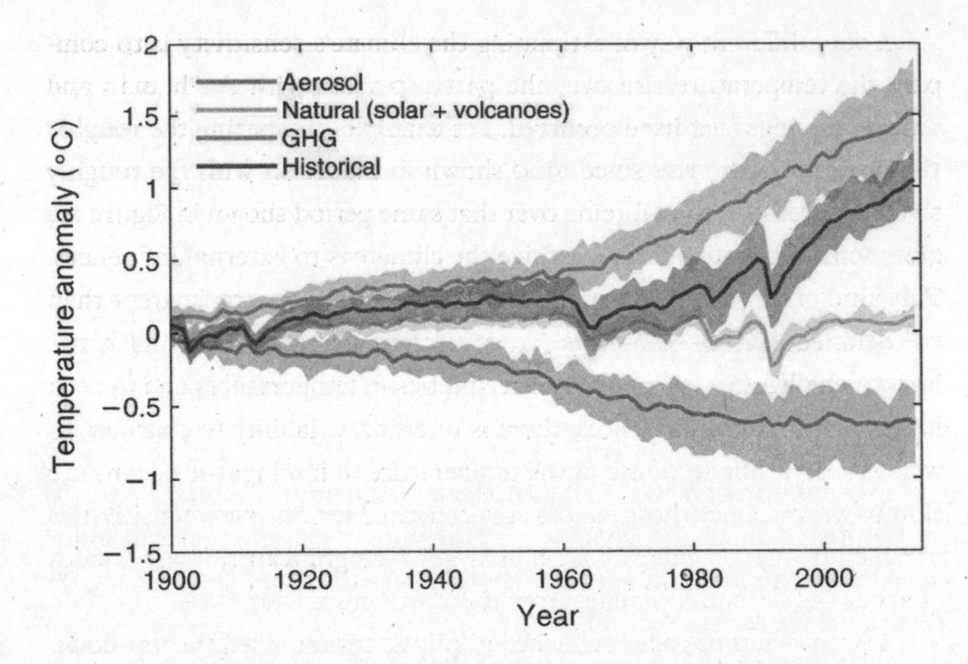

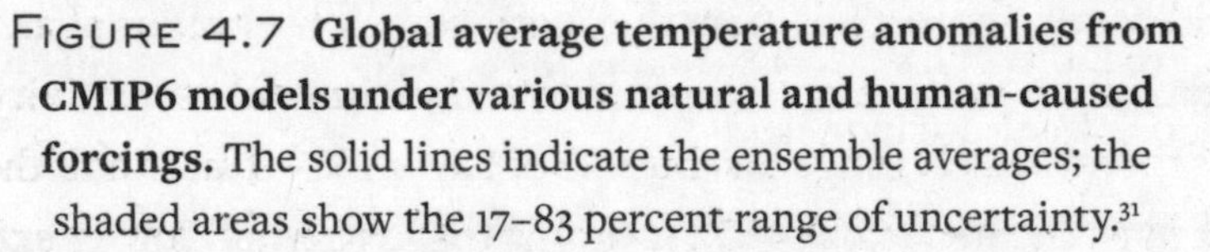
FIGURE 4.7 **Global average temperature anomalies from CMIP6 models under various natural and human-caused forcings.** The solid lines indicate the ensemble averages; the shaded areas show the 17–83 percent range of uncertainty.[31]

and volcanic aerosols don't have much long-term effect). The curve labeled "historical" corresponds to the ensemble with all forcings, natural and human-caused, in place. You can see the cooling impact of especially large volcanic eruptions, especially Agung in 1963 and Pinatubo in 1991; the absence of the observed warming from 1910 to 1940 is also evident in the historical simulations.

Because of this large but uncertain cooling by aerosols, a model with high sensitivities to both aerosols and greenhouse gases could describe the historical record about as well as one with much lower sensitivities. As the influence of greenhouse gases grows and becomes dominant in the coming decades (aerosols are a significant source of local air pollution and reducing them is a priority everywhere), ECS estimates could become more precise.

A very different way of estimating the climate's sensitivity is to compare the temperature rise over the past 140 years with the human and natural forcings that have occurred. For example, comparing the roughly 1°C of temperature rise since 1900 shown in Figure 1.1 with the roughly 2 W/m^2 increase in total forcing over that same period shown in Figure 2.4 gives some estimate of how sensitive the climate is to external influences. This kind of top-down approach is far simpler and more transparent than the detailed grid-based models (no supercomputer required!). But it has its own challenges—apart from uncertainties in temperatures and in both human and natural influences, there is internal variability to consider, as well as a lag in the response of the temperature to forcing (the oceans are slow to warm). Once these factors are accounted for, however, sensitivities obtained from such energy budget analyses are significantly lower (values of about 1.5°C) than the values from the CMIP ensembles.[32]

A twenty-author paper published in July 2020 combined the top-down and grid-based approaches (together with some observational and paleo information) in an attempt to pin down the climate's sensitivity.[33] The authors found a likely range for the ECS of 2.6–4.1°C—this is half the range estimated by the AR5 (1.5–4.5°C), meaning that extremely low or extremely high values are deemed less likely.

There's still much work to do to understand just how sensitive the climate system is to various influences; it would be a big deal if the climate were much less (or much more) sensitive than is currently supposed.

So there's a lot to fret about in the climate modeling business. Apart from the computational challenge of simulations that can take months to run on even the world's fastest computers, there's the ambiguity in tuning, ill-quantified natural variability, and intricacies like the trade-off between greenhouse gas warming and aerosol cooling to contend with. No wonder we've got a poor understanding of how the climate will respond to rising greenhouse gas concentrations. The more we learn about the climate system, the more we realize how complicated it is.

It's uncommon for the popular media to discuss how problematic our climate models are. But you can sometimes get a glimpse of it, if you're paying attention. For example, climate models are also used to assess the effects of various climate response strategies, such as deliberately enhancing the earth's reflectivity (its albedo) to counteract greenhouse gas warming; such "geoengineering" is discussed in Chapter 14. A recent National Academies' report noted:

> The uncertainties in modeling of both climate change and the consequences of albedo modification make it impossible today to provide reliable, quantitative statements about relative risks, consequences, and benefits of albedo modification to the Earth system as a whole, let alone benefits and risks to specific regions of the planet.[34]

If "the uncertainties in modeling" mean these models can't give us useful information about what albedo modification might do, it's hard to see why they would be any better at predicting the response to other human influences. After all, these are the same models discussed in this chapter, with only minor modifications (say, making the sun 1 percent less intense or adding a bit more aerosol in the upper atmosphere) to simulate the change in albedo.[35] Yet it's difficult to imagine a statement like the one above, except about greenhouse gases instead of albedo modification. Here's what it might say:

> *The uncertainties in modeling of both climate change and the consequences of future greenhouse gas emissions make it impossible today to provide reliable, quantitative statements about relative risks and consequences and benefits of rising greenhouse gases to the Earth system as a whole, let alone to specific regions of the planet.*

I don't think you'll ever see a statement like that in the assessment reports.

As the reports themselves always note, the models are the best we've got, and they are becoming more sophisticated all the time. But for now, they're still beset by myriad problems, and I, along with everyone else, sure wish they were better.

MORE ON MODELS—2024 UPDATE

The deficiencies of climate models have become even more evident since *Unsettled*'s initial publication. Perhaps more importantly, they are increasingly being openly acknowledged by the experts, although still rarely in the media and non-expert discussions.

As expected, AR6, the most recent assessment report from the IPCC, had to deal with the oversensitivity of many of the models illustrated in Figure 4.6. That is, they warmed too rapidly under human influences. At issue are the several types of feedbacks discussed in this chapter. One such feedback is that snow and ice disappear as the planet warms, decreasing the albedo and so amplifying the warming. Another feedback is that the greater amount of water vapor in a warmer atmosphere increases its heat-intercepting ability and again amplifies the warming. The aggregate of these and other feedbacks is thought to double or triple the climate's sensitivity.

The AR6 "fix" for the models' oversensitivity was an analysis that claimed to reduce uncertainties by using observations to determine feedbacks, rather than having them emerge from the General Circulation Models (GCMs). (This analysis, by Sherwood et al., was discussed earlier in the chapter, and is cited at note 33.) For example, observing how snow cover changes with temperature can isolate and quantify the ice-albedo feedback. And observing the growth of the atmosphere's moisture with rising temperature can measure the water vapor feedback. Combining such studies for each type of feedback gives an estimate of the total sensitivity.

The result of all of this, as mentioned in the preface, is that for AR6, the final likely range for the Equilibrium Climate Sensitivity is 2.5–4.0°C. That's a greater lower bound and significantly narrower range than had been given in AR5 (1.5–4.5°C), which was based solely on model results.

However, the Sherwood et al. (and hence AR6) estimates of climate sensitivity have been shown[36] to be biased by the statistical methods used, and a more objective analysis of the same data lowers the likely range of sensitivity to 1.75–2.7°C. So this issue remains to be settled.

Quite apart from the numbers, it is disturbing that a fair fraction of the allegedly best models produced by the world's best modelers are far more sensitive than the aggregate of the individual feedbacks, since the models supposedly embody that aggregation. To "fix" that, there have been calls[37] to abandon "model democracy"—the practice of treating all models equally—and utilize results only from those models deemed to be more plausible.

The deficiencies of climate models that I wrote about in this chapter have been acknowledged more publicly and explicitly. For example, two senior climate researchers, Professor Tim Palmer at the University of Oxford and Bjorn Stevens, Director of Germany's Max Planck Institute for Meteorology, have written:

> [F]or many key applications that require regional climate model output or for assessing large-scale changes from smallscale processes, we believe that the current generation of models is not fit for purpose.[38]

That's particularly important because adaptation measures depend upon regional model projections. Stevens has gone so far as to describe the gaps in knowledge as "frightful":

> [I]t is difficult, and in many places impossible, to scientifically advise societal efforts to adapt in the face of unavoidable warming. Our knowledge gaps are frightful because they make it impossible to assess the extent to which a given degree of warming poses existential threats.[39]

Users of the models' results are similarly cautioned:

> [T]he use of these [climate] models to guide local, practical adaptation actions is unwarranted. Climate models are unable to represent future conditions at the degree of spatial, temporal, and probabilistic precision with which projections are often provided, which gives a false impression of confidence to users of climate change information.[40]

Unsettled has been criticized for giving too little credence to climate model projections.[41] But the meaning of these statements is plain enough:

Our understanding of how the climate responds to human influences—especially at the regional scale—is too poor to be of much use. How can a country adapt if it doesn't even know the sign of precipitation change with any confidence, as the AR6 figures show?

On a more optimistic note, there are recent efforts to apply machine learning and artificial intelligence; the National Science Foundation has funded a center[42] toward that end. While a currently promising avenue of research, it remains to be seen whether these methods will significantly reduce uncertainties in climate projections.

HYPING THE HEAT

Today, TV weather presenters have morphed into *climate* and weather presenters, blaming a "broken climate" for many of the severe weather events that they cover. Indeed, it has become *de rigueur* for the media, politicians, and even some scientists to implicate human influences as the cause of heat waves, droughts, floods, storms, and whatever else the public fears. It's a pretty easy sell: the on-the-scene reporting is powerful—and often moving—and our poor memories of past events can make "unprecedented" quite convincing.

But the science tells a different story. Observations extending back over a century indicate that most types of extreme weather events don't show *any* significant change—and some such events have actually become less common or severe—even as human influences on the climate grow. In general, there are high levels of uncertainty involved in detecting trends in extreme weather. Here are some (perhaps surprising) summary statements from the IPCC's AR5 WGI report, indicating what we know (or don't know) about a few such trends:

- " . . . *low confidence* regarding the sign of trend in the magnitude and/or frequency of floods on a global scale."[1]
- ". . . *low confidence* in a global-scale observed trend in drought or dryness (lack of rainfall) since the middle of the 20th century . . ."[2]
- ". . . *low confidence* in trends in small-scale severe weather phenomena such as hail and thunderstorms . . ."[3]
- ". . . *confidence* in large scale changes in the intensity of extreme extratropical cyclones [storms] since 1900 is *low*."[4]

There are many reasons for science's generally low confidence in detecting changes in extreme weather events and then attributing them to human influences, all of which we've discussed earlier: short and low-quality historical records, high natural variability, confounding natural influences, and disagreements among the many models used. Yet even though we've little evidence of much changing, the media maintains a flow of "news" connecting weather events to climate—in part, by relying on what are called "event attribution studies," which have become a growing branch of climate science.[5]

Here's how attribution studies work: After some extreme weather event such as a storm or flood or drought or heat wave, researchers combine climate modeling and historical observations to attempt to determine the role human influences (usually warming) played in its occurrence or severity. Any time you see an article after a hurricane or similar claiming that climate change (by which the authors mean human-caused climate change) made the event XX percent more likely or YY percent more severe, you're seeing the results of an attribution study.[6, 7] As might be expected, attribution studies almost always focus on weather disasters, not benign weather occurrences.

By this point, I probably need hardly tell you that such studies are rife with issues. Many factors contribute to extreme events, and teasing out the result of human influences in any one event is surely a challenge. The Executive Summary of Chapter 3 of IPCC's 2012 Special Report on Extreme Events (SREX) states the problem well:

> Many weather and climate extremes are the result of natural climate variability (including phenomena such as El Niño), and natural decadal or multi-decadal variations in the climate provide the backdrop for anthropogenic [human-caused] climate changes. Even if there were no anthropogenic changes in climate, a wide variety of natural weather and climate extremes would still occur.[8]

The World Meteorological Organization goes even further, saying:

> . . . any single event, such as a severe tropical cyclone [hurricane or typhoon], cannot be attributed to human-induced climate change, given the current status of scientific understanding.[9]

Practitioners argue that event attribution studies are the best climate science can do in terms of connecting weather to changes in climate. But as a physical scientist, I'm appalled that such studies are given credence, much less media coverage. A hallmark of science is that conclusions get tested against observations. But that's virtually impossible for weather attribution studies. It's like a spiritual adviser who claims her influence helped you win the lottery—after you've already won it. The only way to test that extraordinary claim would be to play the lottery many times with her help (no doubt at considerable cost!) and see if you win more than expected. Data is the touchstone of science; the only solid way to test weather event attribution is to see whether the statistical properties of extreme events have changed—which would eliminate the need for attribution studies in the first place.

The bottom line is that the science says that most extreme weather events show no long-term trends that can be attributed to human influences on the climate. (What models might project for future extremes is quite a different matter, though it's often conflated with what the observational record shows.) Yet the popular perception that extreme events are becoming more common and more severe remains. This is not solely the result of the increased prevalence of event attribution studies, or even of the media garbling the message—it is also due to the failure of the official assessment reports to be transparent, or occasionally even correct, about what the science actually says.

In the next few chapters, we'll discuss the specific evidence for several phenomena (some extreme weather events among them) commonly attributed to human-caused climate change. In this chapter, we'll begin with a topic that offers an illuminating look at how the disconnect between the popular perception of weather extremes and the science involved arises. While only a small piece of climate science, it illustrates many of the problems in how the science is portrayed to non-experts, including contrived analyses, misrepresentation of results, failure of review processes, and media exaggeration. It's a topic that garners a great deal of attention: record temperatures.

We can all agree the globe has gotten warmer over the past several decades. Here's another summary statement from the IPCC's AR5:

> [S]ince about 1950 it is *very likely* that the numbers of cold days and nights have decreased and the numbers of warm days and nights have increased . . . there is *medium confidence* that globally the length and frequency of warm spells, including heat waves, has increased since the middle of the 20th century.[10]

The science pointing to longer or more frequent warm spells like heat waves inspires a lukewarm "medium confidence," but that of a general trend toward warmer temperatures gets the IPCC's "very likely" designation, meaning there is a mere 10 percent chance of it being mistaken.

Yet the public perception that *extreme* high temperatures are on the rise—fostered by headlines like "Daily temperature records run rampant as the globe roasts!"—is simply incorrect. In the US, which has the world's most extensive and highest-quality weather data, record low temperatures have indeed become less common, but record daily high temperatures are no more frequent than they were a century ago.

Headlines about record highs (often accompanied by visuals of red thermometers and barren desert vistas) don't come out of nowhere, however. The US government's most recent assessment report, the 2017 Climate Science Special Report (CSSR), is not just misleading on this point—it's wrong. I say that, to use the assessment reports' lingo, with *Very High Confidence* because of some sleuthing I did in the spring of 2019. What emerged is a disturbing illustration of how non-experts are misled and science is spun to persuade, not inform. In fact, page 19 of the CSSR's Executive Summary says (prominently and with *Very High Confidence*):

> There have been marked changes in temperature extremes across the contiguous United States. The number of high temperature records set in the past two decades far exceeds the number of low temperature records.

It offers Figure 5.1 (their Figure ES.5) in support:

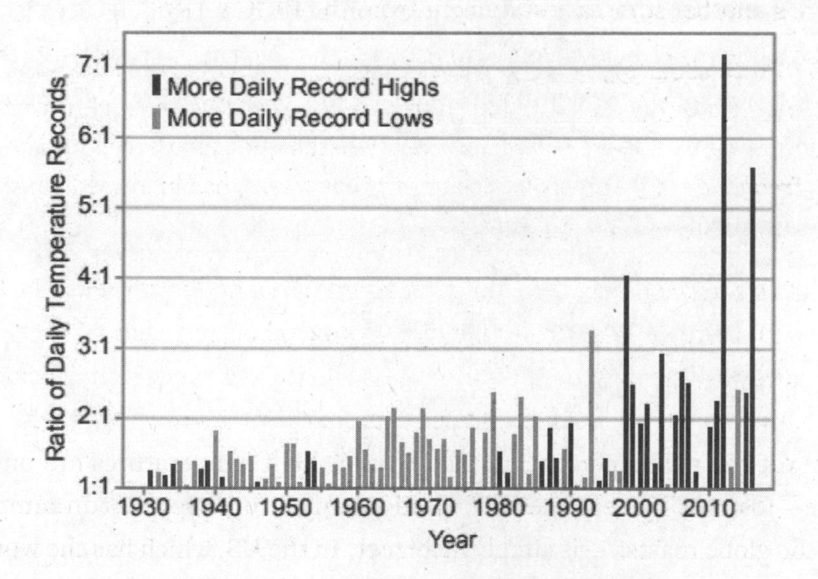

FIGURE 5.1 **The ratio of the number of daily record high temperatures to daily record low temperatures for stations across the forty-eight contiguous states from 1930 to 2017.**[11] (CSSR Figure ES.5.)

The darker bars show a year in which there were more record highs than lows, while the lighter bars show years in which there were more record lows than highs. The height of the bar indicates the ratio of record highs to record lows for each year. For example, a ratio of 2:1 for a light bar means that the stations had twice as many record daily lows that year as they had record daily highs.

I suspect that most readers were shocked by that figure, as I was when I first saw it. Who wouldn't be? An attention-grabbing title ("Record Warm Daily Temperatures Are Occurring More Often") backed up by data with a hockey-stick shape veering sharply upward in recent years (and, in the original, years with more "highs" portrayed in an alarming scarlet). It sure looks like temperatures are going through the roof.

But I was disturbed by an apparent inconsistency between that figure and some others deeper in the report, particularly the figure reproduced as our Figure 5.2. It shows that the average coldest temperature of the year has clearly increased since 1900, while the average warmest temperature has hardly changed over the last sixty years and is about the same today as it was in 1900. (If you look at the graph of highest temperatures in the right-hand panel below, you can also see the "warm" 1930s of the Dust Bowl, when agricultural practices such as overplowing of greatly expanded cropland in the Great Plains enhanced natural variability.)

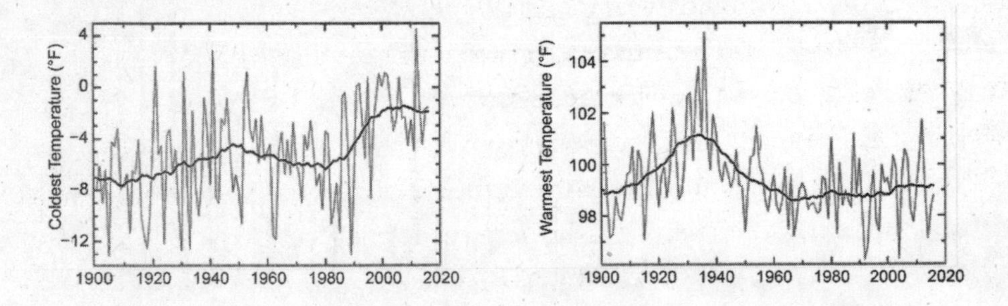

FIGURE 5.2 **The coldest (left) and warmest (right) temperatures of the year since 1900, averaged across the forty-eight contiguous US states.** The spiky light lines show the year-by-year values, while the darker lines show the smoothed behavior. (CSSR Figure 6.3.)[12]

Of course, these average annual record temperatures are not the same thing as the daily record temperatures at individual stations used to construct Figure 5.1. But it sure seemed possible that the ratio of record highs to record lows shown in Figure 5.1 goes up not because record highs are becoming more common, but because as the coldest temperatures warm, the ratio's denominator (number of daily record colds) is getting smaller, while its numerator (number of daily record warms) has hardly changed in recent decades.

Inconsistencies are red meat to a scientist. Resolving them can lead to major insights—and I was determined to get to the bottom of this one. To do that, I first looked up the research papers describing how the CSSR determined the incidence of daily temperature extremes.[13,14] Those papers describe a method that counts "running records," tallying a record daily high for a given station and a given day of the year if the high temperature that day exceeds that of all prior years. (Similarly, a record low is tallied if the low temperature is less than the low of that day in all prior years.) This is the sort of record that's announced in your daily weather reports.

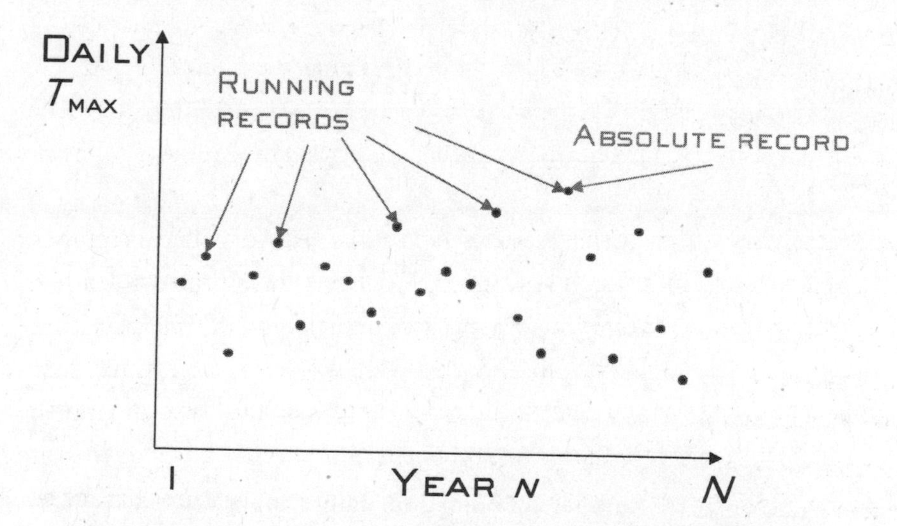

FIGURE 5.3 **The difference between running records and the absolute record, illustrated for daily high temperatures at a single station.** Each of the running records is higher than all years before it, while the sole absolute record is the highest of all years.

But there's another way of counting records, as illustrated for high temperatures in Figure 5.3. For a given station on a given day of the year, a "running record" high temperature occurs in any year where the maximum temperature exceeds that of all the years before it. So there are many of these records over a period of observation (for instance, in Figure 5.1, the period from 1930 to the present). In contrast, an "absolute record" high temperature occurs only once over the period of observation, in the year that has the highest daily maximum temperature.

I quickly understood that there's a big problem with running records—they tend to become less frequent as the years go on because each new record "raises the bar" and makes it more difficult to achieve a subsequent record. Think about it: two years in from the start of the period of observation, a daily temperature only had to be higher than that in 1930 and 1931 to be a "record high." In 1980, to be a record high, the temperature would have to be higher than any on that day over the past fifty years. And in

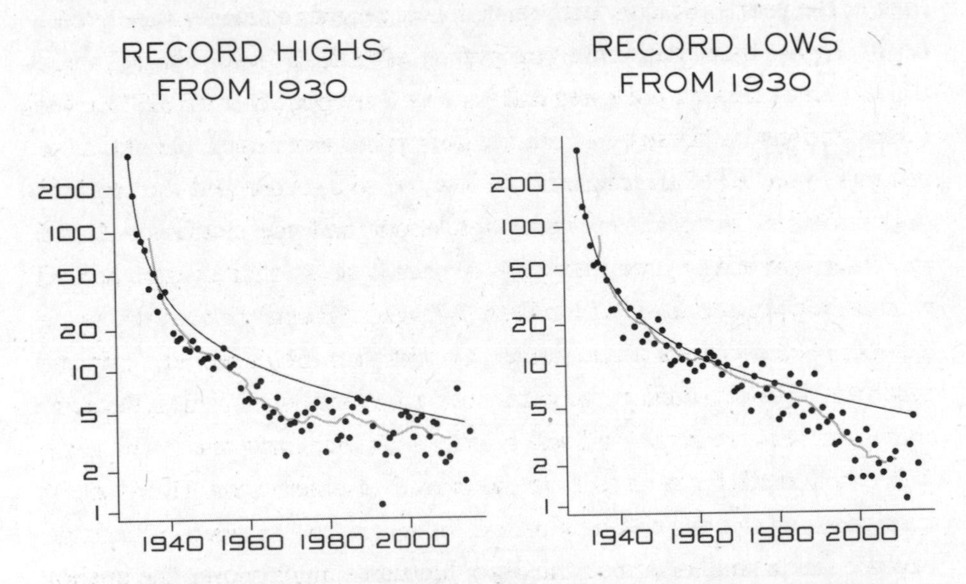

FIGURE 5.4 **Numbers of record US daily temperature extremes calculated by the "running" method used in the CSSR.** The dots show the incidence of record temperatures (highs in the left panel, lows in the right panel), while the gray line is the eleven-year running average. The black line shows the expected decline if there were no trend in temperatures.

fact, the decline in the number of running records is shown in Figure 5.4, which appears in a reference cited by the CSSR.[15] The black line shows the expected decline even if there were no trend in the data—for instance, if the lowest temperatures weren't warming, the yearly incidence of running records (the dots) would be expected to align closely with that descending curve (if low temperatures were strongly cooling, on the other hand, they would tend to be above the curve). As you can see, the numbers of record highs and lows both decline precipitously between 1930 and the present day, with the number of lows declining more rapidly—explaining why the ratio of record highs to record lows is larger in recent years, even as both types of records are becoming much less frequent.

But there's another problem with CSSR's ratio graph (our Figure 5.1): the method of running records guarantees that it will show a flat trend in early years followed by dramatic fluctuations later on.

To see that, think about the second year (1931) of the analysis. On each of the year's 365 days, each station that records a temperature even a degree higher than that of the year before will tally a "record high," while those with a temperature lower will not. We'll suppose that the CSSR used 1,400 stations (it doesn't say exactly how many were used, but the 2016 analysis[16] used 1,408); because of the low bar to a record and the random fluctuations of temperature variations, about half the stations will fall on either side of the previous year's temperature, so that 255,500 record warms (365 × 1,400/2) will be tallied that year. Similarly, about the same number of record colds will be tallied, so that the ratio of record warms to record colds in 1931 (and in the years soon after) will be close to 1. Because these numbers are large (and will remain so in the early years), the ratio won't vary much from 1:1 early in the period of observation. However, in later years, as the bar to a new record increases, the numbers of records become much smaller, and so the ratio fluctuates much more. The upshot is that by using the running records method, the ratio graph is guaranteed to show a long period of values around 1 at the start of the record, followed by dramatic variations toward the end, creating the impression of large changes in recent decades, even if they aren't present. While it produces a scary visual, this ratio has almost nothing to do with how temperatures are actually changing.

Having understood that the CSSR presentation of US daily record temperatures was badly misleading, I naturally wanted to know what a proper analysis—one using absolute records—would show. I was also interested to see what was happening with record temperatures before 1930, since US temperature observations were certainly available even before 1900.

To answer those questions myself, I would have had to start from scratch—download a great quantity of US weather station data, clean it up, and write code to analyze the data. But an advantage in having a wide network of scientific acquaintances is that I can almost always find someone who can do an analysis more quickly (and better) than I could myself: I contacted Professor John Christy at the University of Alabama at Huntsville, whom I had met when he participated in the 2014 APS workshop. John has access to a vast trove of US meteorological records and is adept at analyzing them in different ways. In short order, he was on the case.

Careful researchers always make sure that they can reproduce existing results before attempting something new. That was particularly important in this instance, as John at first didn't believe me when I told him about the problems with the running records method used in the CSSR. So he first demonstrated that he could reproduce the CSSR's Figure ES.5 with his own data set. Then he went on to do an analysis using the standard method of "absolute records." Tallies of these absolute records do not have the structural decrease of the running records method; if the number of these records changes significantly over the period of observation, it is because of a trend in the temperatures themselves.

Christy's analysis of absolute record warms and colds used the data from 725 US stations beginning in 1895. That's about half the number of stations used in the CSSR analysis from 1930, because fewer stations have quality records extending back to that earlier date. Nevertheless, his results are compelling, as shown in Figure 5.5.

The record highs clearly show the warm 1930s, but there is no significant trend over the 120 years of observations, or even since 1980, when

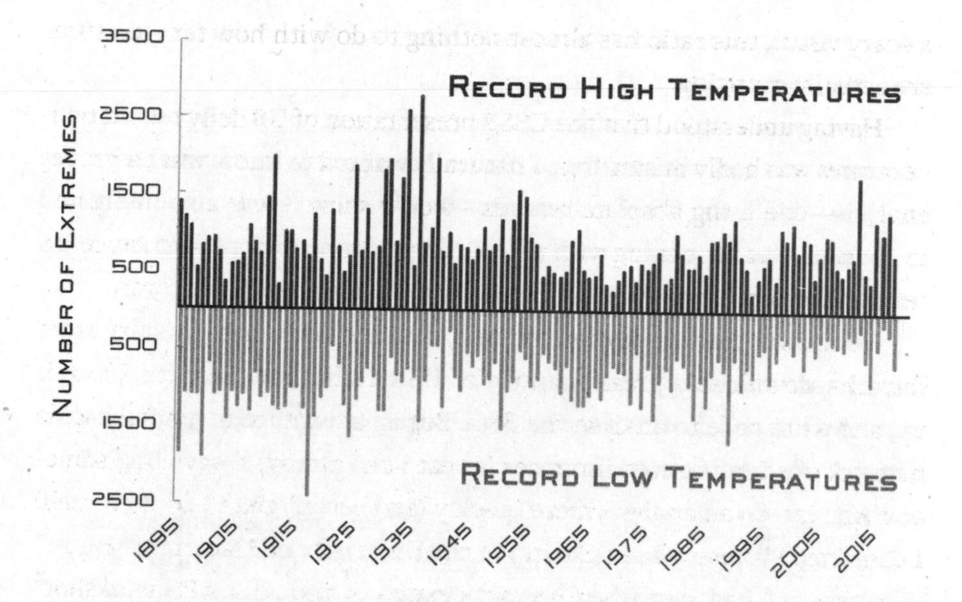

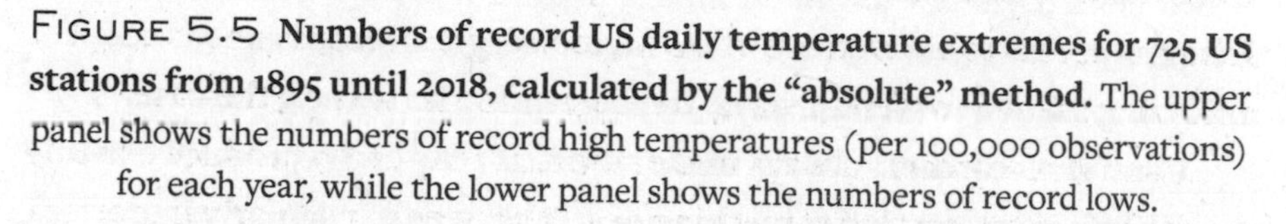

FIGURE 5.5 **Numbers of record US daily temperature extremes for 725 US stations from 1895 until 2018, calculated by the "absolute" method.** The upper panel shows the numbers of record high temperatures (per 100,000 observations) for each year, while the lower panel shows the numbers of record lows.

human influences on the climate grew strongly. In contrast, the numbers of record daily cold temperatures decline over more than a century, with that trend accelerating after 1985. These two panels together show something that is completely contrary to common perception—that temperature extremes in the contiguous US have become less common and somewhat milder since the late nineteenth century.

Yet the Executive Summary of the CSSR prominently features the faulty ratio graph (our Figure 5.1) with the legend "Record warm daily temperatures are occurring more often." Even if you charitably argue that the authors "forgot" to add ". . . compared to record cold temperatures," there is no arguing that it is shockingly misleading, especially when taken along with the rest of the report's material on temperature extremes. When the graph appears again in Chapter 6 of the report, it is beside text that reads "the number of record lows has been declining since the late-1970s while the number of record highs has been rising." But by the CSSR's own definition, the number of record highs has been falling.

How could a report that proclaims itself "designed to be an authoritative assessment of the science of climate change" so mischaracterize the data?[17] After all, the CSSR was subject to multiple reviews, including one by an expert panel convened by the National Academies of Science, Engineering, and Medicine (NASEM).

When I dug a bit further, I found that in fact the National Academies' expert review panel had criticized[18] the discussion of temperature extremes in a late draft[19] of the CSSR. Here is the Key Finding 2 of Chapter 6 in that draft:

> Accompanying the rise is [*sic*] average temperatures, there have been—as is to be expected—increases in extreme temperature events in most parts of the United States. Since the early 1900s, the temperature of extremely cold days has increased throughout the contiguous United States, and the temperature of extremely warm days has increased across much of the West. In recent decades, intense cold waves have become less common while intense heat waves have become more common. (*Extremely likely, Very high confidence*)

The Academies' review panel, in the diplomatic language of an academic review, criticized that finding as follows:

> Further, it is difficult to understand how a statement that includes increases in extreme warmth can be associated with a high confidence or extremely likely statement, given that most of the graphics in this chapter show a decrease in extreme warmth in the historical record.

This is, of course, the same inconsistency that had raised my own suspicions.

The federal official then in charge of the CSSR (Michael Kuperberg, then executive director of the US Global Change Research Program) responded to the Academies' review of Chapter 6 by noting:

> Almost every recommendation from the NAS was incorporated [in the final version] . . . and a new figure was added on changes in record high and low temperatures.[20]

The "new figure" referred to is the final report's infamous ratio graph, Figure ES.5 (our Figure 5.1). It seems unlikely that the Academies panel ever saw that figure after it was added—they surely would have commented if they had, and I doubt it would have then survived to publication. The figure's problems were evidently not flagged (or were ignored) in the subsequent internal government reviews.

So those are the reasons I have *Very High Confidence* in identifying and correcting a prominent misrepresentation of climate science in an official government report. This isn't picking at a nit; it really does matter. The false notion of more frequent US high temperature records is likely to pollute subsequent assessment reports, which invariably cite prior reports. More generally, it matters for those who care about the quality of scientific input to societal decisions and the integrity of the processes by which it's generated. It should also matter to those who proclaim the unimpeachable authority of assessment reports. And it matters for media representations of climate science, which give voice to such misleading "conclusions."

The CSSR's failures on the subject of record temperatures might be due to incompetence, but I suspect otherwise. It would have been more natural for the report to have presented separately the numbers of record highs and record lows, instead of their ratio, which, as we've seen, is rather contrived. But that would have looked something like the declining curves in Figure 5.4, making it difficult to offer increasing temperature extremes as evidence of a broken climate. I would be delighted if someone could explain how this misleading analysis was in the service of informing rather than persuading.

It's no surprise that the media breathlessly spreads the CSSR's disinformation about record temperatures. For example, in March 2019 the Associated Press published a widely syndicated article under the headline "Heat records falling twice as often as cold ones, AP finds."[21] The reporters used data from 424 stations to extend the CSSR running records analysis back to 1920. Interestingly, the figure they show starts only in 1958, omitting more than one-third of the years they analyzed (1920–1957)—perhaps because, as Christy's analysis shows, the 1930s and 1940s would be inconvenient for their thesis.

Nowhere in the AP article (except for the figure heading) do they mention that the numbers of *both* hot and cold records are declining. Incredibly, they even include a quote from a former Weather Channel meteorologist that's directly contradicted by their own figure: "You are getting more extremes. Your chances for getting more dangerous extremes are going up with time."

In short, I would summarize the data on extreme temperatures with the following statement. It has far less headline potential than that from the CSSR quoted on page 105, but has the advantage of being correct:

> *There have been some changes in temperature extremes across the contiguous United States. The annual number of high temperature records set shows no significant trend over the past century nor over the past forty years, but the annual number of record cold nights has declined since 1895, somewhat more rapidly in the past thirty years.*

Of course, temperatures getting milder in this way (fewer harsh winters and cold evenings) makes for a very different (and less alarming) story than torrid summers and blazing afternoons becoming more common. As it happens, the evidence of a rise in the coldest temperatures is perfectly consistent with a warming globe—just not a "roasting" one that lends itself to graphics of bursting thermometers.

HEAT HYPE HEATS UP—2024 UPDATE

Observations from surface weather stations unambiguously show that the globe is warming, although as I discussed in Chapter 1, there are challenges in putting a precise number on that increase. Satellite observations provide an alternative estimate of the warming. They have their own challenges: Observations extend back fifty years at best, and satellite orbits and sensors change over the years. But they offer global coverage in near real time and

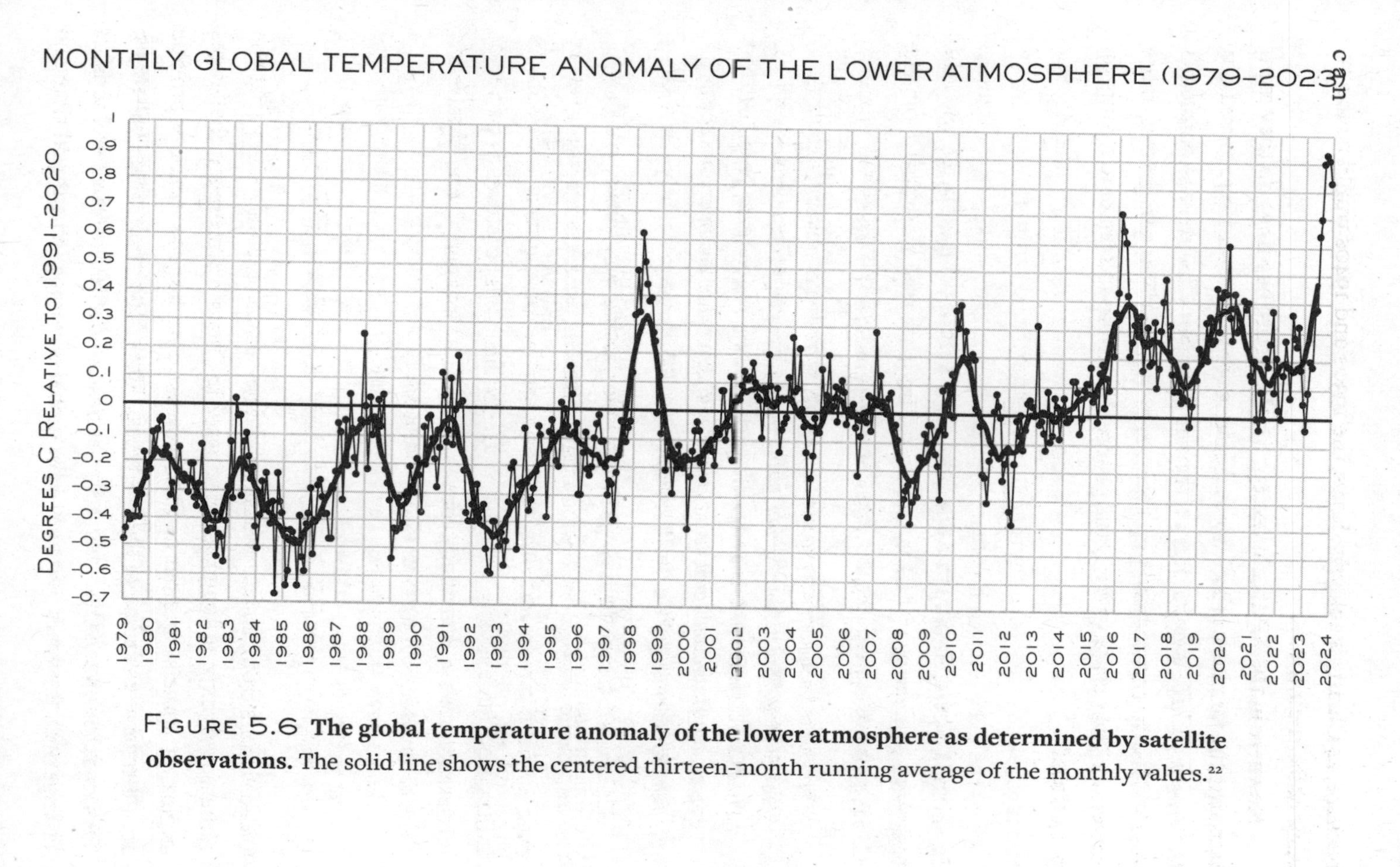

FIGURE 5.6 **The global temperature anomaly of the lower atmosphere as determined by satellite observations.** The solid line shows the centered thirteen-month running average of the monthly values.[22]

measure temperatures not only at the surface but also through the height of the atmosphere.

Figure 5.6 shows the satellite-determined monthly global temperature anomaly of the lower atmosphere. There is warming of about 0.6°C over the forty-five years of data available, with month-to-month and year-to-year fluctuations superimposed on the longer-term trend.

Some part of that longer-term trend can be attributed to human-caused greenhouse gases, which exert a steadily growing warming influence. But the few-year changes are another story. Some are just minor natural wobbles of the ever-changing climate system. The bigger spikes (for example, those in 1998 and 2016) are connected with El Niño events that warm the tropical eastern Pacific Ocean and have a global effect on weather.

This data helps put the most recent temperature rise into context. Although it's drawn much press coverage due to human influences, the strong El Niño brewing in the latter half of 2023 gives us good reason to believe that the temperature will revert to the longer-term trend within a year or so. But that's not guaranteed. The world has been reducing its aerosol emissions, particularly from ships. That's been slightly decreasing the albedo and, hence, increasing the warming influence,[23] as I described in Chapter 2. The next year or two will tell.

As the average temperature increases, for whatever reasons, there will be changes in phenomena directly related to the temperature. As AR6 put it:

> In summary, it is virtually certain that there has been an increase in the number of warm days and nights and a decrease in the number of cold days and nights on the global scale since 1950. Both the coldest extremes and hottest extremes display increasing temperatures. It is very likely that these changes have also occurred at the regional scale in Europe, Australasia, Asia, and North America. It is virtually certain that there has been increases in the intensity and duration of heatwaves and in the number of heatwave days at the global scale.[24]

But as I teach my students, it's important to look at the details of these carefully worded statements—in this case, "What about before 1950?" While global data is poor before then, the US data shows that heat waves

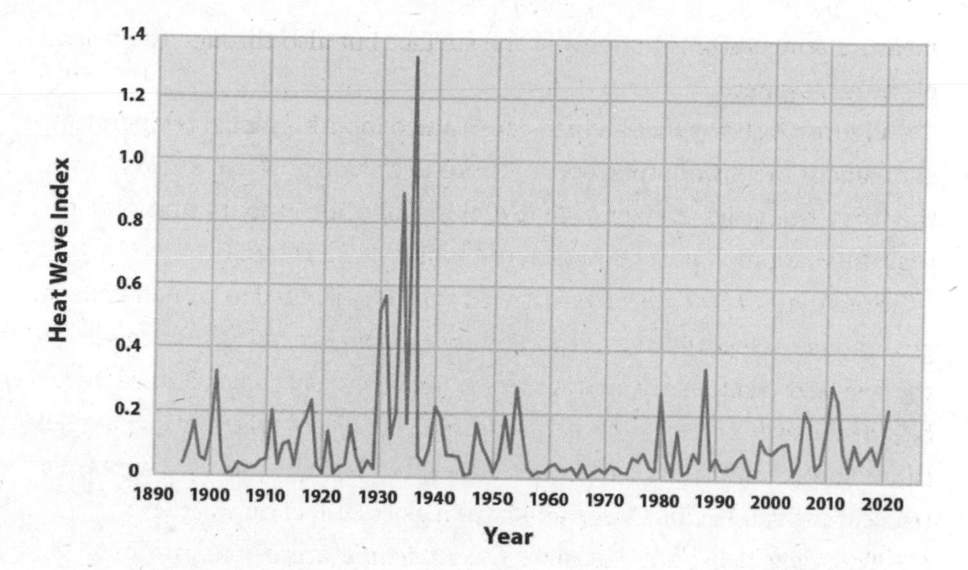

FIGURE 5.7 **The annual values of the US Heat Wave Index, which covers the contiguous forty-eight states** (1895–2021). An index value of 0.2 (for example) could mean that 20 percent of the country experienced one heat wave, 10 percent of the country experienced two heat waves, or some other combination of frequency and area resulted in this value.[25]

have indeed been growing since the 1950s. But the EPA figure above (Figure 5.7) shows that they are no more common recently than in the early twentieth century and are still far below the peak in the 1930s, when human influences on the climate were much smaller.

As these examples show, context is crucial to understanding "unprecedented" weather and climate phenomena. Alas, this context is almost always absent from media or political discussions.

TEMPEST TERRORS

THIS ERA OF DEADLY HURRICANES WAS SUPPOSED TO BE TEMPORARY. NOW IT'S GETTING WORSE.

—*Forbes*, October 7, 2020[1]

Like everyone else, I've heard some version of the above shouted by the media each time a hurricane strikes the US. The message is clear: Storms are becoming more common and more intense, and rising greenhouse gas emissions are going to make it all a lot worse. But the data and research literature are starkly at odds with this message. At the center of this confusion are the assessment reports, which present summary "spin" inconsistent with their own findings. This chapter will chase down the facts about climate and storms, teasing out the truth from the tempest and demonstrating that hurricanes and tornadoes show no changes attributable to human influences.

First, a little background. Technically, "hurricane" is the term for a tropical cyclone in the Atlantic or eastern Pacific; these storms are called "typhoons" in the western Pacific and just "cyclones" in the Bay of Bengal and northern Indian Ocean. I won't make those distinctions and will generally use the US term "hurricanes" for all of them. Up to a few hundred miles in extent, these storm systems feature a low-pressure center (the eye) surrounded by a spiral arrangement (counterclockwise in the Northern Hemisphere, clockwise in the Southern Hemisphere) of thunderstorms and tornadoes producing heavy rain. The lower the eye pressure, the stronger the winds surrounding it. A hurricane has winds greater than 119 km (74 miles) per hour; if weaker, it's called a tropical storm and, weaker still, a tropical depression. Hurricanes are categorized by intensity via the Saffir-Simpson scale, which runs from 1 to 5.[2] Major storms (those in Categories 3 to 5) have winds in excess of 179 km (111 miles) per hour.

Hurricanes grow from tropical depressions (low-pressure areas) born over the oceans alongside the equator. They then move toward the poles, their precise path depending upon the regional winds; most never make it to land. There are about forty-eight hurricanes each year across the globe. Two-thirds of them are in the Northern Hemisphere (where hurricane season is June though November) and one-third is in the South (where the season is November through May). In round numbers, about 60 percent are in the Pacific, 30 percent in the Indian Ocean, and 10 percent in the North Atlantic; they are very rare in the South Atlantic.

Along with tracking the number of hurricanes each year in each category and location, scientists have developed other measures of storm activity. One is the Accumulated Cyclone Energy (ACE), which combines the number of storms with their intensities, weighting each according to strength (storms are weighted by the square of their wind velocity).[3] Another measure is the Power Dissipation Index (PDI), which is similar to

ACE but gives even greater weight to the most intense storms (each storm is weighted by the cube, or third power, of its wind velocity).[4]

There are good records of hurricanes going back to the advent of satellite-based observations in 1966. Who hasn't seen awesome pictures of an eye surrounded by a circular cloud pattern? And while far from complete, aircraft observations extend back to about 1944. Before then, however, there are only records of those storms that made landfall or, occasionally, reports from ships unlucky enough to have encountered one. Going back even further means relying on historical reports and various proxies (*paleotempestology* is the wonderful name for that field of study). So to understand trends over more than seventy years (before the onset of significant human influences), we have to correct for imprecise and incomplete observations, unfortunately all too common across climate science.

The low-pressure areas that become hurricanes arise from evaporation of water from a warm sea surface; that water then releases heat as it condenses high in the atmosphere. This is the same process that grows and sustains a hurricane once it's born. So you might expect to see a steady increase in hurricane activity as the sea surface has warmed. Unfortunately, it's not that simple, as can be seen in the long-term record of the annual number and Accumulated Cyclone Energy of hurricanes in the North Atlantic shown in Figure 6.1. The long-term swings of the Atlantic Multidecadal Oscillation (AMO—as discussed back in Chapter 4) affect the sea surface temperature in the region where hurricanes form and so can enhance or suppress hurricane activity.

Even with an obligingly warm ocean temperature, atmospheric conditions have to be just right for a hurricane to form. Figure 6.2 shows the variation of the Power Dissipation Index along with sea surface temperature. As you can see, strong hurricane activity closely tracks the sea surface temperature until about 2008 . . . and then it doesn't. That's because there are a number of other environmental factors[5] at play, including the amount of wind shear (variation of wind speed or direction with altitude)[6] and the presence of dust from the Sahara (neither of which is well described by climate models).[7,8]

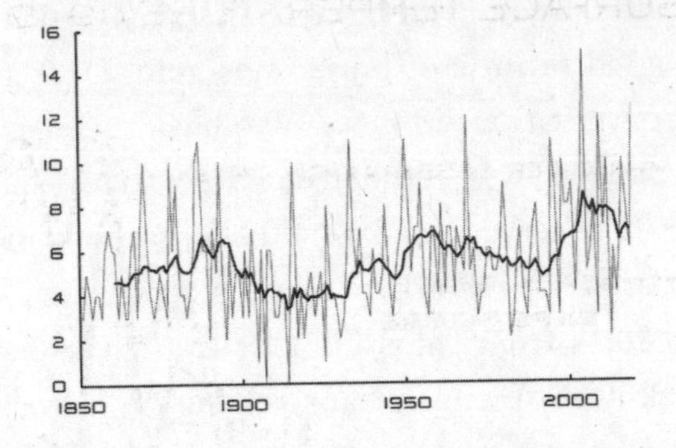

ACCUMULATED CYCLONE ENERGY (1851-2020)

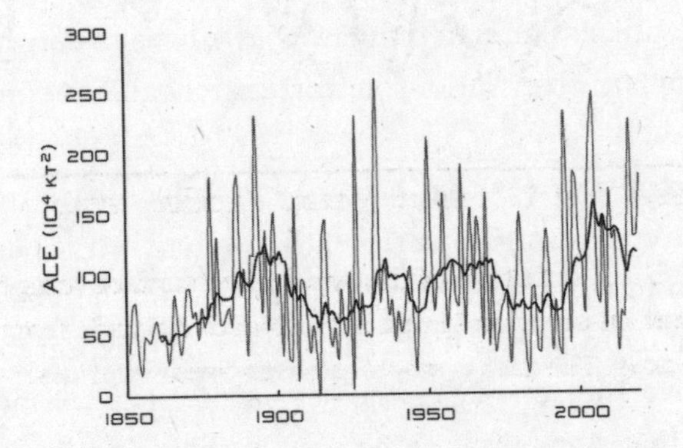

AMO INDEX (1856-2020)

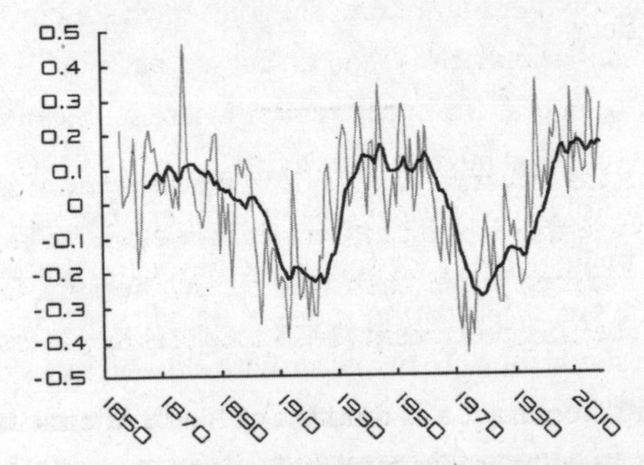

FIGURE 6.1 **Annual number of hurricanes (upper) and Accumulated Cyclone Energy (middle) in the North Atlantic for each year from 1851 to 2020.** The lower panel shows the AMO index from Figure 4.4. In each panel, light lines show the year-to-year variation, while the black line is the ten-year trailing average.[9]

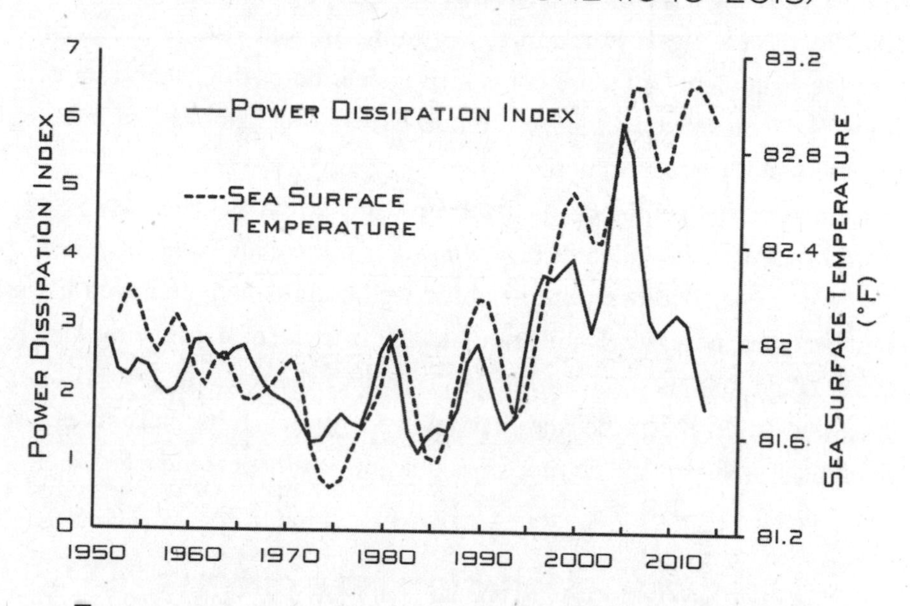

FIGURE 6.2 **Variation of the annual sea surface temperature and Power Dissipation Index in the North Atlantic from 1949 to 2015.** The data is smoothed over 5-year intervals.[10]

Of course, just because temperature isn't the only factor in hurricane formation, that needn't mean warming, whether natural or human-caused, hasn't had an effect.

While advising a US government agency in the summer of 2016, I had reason to look into whether human influences have made hurricanes worse in recent decades. I turned to the (then most recent) National Climate Assessment issued by the US government (NCA2014). Its Key Message 8 reads:

> The intensity, frequency, and duration of North Atlantic hurricanes, as well as the frequency of the strongest (Category 4 and 5) hurricanes, have all increased since the early 1980s. The relative contributions of human and natural causes to these increases are still uncertain. Hurricane-associated storm intensity and rainfall rates are projected to increase as the climate continues to warm.[11]

The report backs up that statement with the graph reproduced in Figure 6.3, showing a seemingly alarming increase in the North Atlantic PDI (that is, the strongest hurricanes) beginning in 1981. While the decline after 2005 evident in Figure 6.2 is also visible here, the general upward trend is emphasized, so that to the non-expert eye, it looks like we're in trouble—and headed for more.

But since the graph began only in 1970, my scientist's curiosity kicked in and I naturally wondered: *How unusual is this trend? What happened in earlier years?* Hurricanes were recorded well before then, and even if the data becomes increasingly uncertain deeper into the past, knowing what it shows could help us shed light on the trends of the present. For example, does the rise in recent decades have precedents when human influences were much smaller? And do those straight lines really portend what's going

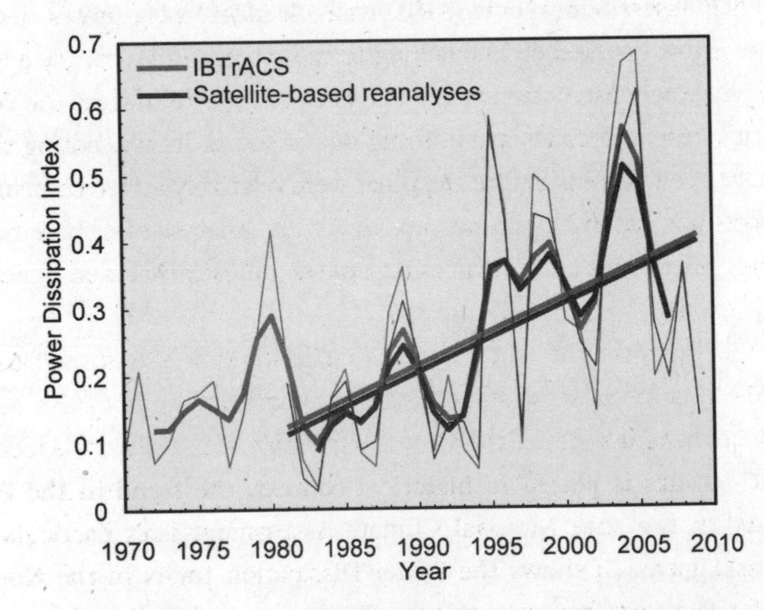

FIGURE 6.3 **Power Dissipation Index in the North Atlantic Ocean.** Two different analyses of the data are shown, along with straight lines indicating the trend in each. (NCA2014, Figure 2.23.)

to happen in the future? If in nothing else, climate data is usually consistent in showing an awful lot of ups and downs.

So I dug into the main research paper cited by the assessment. To my surprise, I found it stated quite explicitly that there are *no* significant trends beyond natural variability in hurricane frequency, intensity, rainfall, or storm surge flooding.[12]

This seemed directly at odds with the National Climate Assessment's alarming figure, so I went back and searched the NCA more thoroughly. On page 769, buried in the text of Appendix 3, I found this statement:

> There has been no significant trend in the global number of tropical cyclones nor has any trend been identified in the number of US landfalling hurricanes.[13]

Wow! I thought to myself. *That's surprising and pretty important. How come this isn't up front as a Key Message?*

The absence of significant trends in hurricane data was hardly unknown to experts at the time the 2014 NCA was being prepared. The IPCC's Fifth Assessment report (AR5), available in late 2013, states clearly that there is low confidence in any long-term increase in hurricane activity. And a 2012 reconstruction of the PDI back to 1880 reinforces the conclusion that recent decades are nothing out of the ordinary, noting that "there have been periods before 1949 that were relatively active compared to the post-1995 era of heightened activity."[14] In other words, there have been times before human influences became significant that were at least as active as today.

Whether because the AMO cycle is currently "high" (as seen back in Figure 6.1) or because of something else, hurricane activity has been higher since the mid-nineties than it was in the 1970s. But when the record of recent decades is placed in historical context, the trend in the PDI highlighted in the 2014 National Climate Assessment isn't particularly surprising. Figure 6.4 shows the Power Dissipation Index in the North Atlantic for each year from 1949 to 2019. The heavy straight line shows the trend that NCA2014 featured so prominently. Given the large year-to-year variations in the data, one could plausibly draw a line from 1960 to 1985

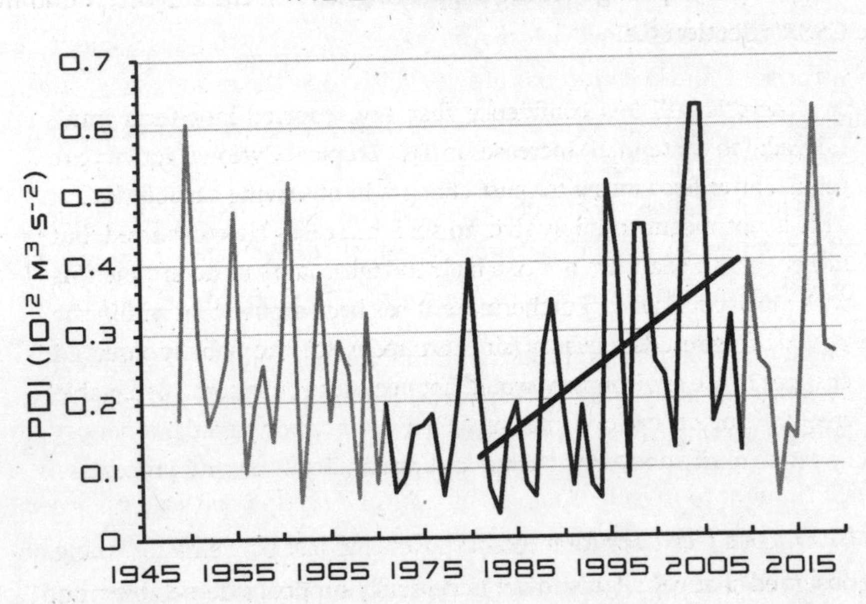

FIGURE 6.4 **Power Dissipation Index in the North Atlantic from 1949 to 2019.** The black data and trend line were highlighted in NCA2014 (our Figure 6.3), while the gray data shows years prior to 1971 and after 2009.[15]

with a comparably rapid *negative* trend. In other words, NCA2014 failed to include data from earlier years that makes the highlighted trend seem much less significant.

The subsequent National Climate Assessment, issued in 2017 as the Climate Science Special Report (CSSR), continued the practice of "burying the lede" on hurricanes. Key Finding 1 of its Chapter 9 reads:

> Human activities have contributed substantially to observed ocean-atmosphere variability in the Atlantic Ocean (*medium confidence*), and these changes have contributed to the observed upward trend in North Atlantic hurricane activity since the 1970s (*medium confidence*).[16]

That's a pretty weak statement for a Key Finding—medium confidence that human activities contributed to variability (*by how much?*), followed in turn by medium confidence that the variability contributed to "an observed upward trend" (*by how much?*). Even so, it is a good deal stronger than you might expect given the supporting text on the subject, found in the CSSR's Section 9.2:

> . . . there is still low confidence that any reported long-term (multidecadal to centennial) increases in TC [Tropical Cyclone] activity are robust, after accounting for past changes in observing capabilities . . . This is not meant to imply that no such increases have occurred, but rather that the data are not of a high enough quality to determine this with much confidence. Furthermore, it has been argued that within the period of highest data quality (since around 1980), the globally observed changes in the environment would not necessarily support a detectable trend in tropical cyclone intensity. That is, the trend signal has not yet had time to rise above the background variability of natural processes.

Wow! Let's step back for a second to let that last part sink in: "the globally observed changes . . . would not necessarily support a detectable trend . . . the trend signal has not yet had time to rise above the background variability of natural processes." If we can't have confidence that a durable trend exists, we certainly can't confidently attribute one to human influences.

The National Academies' review of the CSSR doubled down on burying the lede.[17] On its page 38 is a recommendation that the CSSR emphasize the recent upward trend in PDI (a phenomenon that we saw in Figure 6.4 is not unusual), even if its origins are not understood.

The discussion of hurricanes in the 2017 CSSR is a profound violation of Feynman's Wesson Oil caution, that a scientist must "try to give all of the information to help others to judge the value of your contribution; not just the information that leads to judgment in one particular direction or another."

Recent research only reinforces the lack of "news" about tropical cyclones. A landmark paper in 2019 co-authored by eleven tropical cyclone experts was unusual for presenting the diversity of expert opinion.[18] Those authors found that the *strongest* case for any detectable change in tropical cyclone activity was a very slow northward shift of the average track of storms in the northwest Pacific (0.19° ± 0.125° latitude per decade over the past seventy years, a $1\frac{1}{2}\sigma$ result). Moreover, even for that slow, small change (21 km or 13 miles per decade), eight of the eleven authors had only low to medium confidence. Most significantly, the majority of the authors had only low confidence that any other observed tropical cyclone changes were beyond what could be attributed to natural variability.[19]

I've been unable to find any media coverage (or even a press release) of that paper. Instead, the media continues to promulgate unsupported alarm. For example, to herald the publication of a different study, *USA Today* used the headline "Global warming is making hurricanes stronger, study says."[20] In that study, the researchers used a new method to analyze satellite imagery of tropical cyclones to determine storm intensity.[21] They found a short-term trend toward more intense storms in the North Atlantic basin and linked it to a multidecadal variability that "complicates detection because the climate drivers of that variability are not fully understood." Their bottom-line conclusion reads:

> Ultimately, there are many factors that contribute to the characteristics and observed changes in TC intensity, and this work makes no attempt to formally disentangle all of these factors. In particular, the significant trends identified in this empirical study do not constitute a traditional formal detection, and cannot precisely quantify the contribution from anthropogenic factors.

Yet the second sentence of the *USA Today* article says unequivocally: "Human-caused global warming has strengthened the wind speeds of hurricanes, typhoons and cyclones around the globe."

This is more than the imprecision that is so often a hallmark of climate reporting. The fact is that, while it is not unreasonable to think that warming might indeed lead to some kind of change in hurricane activity at some point, right now there simply isn't evidence that this is happening. Yes, economic damages from hurricanes are increasing, but that's because there are more people and more valuable infrastructure near the coasts, not because storm characteristics are changing long-term.[22] And while it is possible that storms might get worse in the future, an assessment of model-based storm projections under 2°C (3.6°F) of human-caused warming shows changes that can hardly be characterized as catastrophic—medium to high confidence in a 10 to 20 percent increase in many, but not all, measures of storm activity.[23, 24]

Whatever the future holds, the descriptions of hurricane data in the assessment reports mislead by omission. They violate Einstein's famous dictum prominently displayed on the National Academies building in Washington, DC: "The right to search for truth implies also a duty; one must not conceal any part of what one has recognized to be true." As for the media, pointing to hurricanes as an example of the ravages of human-caused climate change is at best unconvincing, and at worst plainly dishonest.

Of course, hurricanes aren't the only storms that wreak havoc and garner headlines. Although tornadoes occur all over the globe, the United States has the greatest number of twisters of any country. US tornadoes are most frequent in the spring, along a swath called Tornado Alley that extends from North Texas up to South Dakota. Tornadoes appear unpredictably and travel along a seemingly random track for an average of 8 km (5 miles). They're highly localized (typically 160 meters, or 500 feet, across, although some can be several times larger), but along that slender path the damage they cause is severe; after lightning strikes, they are perhaps the most "personal" of extreme weather events.

It's natural to ask whether tornadoes have changed in response to changes in climate—and to wonder how they might change in the future, as human influences on the climate grow. For a scientist, answering these questions begins with a look at the data.

Figure 6.5 shows the annual number of tornadoes in the US. It certainly doesn't look like good news: In the past two decades, tornadoes have been more than twice as frequent as they were in the twenty years after 1950, the change occurring over a period when the globe warmed notably.

But this is a perfect illustration of the perils of correlation. A quick Google search reveals that both the number of fishing boats and the amount of movie violence are also among the things that have doubled since 1950, and certainly neither of these trends are due to changes in climate. In the case of tornadoes, the key to this "trend" lies in understanding how the data is compiled—which is often just as important as the data itself. So how are tornadoes counted?

Today, weather radar can detect even very weak tornadoes from distances of more than 160 km (100 miles). Before radar was widely deployed, however, weak tornadoes didn't always make it into the record. While strong tornadoes leave an evident trail of destruction, weaker tornadoes can come and go without a trace, particularly in sparsely populated areas. To see if there's been a real change in the number of tornadoes over the past seventy years, we've got to correct for the observing bias in favor of strong storms early in the record.

Tornado strength is measured according to the Enhanced Fujita Scale; the original Fujita Scale was developed in 1971, with the enhanced version adopted in 2007. Strength categories range from EF0 for the very weakest storms to EF5 for those with winds above 260 mph. In the US today, 60 percent

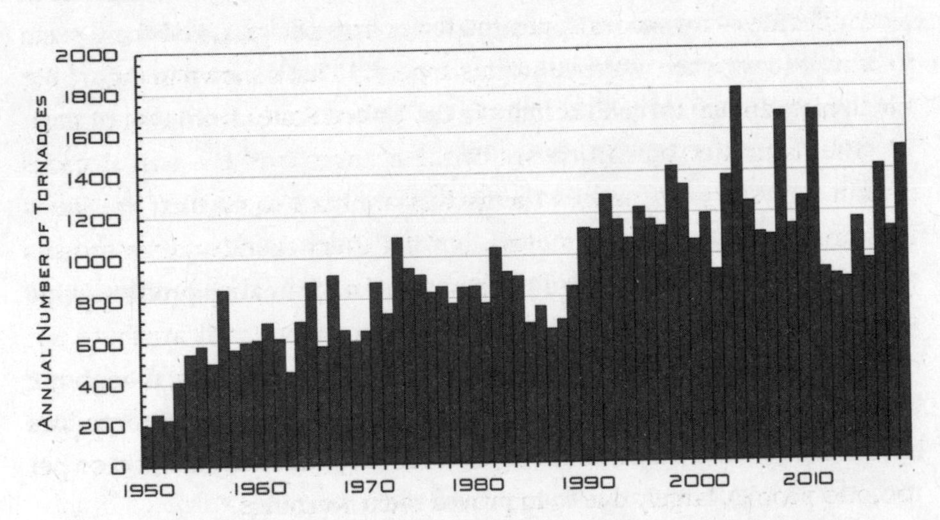

FIGURE 6.5 **Number of tornadoes recorded by NOAA each year from 1950 to 2019 in the contiguous forty-eight states.**[25]

of recorded tornadoes are category EF0, while back in 1950 such storms made up only about 20 percent of the recorded total. This suggests that the growth in the number of recorded tornadoes is due to counting more weak storms in recent decades, which according to NOAA is, in fact, the case.[26]

We can correct for the past observation bias against counting weak storms by looking only at storms in category EF1 and stronger (those most likely to cause destruction). This gives us the two graphs in Figure 6.6. The upper graph, an annual count of US tornadoes strength EF1 or greater, shows no trend over the past sixty years, although there is a hint of a forty-year cycle. The lower graph looks at only the strongest tornadoes (EF3 or above) and shows that their number *decreased* by about 40 percent during the sixty years following 1954. In other words, as human influences have grown since the middle of the twentieth century, the number of significant tornadoes hasn't changed much at all, but the strongest storms have become less frequent.

Even as the total number has declined, there has been a shift in tornadoes away from the Central and Southern Great Plains and toward the Midwest and Southeast.[27] Trends in other tornado properties are less certain. As the CSSR notes, there has been more variability in tornado occurrence in recent decades—tornadoes appear on fewer days per year, and there seem to be more days when multiple storms appear.[28] That's shown in Figure 6.7, which plots annual tornado activity in the United States from 1955 to 2013.

The natural or human causes of the changes over the past decades remain a mystery. Tornadoes themselves are hard to predict: We know they are formed by thunderstorms, but not every thunderstorm spawns tornadoes, even if the storm's temperature and humidity profiles, wind shear, and "spin" are favorable. However, we *can* confidently attribute one dramatic tornado-related change primarily to human actions, though not in the way that usually comes to mind—annual US deaths from tornadoes have fallen by more than a factor of ten since 1875 (currently about 0.02 per 100,000 people), largely due to improved radar warnings.[30]

As for how tornadoes might change in the future, our discussion of both climate models and the factors involved in creating tornadoes makes it clear

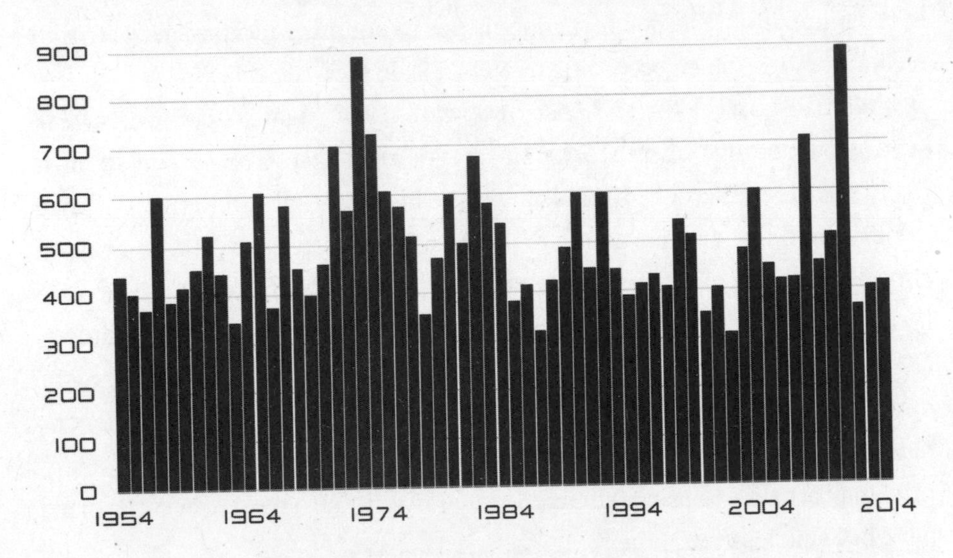

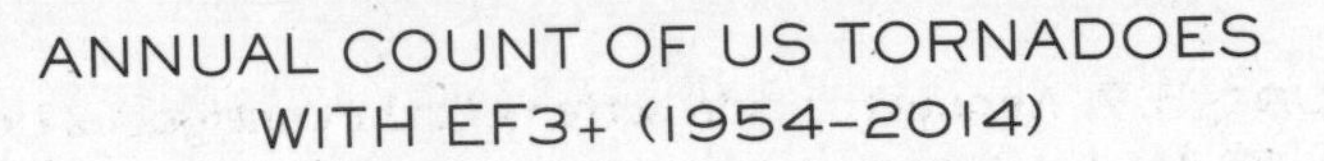

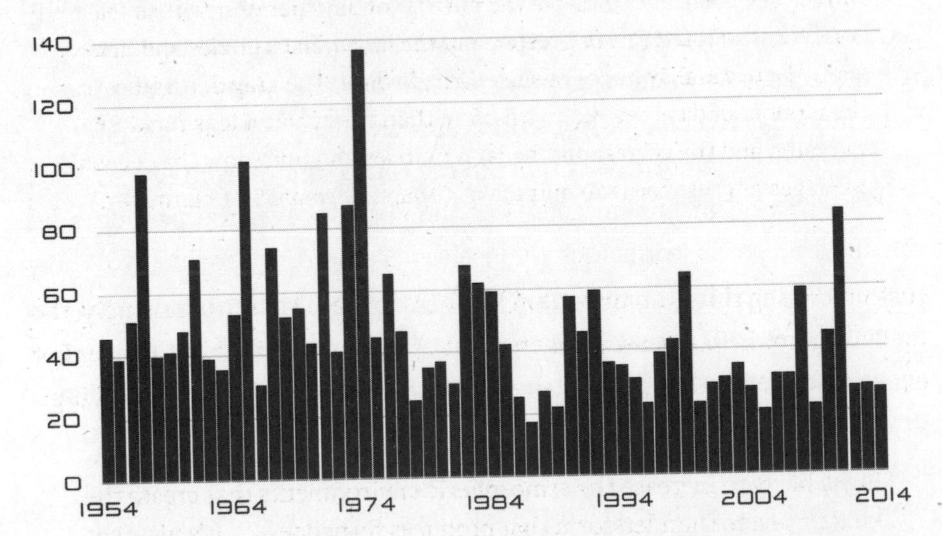

FIGURE 6.6 **Number of tornadoes recorded by NOAA each year from 1954 through 2014 in the forty-eight contiguous states.** The upper panel shows tornadoes in category EF1 or stronger, while the lower panel shows only the strongest tornadoes, which have strengths EF3 or greater.[29]

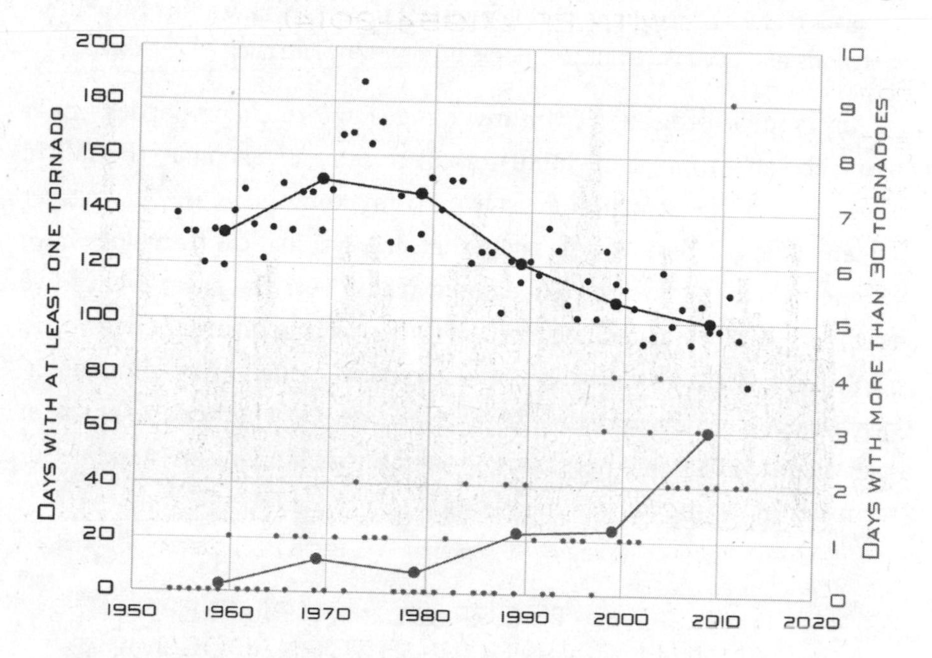

FIGURE 6.7 **Annual tornado activity in the contiguous United States.** The black dots indicate the number of days per year with at least one tornado rated EF1 or greater, and the larger black circles and line show the decadal averages of such tornado days. The gray dots indicate the number of days per year with more than thirty tornadoes rated EF1 or greater, and the corresponding larger circles and line show the decadal averages of these tornado outbreaks. (Adapted from CSSR Figure 9.3.)

that predicting this is an uncertain business indeed. Yet unsurprisingly, the media cannot help at least implying that things are going to get worse. For example, the *New York Times* quotes a Stanford climate scientist as saying:

> We do have strong evidence that at the large scale that global warming is likely to increase the atmospheric environments that create the kind of severe thunderstorm that produces tornadoes . . . It's just that we can't distinguish the signal from the noise.[31]

At the same time, the IPCC's 2012 Special Report on Extreme Events states in the Executive Summary of its Chapter 3:

> There is low confidence in projections of small-scale phenomena such as tornadoes because competing physical processes may affect future trends and because climate models do not simulate such phenomena.

Any credible projection of future changes in tornado properties would have to be able to explain the historical trends, for example, the fall in the number of the strongest twisters. To my knowledge, that hasn't yet happened. So the best we can say is that, if anything, US tornadoes have become more benign as the globe has warmed over the past seventy-five years, and we have no credible method for projecting future changes.

Unfortunately, that last is not an unusual situation in climate science. Remember that next time you hear someone, whether a scientist or a weather presenter or a politician, proclaim that humans are making our storms stormier and our weather worse.

WILD WEATHER—2024 UPDATE

Today's everywhere-all-the-time global media coverage distorts our perception of extreme weather. More and more one hears, as though established fact, that "extreme weather is becoming more common." To judge from media coverage, extreme weather events such as heat waves, storms, droughts, floods, and wildfires are increasingly ravaging the planet. Not a surprise, given the need to attract clicks and eyeballs, the impact of repetition, and how common "extreme" events have always been on a global scale—much more common than you'd think.

Here's why. Suppose an extreme weather event, like a flood in the American Midwest, takes place in a one-hundred-miles-sided square. There are six thousand such squares covering the land area on the earth. A "once in a century" flood will occur in each square about once every one hundred years—meaning that, on average, there would be sixty extreme floods around the globe each year. That is more than one "once in a century" flood every week!

One can quibble with the assumptions that go into this estimate, and of course we would also have to count other types of events. But the upshot is that although we can't say exactly how often a hundred-year weather event might occur somewhere around the globe, it's clear that they're far more common than once per year, as they have been for a long time. What has increased is simply our awareness of them.

Observations of the climate over the past many decades tell pretty much the same story. This table, adapted from one in AR6, reaffirms that trends have not (yet?) been observed in most weather extremes:

Impact-driver Type	Impact-driver Category	Observed trend
Heat and Cold	Mean air temperature	+
	Extreme heat	+
	Cold spell	-
	Frost	
Wet and Dry	Mean precipitation	
	River flood	
	Heavy precipitation and pluvial flood	
	Landslide	
	Aridity	
	Hydrological drought	
	Agricultural and ecological drought	
	Fire weather	
Wind	Mean wind speed	
	Severe wind storm	
	Tropical cyclone	
	Sand and dust storm	

Impact-driver Type	Impact-driver Category	Observed trend
Snow and Ice	Snow, glacier, and ice sheet	
	Permafrost	-
	Lake, river, and sea ice	- (Arctic Sea ice only)
	Heavy snowfall and ice storm	
	Hail	
	Snow avalanche	
Coastal	Relative sea level	
	Coastal flood	
	Coastal erosion	
Open Ocean	Mean ocean temperature	+
	Marine heat wave	
	Ocean acidity	
	Ocean salinity	+
	Dissolved oxygen	-
Other	Air pollution weather	
	Atmosphere CO_2 at surface	+
	Radiation at surface	

FIGURE 6.8 **Trends detected in various drivers of climate impact.** The degree of shading indicates the confidence with which a trend has been detected in the historical period, while the + or – indicates the sign of the trend. No shading indicates low confidence in any trend.[32]

The *projection* of changes in these phenomena, which is quite a separate matter from whether changes have already been observed, is a highly uncertain business, because of their small extent, the deficiencies of the models, and how infrequent any particular type of extreme event is in any particular region.

HURRICANES—2024 UPDATE

This chapter discussed the fact that studies had found no long-term trend in the properties of hurricanes (technically called "tropical cyclones"). Since the first edition, AR6 has reaffirmed the finding that

> There is low confidence in most reported long-term (multidecadal to centennial) trends in TC [tropical cyclone] frequency- or intensity-based metrics.[33]

but then noted

> It is likely that the proportion of intense tropical cyclones has increased over the last four decades and that this cannot be explained entirely by natural variability.[34]

and that

> Observational evidence for significant global increases in the proportion of major TC intensities is consistent with both theory and numerical modeling simulations.[35]

These latter two findings are based upon general physical principles of hurricanes and the Kossin et al. 2020 observational paper[36] I discussed earlier in the chapter.

Unfortunately, a published correction to the Kossin et al. paper eliminated the statistical significance of the trends it had claimed in all but the Atlantic and southern Indian Ocean basins.[37] And for the remaining trend in the North Atlantic, a historical analysis suggests that the increase since 1980 in the fraction of storms in the strongest category is merely a return to normal.[38] As the authors of the analysis note:

> We hypothesize that these recent [late 20th century] increases contain a substantial, even dominant, contribution from internal climate variability, and/or late-20th century aerosol increases and subsequent decreases, in addition to any contributions from recent greenhouse gas-induced warming.

Overall, the hurricane story remains quite unsettled. While there are continuing hints of a greater proportion of more intense storms or that storms are intensifying more rapidly, the confident detection of such trends over decades, much less their attribution to human influences, remains a work in progress.

PRECIPITATION PERILS—FROM FLOODS TO FIRES

When I joined the Obama administration in May 2009, my wife and I moved to Chevy Chase, a Maryland suburb just north of Washington, DC. The lovely spring of that year, the not-so-lovely summer, and the pleasant fall gave way to the snowiest winter ever recorded in the Capital area—the largest storm, dubbed "Snowmageddon" by the local media, dumped twenty-eight inches over two days. On two separate occasions, unable to shovel our way out the front door, we were trapped in our house for several days with no electricity and no heat save for that most ancient of energy sources, wood in the fireplace. The federal government shut down for days as well.

As is the custom these days whenever weather surprises us, the term "climate change" was bandied about in describing the event. While some asserted that our changing climate had clearly played a role, others—particularly those who insist that any evidence of human-caused climate change is part of a "hoax"—touted Snowmageddon as "proof" that the globe could not be warming after all.

Psychology researchers at Columbia University have found that our views of changes in the climate are influenced by the weather in a fairly simplistic way: When we think it's warmer than usual, we're more likely to be concerned about a warming climate, and vice versa—despite the fundamental fact that, as we've seen throughout this book, weather and climate are not at all the same thing.[1] The relationships between the two are complicated, especially for weather phenomena related to precipitation, otherwise known as rain and snow. For example, though it may seem counterintuitive, rising temperatures can indeed lead to more snow—for instance, if a rise in low temperatures keeps the Arctic Ocean from freezing in winter, more water will evaporate into the atmosphere.

Since climate is a statistical concept over decades, no individual weather event can ever be firmly attributed to human influences, but it's certainly possible that human influences made Snowmageddon, well, snowier. Yet such a statement, as always in science, is ultimately judged true or false by comparison with the data—in this case, long-term changes in the average weather.

Luckily, it's easy to check long-term trends in the weather for almost any location in the developed world. Figure 7.1 shows annual snowfall totals in the Washington, DC, area. The trend has been toward declining total snowfall, with about a 40 percent drop over the 130-year period covered (1889 to 2018). But the fifteen-year trailing average shows ups and downs, and the year-to-year variation is even more dramatic.

So just how unusual was Snowmaggedon? We can judge that from a list of the snowiest and least snowy DC winters.[2] The Snowmaggedon winter of 2009–10 was indeed the snowiest since 1888, as is evident in Figure 7.1. But the (barely) second snowiest winter was 1898–99, more than a century before that, and well before human influences on the climate were significant. Seven of the fifteen snowiest years, about half, occurred after 1950, which is what you'd expect if there were no significant trends (the sixty-seven years from 1950 to 2017 are about half of the 130-year history).

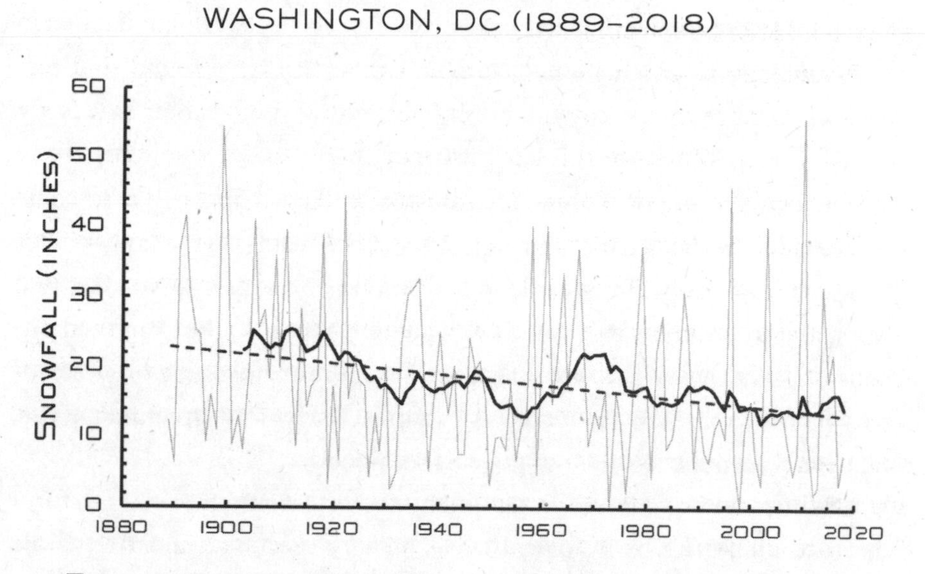

FIGURE 7.1 **Annual cold season snowfall totals for Washington, DC, from 1889 to 2018.** The dotted line shows the trend, while the solid line shows the fifteen-year trailing average. Data points are labeled by the year in which January occurred.[3]

On the other hand, five of the fifteen *least* snowy years occurred just in the eighteen years after 2000, whereas if there were no trend, one would have expected (18 × 15)/130, or about 2. With no trend, there would be less than a 3 percent chance of finding five or more of the fifteen least snowy winters during those years. So the record of extreme snowfalls agrees with what the trend of annual totals suggests—if anything, human influences are making DC less snowy, rather than more.

Of course, eighteen years is barely enough to say anything about climate, let alone changes in it. And DC is only one location on a much larger globe. To better judge possible changes in events related to precipitation—snowfall and rainfall, droughts and flooding, wildfires—we need to look at the bigger picture of precipitation as the globe has warmed during the past century. Are droughts becoming more or less severe? Are floods becoming more or less frequent? Are wildfires becoming more or less common?

We'll turn to the data to answer these questions, and look at what those answers suggest about trickier queries, for instance: What might happen in the future as human influences grow?

The amount of water on the earth is essentially fixed. Almost all of it (some 97 percent) is in the oceans, and almost all of the rest is on the land—in ice and snow (especially the Greenland and Antarctic ice sheets), in lakes and rivers, and in groundwater. But as we saw in Chapter 2, the one hundred-thousandth of the earth's water that resides in its atmosphere plays a central role in climate—water vapor is the most important greenhouse gas, and clouds account for most of the earth's albedo.

The sun's energy moves water among these various reservoirs to form what's termed "the hydrological cycle." The largest and most dynamic part of this cycle is the flow of water from the earth's surface into the atmosphere (85 percent of this flow comes from evaporation of the ocean, the other 15 percent from the land, much of it transpired by plants). That water remains aloft for an average of ten days before condensing and falling back to the surface as rain or snow (77 percent falling on the ocean and 23 percent on the land).

Precipitation depends upon both how much water vapor is in the air and the air's temperature. Water vapor will condense into liquid or ice—and fall out of the air when the temperature drops. That's why you can see your warm, moist breath on a cold day. For that reason, while the average annual precipitation over the globe is 980 mm (38.6 inches) of water—that is, if every location on earth received the same amount of precipitation every year, that's about what we'd all get—in practice it varies greatly with the weather, the season, and, most importantly, location.[4] Globally, rainfall is high right near the equator (where much of the evaporated water comes back down as the warm, moist air rises and cools), but low where dry air descends, creating the bands of deserts that flank the equator. The driest place on Earth is in South America, on the northern edge of the Chilean Atacama Desert: Arica, which averages 0.6 mm (0.02

inches) of rain each year. The wettest is Mawsynram, India, which averages 11,871 mm (467 inches).

All else being equal, the hydrological cycle is expected to intensify as the globe warms: that is, there'll be more evaporation, and the warmer air will be able to carry more water, leading to more precipitation. Precipitation is also expected to become "lumpier," with dry areas becoming drier and wet areas wetter with more periods of intense rainfall. This could lead to an increase in flooding in some areas, but since higher temperatures would also increase evaporation from land, droughts might also increase. There is little consensus among models about exactly how, where, and when these changes would play out.

Unfortunately, not only are these expectations uncertain, but also it's hard to acquire and analyze data that would test them, even to answer a basic question like "How has average precipitation changed?" Unlike temperature anomalies, precipitation can vary greatly over short times and distances—it can be raining somewhere, yet be perfectly dry twenty (or even two) miles away. That's because, as mentioned before, it involves a sudden change in the properties of water: depending upon the temperature and the amount of water vapor, water will either condense and fall out, or it won't. So unlike with temperature, there is no easy way to combine precipitation data from scattered weather stations to get at the bigger picture.

Nevertheless, a combination of ground and satellite observations can give us a long-term global picture, most accurately over land, where weather stations are most numerous. Figure 7.2 shows global precipitation anomalies (deviations from the average) over land have increased at an average rate of 0.2 percent per decade since 1901. But this isn't a very robust statement because of the high variability of precipitation, sparse data, and the smallness of the change; in fact, as you can see, a steady trend is a poor description of the data. As the IPCC's Fifth Assessment Report (AR5) of 2013 noted:

> Confidence in precipitation change averaged over global land areas since 1901 is *low* prior to 1951 and *medium* afterwards. Averaged over the mid-latitude land areas of the Northern Hemisphere, precipitation

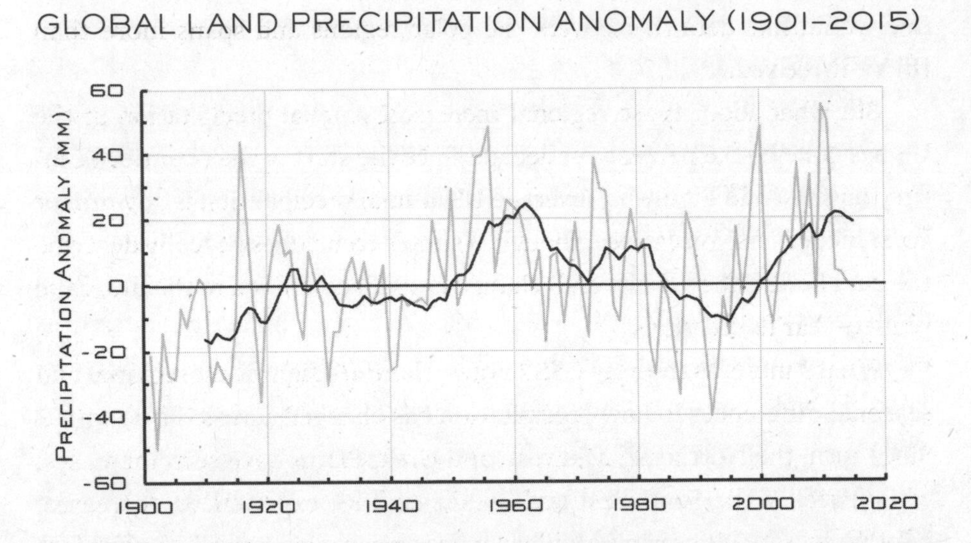

FIGURE 7.2 **Land precipitation anomalies from 1901 to 2015.** The gray line shows the annual values, while the black line is the ten-year trailing average. The average annual precipitation over land is 818 mm.[5]

has increased since 1901 (*medium confidence* before and *high confidence* after 1951). For other latitudes area-averaged long-term positive or negative trends have *low confidence*.[6]

This quote makes plain that there's good evidence for a regional increase in precipitation over land in that part of the Northern Hemisphere that includes the US, Europe, and temperate Asia, especially since 1951, but a more global pattern simply isn't there. A paper published in 2018 reinforces that last conclusion by analyzing more than thirty-three years of high-quality satellite observations covering the globe between 60° S and 60° N (everywhere except the polar regions):

> . . . there seems not to be any detectable and significant positive trends in the amount of global precipitation due to the now well-established increasing global temperature. While there are regional trends, there is no evidence of increase in precipitation at the global scale in response to the observed global warming.[7]

This doesn't accord with the notion that a warming globe will accelerate the hydrological cycle—more rain, more floods. To be certain, though, one would like data that covers the polar regions and spans more than thirty-three years.[8]

But what about those regional increases? Annual precipitation in the US has gone up 0.6 percent per decade since the start of the twentieth century, as shown in Figure 7.3 (average US annual precipitation is 767 mm or 30.21 inches). As you can see, though, a simple trend doesn't really describe the data here, either: the overall change is small compared to the dramatic year-to-year fluctuations.

What's more, as the 2017 CSSR notes, there are significant regional and seasonal differences in how precipitation has changed across the country.[9] Since 1901, the Northeast, Midwest, and Great Plains have seen increases, while parts of the Southwest and Southeast have experienced decreases. In other words, US precipitation has indeed risen a bit overall, but the fact that it varies over both years and location much more than the trend itself makes it hard to draw any solid conclusions about the relative roles of human influences and natural variability.

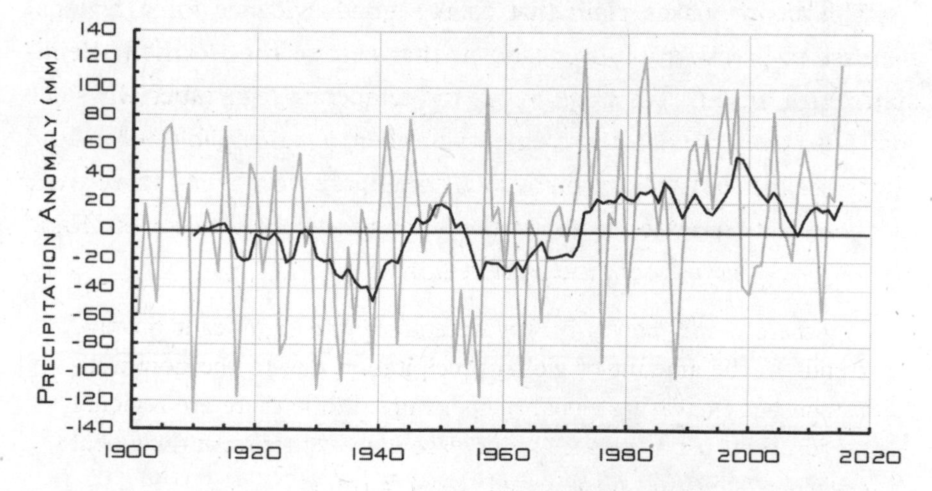

FIGURE 7.3 **Precipitation anomalies for the contiguous United States from 1901 to 2015.** The gray line shows the annual values, while the black line is the ten-year trailing average. The average annual precipitation is 767 mm.[10]

One detectable change that *has* occurred in US precipitation is that it has become lumpier over the past four decades—there are more episodes of intense rainfall, or episodes that account for an outsized portion of the annual total precipitation, as shown in Figure 7.4. The increase has been greatest in the Northeast and Upper Midwest, while it's much smaller in the West. AR5 notes, rather weakly, that something similar holds for the globe as a whole:

> . . . it is likely that since 1951 there have been statistically significant increases in the number of heavy precipitation events (e.g., above the 95th percentile) in more regions than there have been statistically significant decreases, but there are strong regional and subregional variations in the trends.[11]

Another aspect of precipitation of particular importance to the climate system is snow cover, which enhances the land's albedo. It most obviously

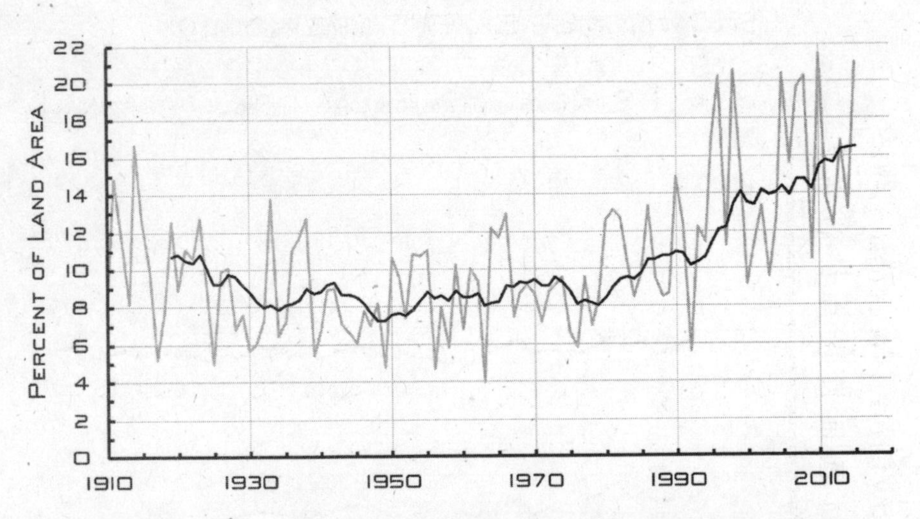

FIGURE 7.4 **The percentage of the land area of the contiguous forty-eight states where a much greater than normal portion of annual precipitation has come from extreme single-day precipitation events.** The gray line represents individual years, while the black line is a ten-year trailing average. (Adapted from EPA data.)[12]

depends upon snowfall, but also upon temperature. Almost all (98 percent) of the earth's snow cover is on the land in the Northern Hemisphere. The amount shows a seasonal cycle—high in the winter and low in the summer. A Key Finding of the CSSR's Chapter 7 is that:

> Northern Hemisphere spring snow cover extent, North America maximum snow depth, snow water equivalent in the western United States, and extreme snowfall years in the southern and western United States have all declined, while extreme snowfall years in parts of the northern United States have increased (*medium confidence*).

Satellite observations of Northern Hemisphere snow cover have been ongoing since 1967—Figure 7.5 shows the data over forty years. Indeed, as the CSSR notes, there is a pronounced decline in snow cover during the spring (and also, to some extent, in summer) as would be expected in a warming globe—especially one in which low temperatures are increasing,

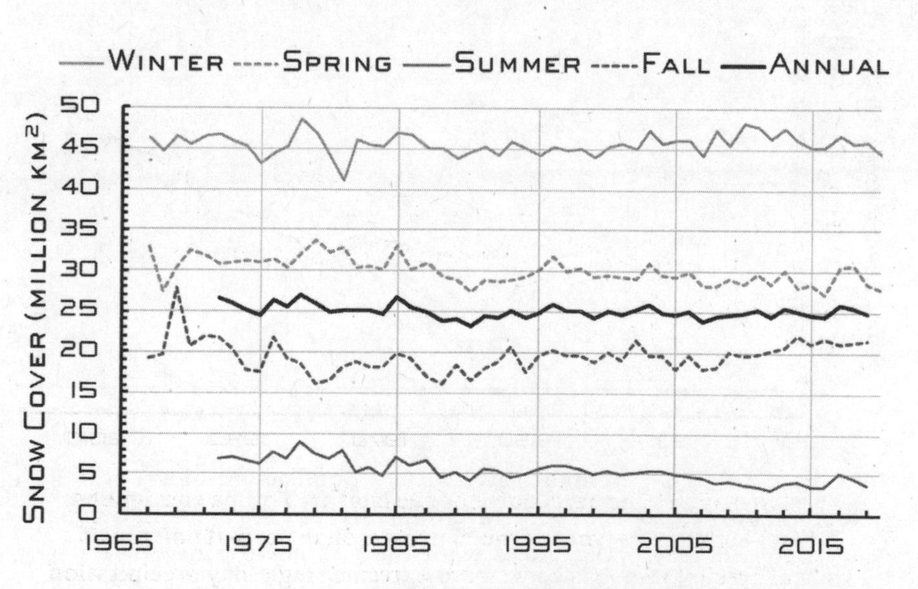

FIGURE 7.5 **Seasonal snow cover extents for land in the Northern Hemisphere from 1967 to 2020 as determined from satellite observations.**[13,14]

as discussed in Chapter 5—while snow cover during the fall and winter has been increasing modestly.

Those contrasting seasonal trends combine to produce a more gradually changing annual snow cover, as shown in more detail in Figure 7.6. There appears to be a distinct drop of about 0.8 million km² (3 percent of the annual average of 25 million km²) between relatively constant values before and after 1989. Declining snow cover is again consistent with a warming globe, though another factor might be dust and soot on the snow, which accelerates melting by absorbing more sunlight. However, a straight-line trend doesn't really describe the data, since there has been no change in the ten-year average during the thirty years since 1990, even as the globe has warmed 0.5°C (0.9°F). I'm surprised not to have seen that in the CSSR's Key Finding (or, in fact, anywhere else in the report).

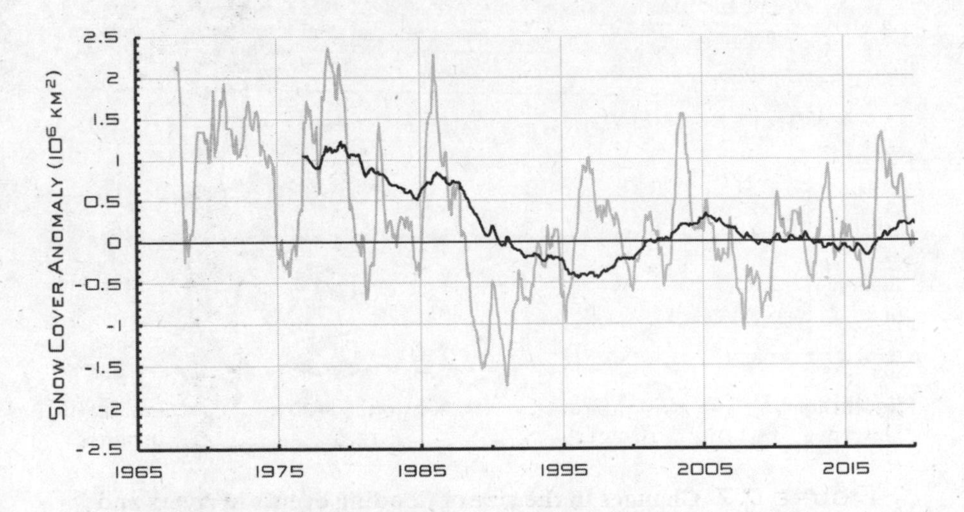

FIGURE 7.6 **Anomalies in Northern Hemisphere snow cover extent, 1967–2020.** The spiky gray line shows the twelve-month trailing average of the monthly values, while the black curve is the ten-year trailing average. Anomalies are relative to the 1981–2010 baseline.[15]

Of course it's the extremes associated with precipitation—floods and droughts—that land "climate" in the news. The modest changes in US rainfall during the past century haven't changed the average incidence of floods. However, trends in flooding vary across the country, with some locations experiencing increases and some decreases, as seen in Figure 7.7, which shows location-specific changes in the size of flooding events in rivers and streams.

CHANGE IN THE MAGNITUDE OF RIVER FLOODING IN THE UNITED STATES (1965–2015)

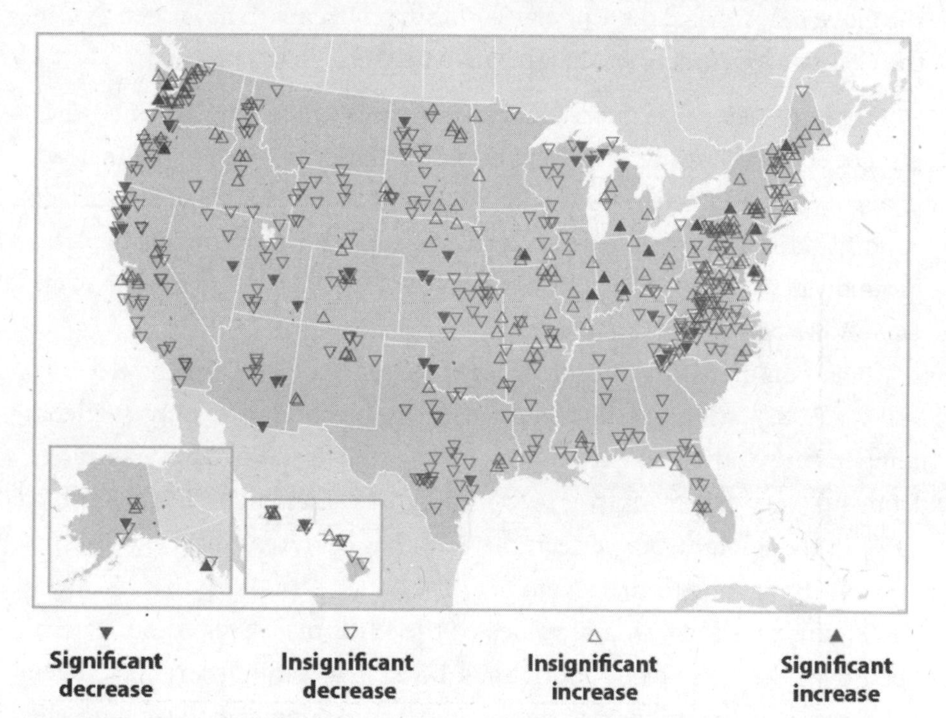

FIGURE 7.7 **Changes in the size of flooding events in rivers and streams in the United States between 1965 and 2015.** Upward-pointing symbols show locations where floods have become larger; downward-pointing symbols show locations where floods have become smaller. The larger, solid symbols represent stations where the change was statistically significant.[16]

As for changes in floods globally, AR5 expresses "*low confidence* regarding the sign of trend in the magnitude and/or frequency of floods on a global scale." In other words, we don't know whether floods globally are increasing, decreasing, or doing nothing at all. The report offers a longer-term perspective in the conclusion of its Section 5.5:

> In summary, there is *high confidence* that past floods larger than those recorded since the 20th century have occurred during the past 500 years in northern and central Europe, western Mediterranean region, and eastern Asia. There is, however, *medium confidence* that in the Near East, India, central North America, modern large floods are comparable to or surpass historical floods in magnitude and/or frequency.

Droughts are even more difficult to assess than floods, since they are not solely the result of precipitation (or rather the lack of it). Instead, droughts involve some combination of temperature, precipitation, surface runoff, and soil moisture. Human activities, such as irrigation that depletes groundwater or the overplowing of the US Great Plains during the 1930s Dust Bowl, can also play a role.

One common measure of drought is the Palmer Drought Severity Index (PDSI), which estimates dryness by combining readily available temperature and precipitation data.[17] The PDSI for a location can range from −10 (very dry) to +10 (very wet), but most values fall in the range of −4 to +4. Although far from perfect, the PDSI has been reasonably successful at quantifying long-term drought.

Figure 7.8 shows annual values of the PDSI from 1895 to 2015, averaged over the forty-eight contiguous US states. While there have been brief regional variations across the country during that time, and the past fifty years have been slightly wetter than average, it's difficult to see much long-term change. The AR5 says pretty much the same thing for the globe as a whole, expressing—no doubt to the surprise of many—"*low confidence* in a global-scale trend in drought or dryness since the middle of the 20th century."

AVERAGE DROUGHT CONDITIONS IN THE CONTIGUOUS 48 STATES (1895–2015)

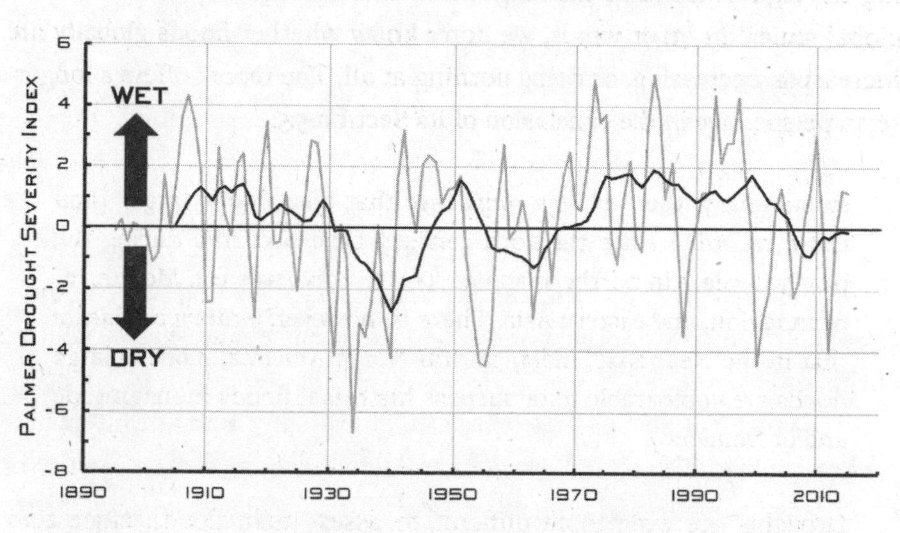

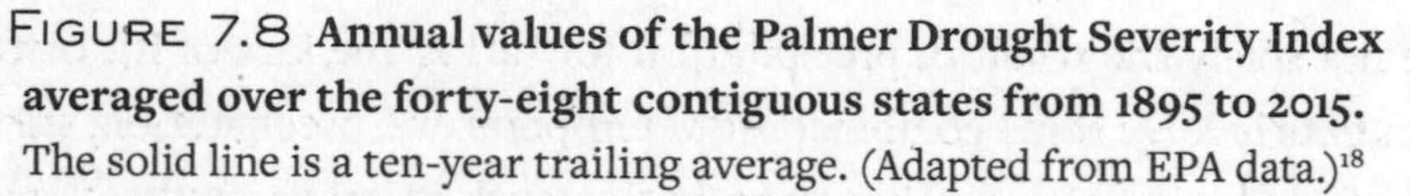
FIGURE 7.8 **Annual values of the Palmer Drought Severity Index averaged over the forty-eight contiguous states from 1895 to 2015.** The solid line is a ten-year trailing average. (Adapted from EPA data.)[18]

But drought *is* a serious issue for regions within the US, as it has been for millennia. You can see that in Figure 7.9, which charts droughts in the Southwest United States over some 1,200 years. The record, derived

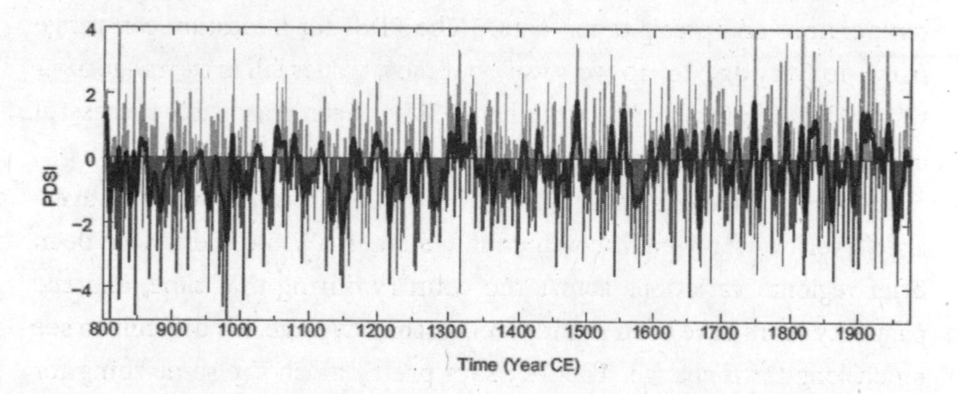

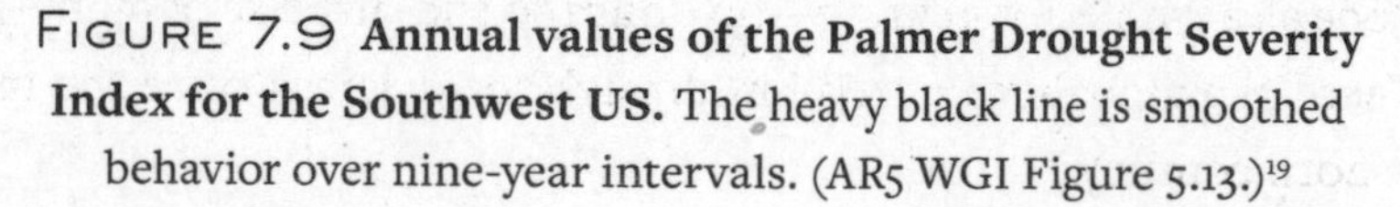
FIGURE 7.9 **Annual values of the Palmer Drought Severity Index for the Southwest US.** The heavy black line is smoothed behavior over nine-year intervals. (AR5 WGI Figure 5.13.)[19]

mostly from tree ring data, shows many droughts that last for decades. Any before 1900 cannot have been due to human influence, and the so-called megadroughts from 900 to 1300 AD are thought to be associated with the naturally warmer globe at that time.[20]

Models project that the Southwest will become steadily drier as the globe warms, but the data shown for the twentieth century are well within the historical context and, as the AR5 notes, the current impact of human influences seems weak in comparison with natural variability.[21] A study published in 2020 confirms the notion that a leading cause of multiyear US droughts over the past millennium has been internal variability of the atmosphere.[22]

The 2009 National Climate Assessment explicitly noted the large natural variability in Southwest droughts by including a graph of the Colorado River flow reconstructed back over 1,200 years.[23] The accompanying text says, "These data reveal that some droughts in the past have been more severe and longer lasting than any experienced in the last 100 years." In 2014, the IPCC's AR5 was similarly straightforward in one of the summary statements for its Chapter 5:

> There is high confidence for droughts during the last millennium of greater magnitude and longer duration than those observed since the beginning of the 20th century in many regions.

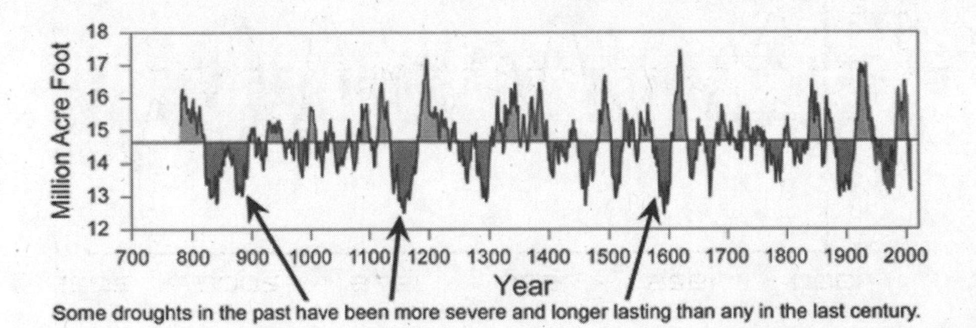

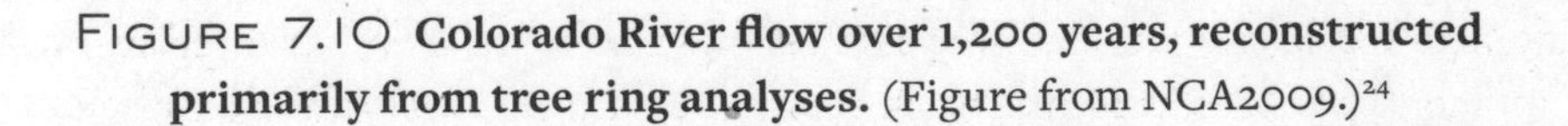

FIGURE 7.10 **Colorado River flow over 1,200 years, reconstructed primarily from tree ring analyses.** (Figure from NCA2009.)[24]

That kind of long-term perspective makes it difficult to attribute recent droughts solely to human influences. But, oddly, it's entirely absent from the 2014 National Climate Assessment. The subsequent CSSR of 2017 doesn't offer any figure like our Figures 7.9 and 7.10, but it does have half a page of text describing the past millennium.[25] Alas, it then spends about twice as long discussing the then most recent six-year California drought, which that state's governor had declared over six months before the CSSR was released.[26] As you can see from Figure 7.11, which shows the Palmer Drought Severity Index for California since 1901, that six-year drought was at its worst during 2014; by 2019 coverage was of the "wet" winter. It is hard to justify a *climate* assessment analyzing any "trend" shorter than ten years, even if that makes its conclusions less newsworthy. On longer timescales, the state has moved toward drought since 2000; it remains to be seen whether that will persist in the coming decades.

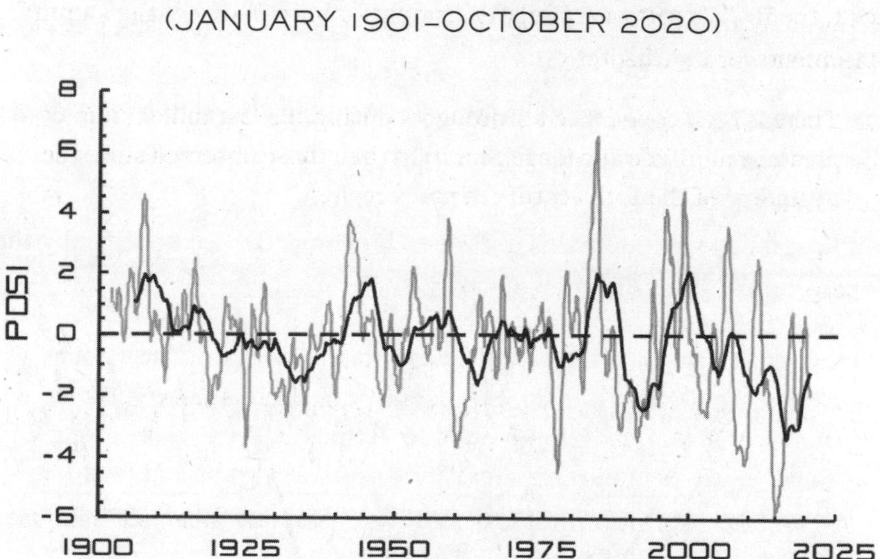

FIGURE 7.11 **Palmer Drought Severity Index for California from January 1901 to October 2020.** The gray curve shows the twelve-month trailing average of monthly values, while the black curve shows the five-year trailing average.[27]

Drought exacerbates wildfires, which garner more dire headlines than any other precipitation-related phenomenon. Media coverage of devastating fires around the world—most recently in Brazil, Australia, and California—portrays them as a horrendous consequence of a warming globe. They can indeed be horrendous: 2020 saw record-breaking fires on the West Coast of the US that destroyed millions of acres, incinerating homes and communities and tragically killing many people, while keeping thousands more trapped inside due to poor air quality. These fires spawned an outpouring of articles with titles like "Western wildfires: An 'unprecedented' climate change fueled event, experts say."[28] And in fact, changes in climate do play a role in the frequency, location, and character of wildfires. But understanding that role, and the part humans play (and might play in the future), requires we dig deeper than headlines.

Sophisticated satellite sensors first began monitoring fires globally in 1998. Unexpectedly, analysis of the images showed that the area burned annually declined by about 25 percent from 1998 to 2015.[29] That's evident in Figure 7.12 from NASA, which shows the global area burned by fires each year from 2003 to 2015, with the straight line indicating the trend. Despite the very destructive wildfires in 2020, that year was among the least active globally since 2003.[30]

Researchers attribute this decline to human activities, specifically the expansion and intensification of agriculture:

> As populations have increased in fire-prone regions of Africa, South America, and Central Asia, grasslands and savannas have become more developed and converted into farmland. As a result, long-standing habits of burning grasslands (to clear shrubs and land for cattle or other reasons) have decreased . . . And instead of using fire, people increasingly use machines to clear crops.[31]

In other words, whatever influence a changing climate might have had on wildfires globally in recent decades, human factors unrelated to climate

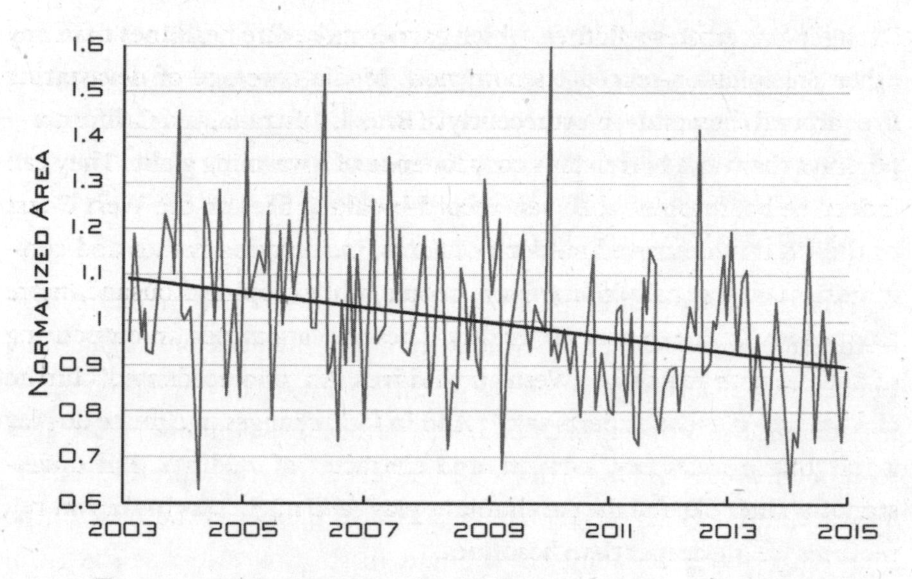

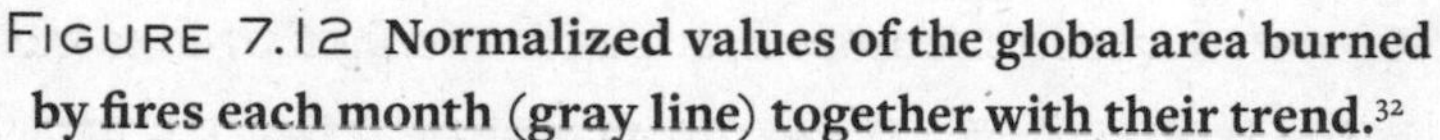
FIGURE 7.12 **Normalized values of the global area burned by fires each month (gray line) together with their trend.**[32]

were dominant. But the satellite data also showed a significant increase in the intensity and reach of fires in the western United States. In fact, Key Finding 6 of the CSSR's Chapter 8 is:

> The incidence of large forest fires in the western United States and Alaska has increased since the early 1980s (*high confidence*) and is projected to further increase in those regions as the climate warms, with profound changes to certain ecosystems (*medium confidence*).

There's a bit more detail in the report's Section 8.3:

> State-level fire data over the 20th century indicates that area burned in the western United States decreased from 1916 to about 1940, was at low levels until the 1970s, then increased into the more recent period.

Climate changes are surely playing a role here. Less rainfall and higher temperatures make for drier "fuel" that's easier to ignite and promotes a fire's rapid spread and intensification. The IPCC's SR15 report of 2018 makes that explicit in its Chapter 3:

> There is additional evidence for attribution of increased forest fire in North America to anthropogenic climate change during 1984–2015, via the mechanism of increasing fuel aridity almost doubling the western US forest fire area compared to what would have been expected in the absence of climate change.

The 2016 study cited to support that statement compared dryness of fuel in climates with and without human influences as simulated by twenty-seven CMIP5 models.[33] Those differences in dryness then result in different fire properties. Of course, attributing a doubling in fire area to human causes assumes that the models correctly reproduce internal variability, which they don't. That's particularly true because of the high longer-term variability of drought in the Southwest as shown in Figure 7.9.

But factors other than climate must also play an important, if not dominant, role, since fires declined in the early part of the twentieth century even as California's drought conditions showed no trend, as seen in Figure 7.11. "Human influences" can take many forms. Forest management (How much fuel is allowed to accumulate? Are fires suppressed or allowed to burn? How much development is permitted in or near forests?) and human-caused ignition (nearly 85 percent of US wildland fires have a human cause) are among the contributors.[34]

While we may not be able to fully quantify, much less control, the many climate-related influences on wildfires, we have significant power to address these human factors. By making the conversation about wildfires only one of unavoidable doom due to "climate change," we miss an opportunity to take steps that would more directly curtail these catastrophes.[35]

Of course, things can (and surely will) change in coming decades. But how they will change is far from certain. We shouldn't place too much confidence in model projections of future changes in precipitation—after all, they come from the models discussed in Chapter 4. And as AR5 notes:

> The simulation of large-scale patterns of precipitation has improved somewhat since the AR4, although models continue to perform less well for precipitation than for surface temperature.[36]

In fact, it's even worse than that. The report states explicitly that, in general, the CMIP5 model ensembles "cannot be taken as a reliable regional probability forecast"[37] for any aspect of the climate. In other words, the models are even worse at describing changes in regional climates than they are at describing changes in global quantities.

Because droughts and floods have such dramatic impact, however, politicians and other officials can't resist citing model results to prophesy future catastrophes. Mark Carney, former head of Canada's central bank and later head of the Bank of England, is probably the single most influential figure in driving investors and financial institutions around the world to focus on changes in climate and human influences upon it. A learned man, with a PhD in economics from Oxford University, he has been an outstanding central banker. Carney is now the United Nations' Special Envoy on Climate Action and Finance. He is also a UK advisor for the 26th annual UN Conference of Parties (COP26), a follow-on to the 2015 Paris climate conference that's due to take place in Glasgow, Scotland, during November 2021. So it's important to pay close attention to what he says.

In a 2015 speech just before the Paris conference, speaking as governor of the Bank of England, Carney laid out many aspects of "the insurance response to climate change."[38] Extreme weather costs insurance companies a lot of money, so perhaps it is no wonder that his appeal included a warning about flooding:

> Despite winter 2014 being England's wettest since the time of King George III; forecasts suggest we can expect at least a further 10% increase in rainfall during future winters.

To support that assertion, he cited Britain's Met Office "research into climate observations, projections, and impacts." These were model forecasts for the next five years, so you might expect they'd be more accurate than those attempting to project climate fifty years out. Let's turn to the data and see.

Figure 7.13 shows the observed winter precipitation (December through February) in England and Wales up through 2020; it's one of the longest instrumental weather series available, beginning in 1766. The average rainfall looks pretty constant over decades from 1780 to 1870 and again from 1920 to the present. A shift occurred somewhere over the fifty years in between, when human influences on the global climate were quite negligible.

Carney was correct that 2014 was a record wet winter (455.5 mm, or 17.9 inches), and it was indeed the "wettest since the time of King George," since George III's reign lasted until 1820. But the Met Office models Carney cited back in 2014 all turned out to be dead wrong. Rainfall during the

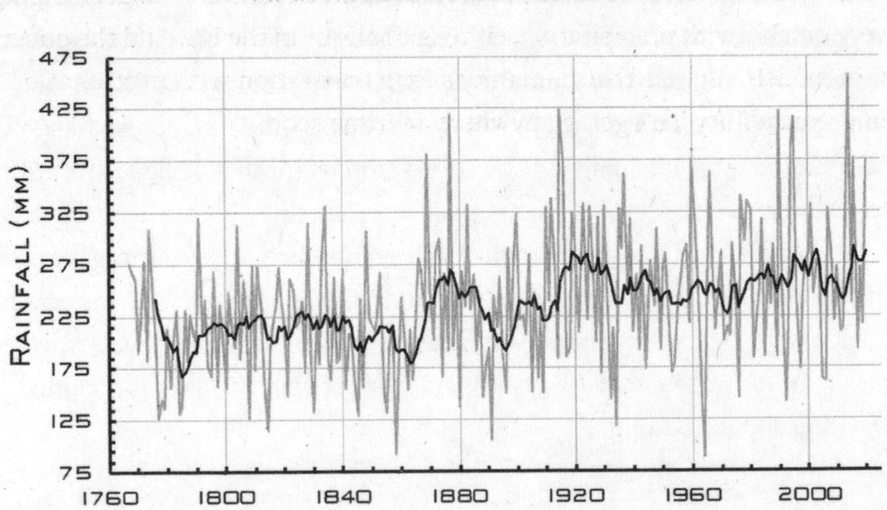

FIGURE 7.13 **Rainfall on England and Wales during December through February from 1767 to 2020.** The spiky gray lines are the values for each year while the dark line is the ten-year trailing average.[39]

six winters after 2014 was well in context with the previous century, and it averaged 278 mm, 39 percent *less* than the 2014 record and nowhere near the "at least" 500 mm implied by the predicted increase. And a Met Office analysis published in 2018 found that the largest source of variability in UK extreme rainfalls during the winter months was the North Atlantic Oscillation mode of natural variability, not a changing climate.[40]

Of course Carney could take refuge in his speech's subjunctive "forecasts suggest" and the indeterminate hedging of "future winters." Nevertheless, it's surprising that someone with a PhD in economics and experience with the unpredictability of financial markets and economies as a whole doesn't show a greater respect for the perils of prediction—and more caution in depending upon models.

Floods, droughts, and fires bring great tragedy and sorrow, and their consequences can be devastating. As the world gets more and more connected through communications, we become more and more aware of these events when they happen. But that does not make them "further proof" of climate change. In the end, the data tells us there's not very much changing very quickly with precipitation, either globally or in the US. And the uncertain models suggest that humanity's long frustration with precipitation's unpredictability isn't going anywhere anytime soon.

SEA LEVEL SCARES

A striking image of a half-submerged Statue of Liberty filled the cover of the September issue of *National Geographic* magazine in 2013, touting its lead story: "Rising Seas—how they are changing our coastlines." Any curious reader could have consulted the record of the tide gauge at The Battery at the tip of Manhattan (less than two miles from the statue) and seen that sea level there has been rising at an average rate of about 30 cm (1 foot) per century since 1855.[1] And a quick calculation would show that—at this rate—it would take more than twenty thousand years for the water to reach the level shown menacing poor Lady Liberty. But along with demonstrating that those who design magazine covers know how to get our attention with nothing but Photoshop and a healthy serving of artistic license (the oldest surviving human structure is less than six thousand years old), the cover makes one thing clear: people are very concerned about rising sea levels.

Should they be? Are sea levels changing in response to our warming globe? And—even if our nation's monuments are safe for another thousand generations—are our human futures in danger as a result?

As with precipitation, sea level is all about where the water is. After the oceans, the earth's largest store of water is in the ice sheets on Greenland and the Antarctic. While many factors contribute to the small changes in sea level we observe over decades, its behavior over geological times depends most importantly upon how much ice is present on the land.

Slow cycles—variations in the earth's orbit and the tilt of its axis over tens of thousands of years—change the amounts of sunlight absorbed by the Northern and Southern Hemispheres. As we've seen in Chapter 1, those changes have caused large swings in the global temperature over the past million years. But they also cause the ice sheets covering the continents to grow or melt (the intervals are technically called "glaciations" and "interglacials," respectively), putting less or more water in the oceans, and so causing sea levels to fall or rise accordingly. Figure 8.1

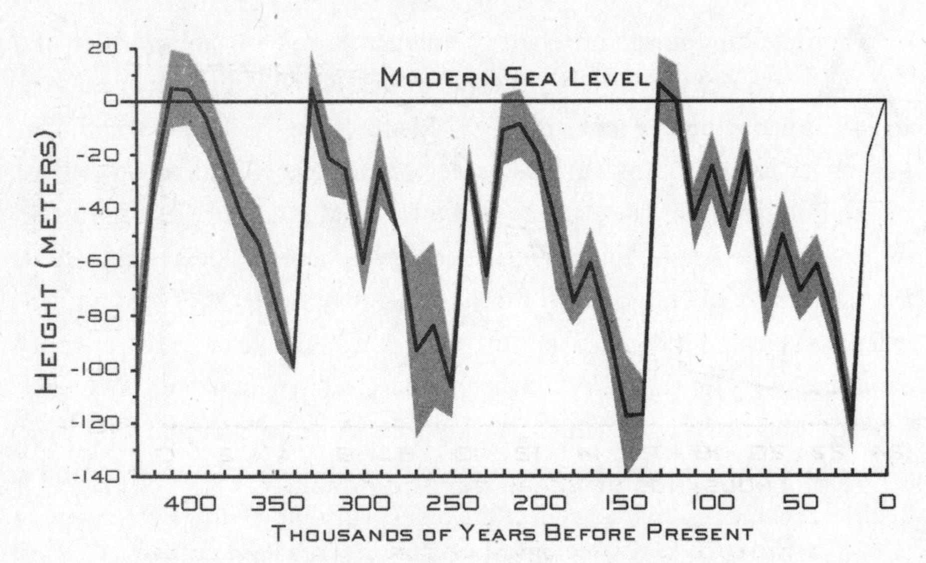

FIGURE 8.1 **Global sea level over the past 400,000 years as estimated from geological proxies.** The typical uncertainty in these estimates is 10 meters.[2]

shows geological estimates of the global sea level extending back more than 400,000 years.

As you can see, the past half-million years tell a story of repeating episodes in which sea level dropped slowly by about 120 meters (400 feet) every 100,000 years as continental-scale glaciers built up, but then rose back up rapidly over about 20,000 years as the glaciers melted again. During the last interglacial (low-ice) period 125,000 years ago, known as the "Eemian," sea levels were some 6 meters (20 feet) higher than they are today!

The Last Glacial Maximum occurred about 22,000 years ago, when the continental glaciers started melting once again. The earth today is in the Holocene interglacial, which geologists deem to have started about 12,000 years ago. According to the geological record, sea level has risen by about 120 meters (400 feet) since the Last Glacial Maximum, as rapidly as 120 mm (5 inches) per decade until about 7,000 years ago, when the rate of rise slowed dramatically; that's shown in Figure 8.2.

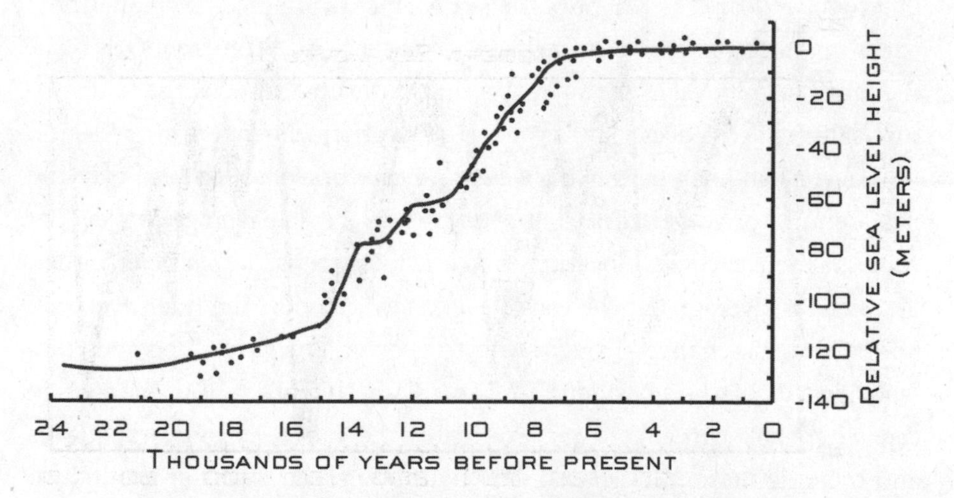

FIGURE 8.2 **Geological estimates of sea level change since the Last Glacial Maximum 22,000 years ago.** The solid curve is the average of estimates from various sites around the globe, shown by the individual representative points.[3]

So the question is not whether sea level is rising—it's been doing that for the past 20,000 years. Instead, what we want to know is whether human influences are accelerating that rise. Since human influences increased dramatically after about 1950, the best way to assess whether sea levels are going up faster than they would without us is to compare measurements since then with those in the more distant past. Shorter timescales mean smaller amounts of rise, so we'll need something more precise than the geological estimates we relied upon for the bigger picture. Luckily, some sea level records are available starting in the eighteenth century from tide gauges at ports in Europe and North America. Today, we have measurements from more than two thousand tide gauges around the globe and, starting in 1992, from satellite observations that use radar to measure the height of the ocean surface.

To talk about the level of water in the ocean, climate scientists use the concept of Global Mean Sea Level (GMSL), which is inferred from measurements over the whole globe; even though "water seeks its own level," there are variations that are important given the fraction-of-a-millimeter precision required—the GMSL is rising today at a rate of only a few millimeters per year. The height of the sea surface is affected by small differences in the earth's gravity from place to place, by ocean currents (think of the lower water level in the swirl around a bathtub drain), by the temperature/salinity of the ocean, and even by the weather-dependent air pressure.

As you can imagine, getting a clear picture of global sea level over the past century or so isn't easy, but even measuring the average sea level at some particular coastal location with a tide gauge isn't so simple, either. You need to average out the waves every few seconds, the tides every six hours, and the changes from season to season. Over time, local sinking (subsidence) of the coast due to natural or human-caused changes in the water table can affect the elevation of the gauge, as can earthquakes and tectonic motion in general. For example, withdrawals of groundwater in the Houston-Galveston area over the past century have caused the ground to compact, lowering the land surface there by as much as 3 meters (10 feet). And of course there are all the usual problems caused by changes in instrumentation or observing protocols. But for better or worse, tide gauge data is what we've got, and some long and precise records are available, for instance through NOAA.[4]

Determining the Global Mean Sea Level requires sophisticated analyses of tide gauge data from many different coastal locations, because just as the ocean isn't "level" across the globe, its rate of rise differs from place to place as well. For example, the local rates of rise along the US coasts vary enormously due to local conditions—Eugene Island on the Gulf Coast is experiencing a rise of 9.65 mm (0.38 inches) per year, while in Skagway, Alaska, the sea is *retreating* at a rate of 17.91 mm (0.71 inches) per year.

At least four independent groups have analyzed tide gauge data to determine the Global Mean Sea Level over more than a century. The results of one such analysis are depicted in Figure 8.3. They show that the GMSL was rising at the end of the nineteenth century—well before there were

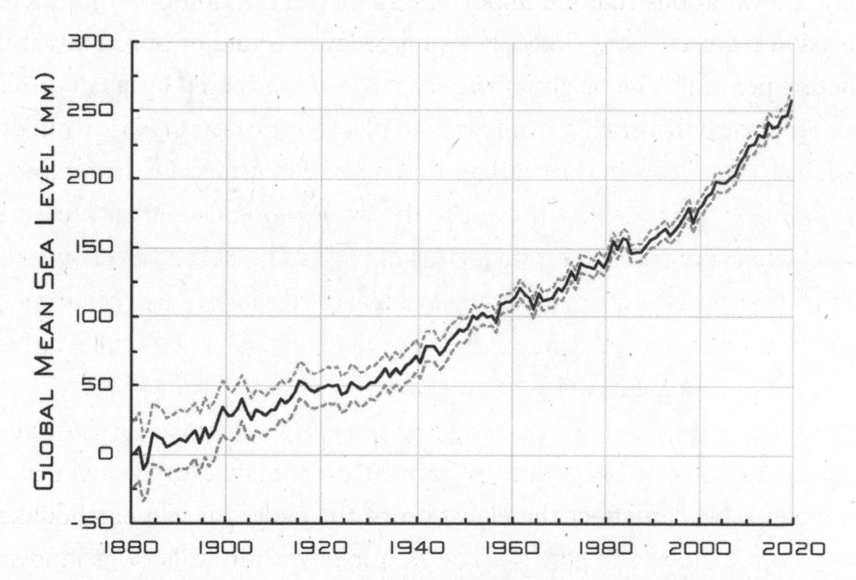

FIGURE 8.3 **Changes in Global Mean Sea Level relative to 1880 as estimated from tide gauge data.** The solid curve indicates the average value and the dashed lines the uncertainty.[5]

significant human influences on the climate—and since 1880 rose by some 250 mm (10 inches), for an average rate of 1.8 mm (0.07 inches) per year. The average rate of rise over shorter periods has fluctuated; for the past three decades it's been about 3 mm (0.12 inches) per year. In the context of the earth's hydrological cycle, these are small numbers—3 mm of sea level rise per year is roughly 0.3 percent of the planet's annual precipitation—so it's not surprising that there are variations from decade to decade.

As mentioned earlier, global sea level has also been measured by satellite since late 1992.[6] In satellite altimetry, a satellite measures its height above a broad patch of the ocean using radar. If the position of the satellite is known precisely, the height of the sea surface can then be determined. Today, radar altimetry provides essentially continuous global coverage of the open ocean, complementary to the coastal measurements of tide gauges.

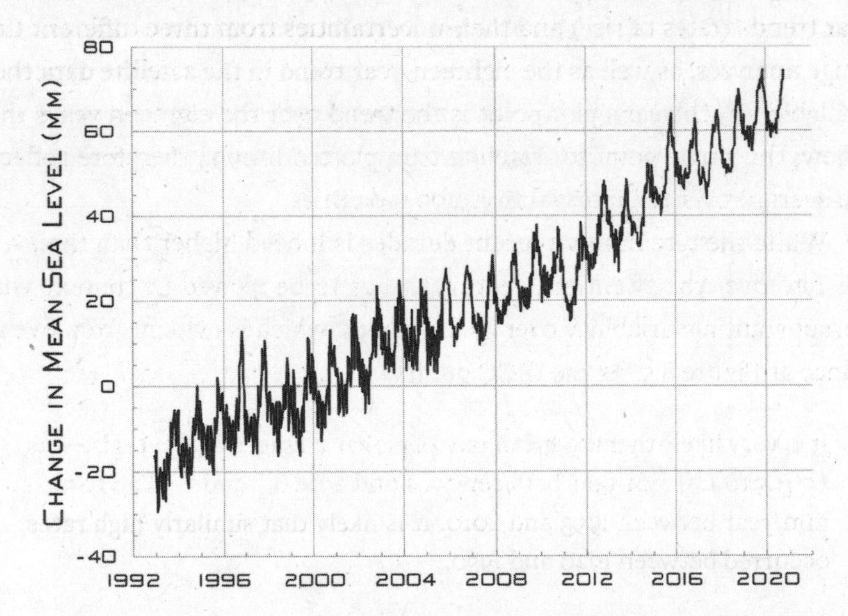

FIGURE 8.4 **Changes in Global Mean Sea Level as measured by satellite altimetry.** A seasonal cycle of about 7 mm (0.2 inches) is superimposed on a trend of 3.0 ± 0.4 mm/yr (0.12 ± 0.02 in/yr).[7]

Four independent groups analyze the data from the eleven satellite altimeters that have flown so far. It's quite a feat to measure the average ocean height to fractions of a millimeter from a satellite 600 km (370 miles) above the surface, and to do so consistently for decades. After some years of refinement, identifying corrections for things like drifts in satellite orbits, results such as those shown in Figure 8.4, from the NOAA group, have emerged.

As measured over twenty-seven years by this series of satellites, the Global Mean Sea Level shows a clear seasonal cycle (ups and downs of about 7 mm or 0.3 inches) superimposed on an average rate of rise of 3.0 ± 0.4 mm (0.12 ± 0.02 in) per year.

So for the past three decades, sea level has been going up by about 3 mm (0.12 inches) each year—higher than the overall average rate (1.8 mm or 0.07 inches per year) since 1880. To judge how the rate of sea level rise might have increased under growing human influences, IPCC's AR5 presents a figure (reproduced as Figure 8.5) that displays year-by-year eighteen-year trends (rates of rise) and their uncertainties from three different tide gauge analyses, as well as the eighteen-year trend in the satellite data then available. (Each year's plot point is the trend over the eighteen years that follow; the single point for satellite data plotted in 1994 therefore reflects the average over the interval from 1994 to 2011.)

While the rate in more recent decades is indeed higher than the average rate over the twentieth century, it has to be viewed in context with the substantial variability over past decades, which is evident from even a glance at Figure 8.5. As the IPCC put it:

> It is very likely that the mean rate of global averaged sea level rise was 1.7 [1.5 to 1.9] mm/year between 1901 and 2010 . . . and 3.2 [2.8 to 3.6] mm/year between 1993 and 2010. It is likely that similarly high rates occurred between 1920 and 1950.

In fact, the rate of rise between 1925 and 1940—a period almost as long as the eighteen-year satellite record then available—was almost the same as that recent satellite value, about 3 mm (0.12 inches) per year.

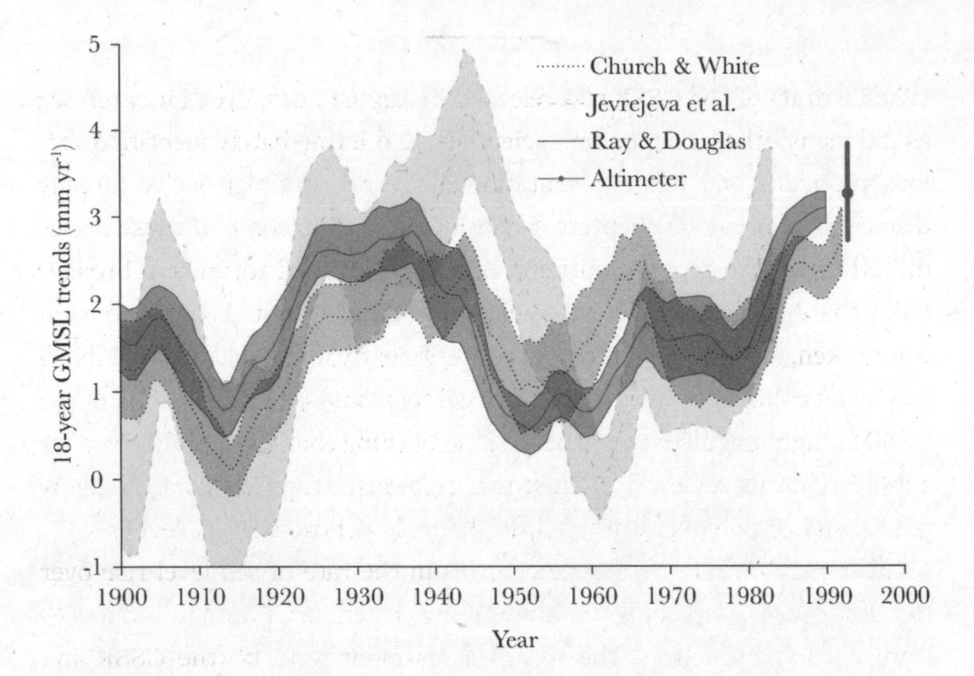

FIGURE 8.5 **Eighteen-year leading trends in Global Mean Sea Level since 1900.** Estimates from three different tide gauge analyses are shown, together with a single value from satellite altimetry. Uncertainties are 90 percent confidence levels; that is, there is only a 10 percent chance that the true values lie outside the shaded area.[8] (AR5 WGI Figure 3.14.)

Since the rate varies so much, it's hard to know for recent years what's human-caused and what's natural. And while the IPCC's 2019 Special Report on the Oceans and Cryosphere in a Changing Climate (or SROCC) expresses high confidence that the satellite data from 1993 to 2015 shows an acceleration (that is, the rate of rise is increasing), the implications are murky both because of the shortness of the record and because there was acceleration well before human influences were significant. The AR5 had this to say:

> It has been clear for some time that there was a significant increase in the rate of sea level rise in the four oldest records from Northern Europe starting in the early to mid-19th century. The results are consistent and indicate a significant acceleration that started in the early to mid-19th century, although some have argued it may have started in the late 1700s.[9]

When a draft of the CSSR was released in August 2017, I read it carefully, as did many other independent scientists, and immediately identified various problems and misrepresentations—several of which we've already discussed in previous chapters. I considered raising some of these issues directly with the report's authors. But I also wanted to make a broader point that had become clear to me: that whether or not the climate itself was broken, the assessment report process clearly was. I decided to publish an Op-Ed calling out one of the more egregious misrepresentations in the CSSR to highlight an example of the kind of thing that showed the need for a more rigorous review. I did just that right after the CSSR was formally published in November 2017, and the example I chose was sea level rise.[10]

Although decade-by-decade changes in the rate of sea level rise over the past century are central to untangling the effect of human influences from natural influences, the recent assessment reports (the CSSR and the IPCC's 2019 SROCC) hardly mention them.[11] There are no graphs like Figure 8.5, where it's very easy to see how that rate changes—sometimes dramatically—over decades. Rather, the reports are filled with graphs of the rising sea level itself, such as Figures 8.3 and 8.4, from which it's almost impossible to judge the variations in, and significance of, how quickly sea level is going up.

All of the assessment reports have plenty of text emphasizing that the rate of sea level rise in the past two decades is higher than the average of the twentieth century. For example, the CSSR offers this on page 16 of its Executive Summary:

> Global mean sea level (GMSL) has risen by about 7–8 inches (about 16–21 cm) since 1900, with about 3 of those inches (about 7 cm) occurring since 1993 (*very high confidence*).

That statement was a red flag for me, because it compares the rise over the past twenty years with that over more than a century. The fact that three of the seven inches of rise since the dawn of the twentieth century occurred in the past twenty-five years does indeed seem alarming—but suddenly less so if you know that it also rose almost as much (6 cm, as opposed to 7 cm) in the twenty-five years between 1935 and 1960. The CSSR makes no mention of that at all, even though its primary reference for sea level rise describes variability like that shown in Figure 8.5.[12] The rate of rise over the most recent twenty-five years should be compared to that over other twenty-five-year periods to understand just how significant the recent rate is.

When I made that point way back in my 2014 *Wall Street Journal* Op-Ed, I was criticized roundly. Here's one example:

> He [Koonin] claims that the rate of sea level rise now is no greater than it was early in the 20th century, but this is a conclusion one could draw only through the most shameless cherry-picking. In reality, according to the data, the sea level trend was .8 millimeters of rise per year from 1870 to 1924, 1.9 millimeters per year from 1925 to 1992, and 3.2 millimeters per year from 1993 to 2014—i.e., the rate has actually quadrupled since preindustrial times.[13]

Notice that the historical intervals cited are fifty-four years (1870–1924) and sixty-seven years (1925–1992), while the recent interval (1993–2014) is twenty-one years. This disingenuously obscures the higher rates in the twenty years from 1925 to 1945, which can be seen quite clearly in Figure 8.5. In the spirit of fruit metaphors, I wasn't cherry-picking, but rather comparing apples to apples. And, at any rate, I was only quoting what the IPCC itself said.

The CSSR, on the other hand, follows the lead of some prominent climate scientists in hiding the large fluctuations in the rate of sea level rise over the past century, presumably because they make the past three decades seem less unusual. The report misleads by omission in not mentioning either the strong decadal variability of sea level rise during the twentieth century or the fact that the then most recent values of the rate were statistically indistinguishable from those during the first half of the twentieth century.

Before publishing the Op-Ed pointing this out, I sent a more technical discussion of the issue to Don Wuebbles of the University of Illinois (the CSSR's senior lead author) and to Robert Kopp of Rutgers University (the main author of the CSSR chapter on sea level rise). Both agreed that my criticism was valid; an October 15, 2017, email to me from Kopp read, in part:

> I think the point about interdecadal variability is a useful one, and had it come during the public review period for the draft, I'm sure it would have been taken into account.

Both Kopp and Wuebbles said that I was the first to raise the point—surprising, since the problem is elementary and many eyes had already been on the draft. Both also said that they would have added a discussion of sea level rise variability in the twentieth century, but that it was then too late (the draft was in final copyedits) and the report was already too long. (That last was another surprise, given the length already and how little space it would have taken to fix the problem.) Wuebbles also said he'd see about including text in the second part of NCA2018 that would remedy the omission; I've not been able to identify any such text in the released version.

To be clear, sea levels do rise as the globe warms. When the earth's surface temperature increases, land ice melts—and as oceans warm, the water in them expands. Levels rise and fall seasonally, over the longer term in response to the orbital cycles we discussed earlier, and in response to other natural or human influences. While rates of rise over the past century have had significant ups and downs, a warming globe does indeed put more water into the oceans. What, then, of future sea level rise? The answer largely depends upon how much of the ice on the land melts as temperatures increase, together with the expansion of the warming oceans.

Insights into why the rate of sea level rise has changed during the past century come from a recent paper that successfully closed the sea level rise "budget."[14] Scientists have struggled in recent years to balance that budget—that is, to reconcile the observed rise in sea levels with what is known about the various factors contributing to it. The new work tallies all of the observed changes since 1900 in land ice (Greenland, Antarctica, and mountain glaciers) and in liquid water stored on the land in aquifers and behind dams, and combines them with estimates of the thermal expansion of the ocean. The paper then compares those contributions with changes in sea level as measured by tide gauges and satellites.

The results are shown in Figure 8.6. The top panel shows how mountain glaciers, land water storage, and the Greenland and Antarctic ice sheets have each contributed to changes in the global sea level. There are a few surprises here, given that the globe has warmed since 1900: the contribution from glacier melting has slightly declined since 1900 and is the same now as it was fifty years ago; the contribution from Greenland went through a minimum around 1995 and is now no higher than it was in 1935; and changes in terrestrial water storage were an important (negative) contributor during the flurry of dam building in the 1970s.

So future global sea level rise is uncertain not only because of all of the model uncertainties in the global temperature rise discussed in Chapter 4, but also because the dynamics of the Greenland and Antarctic ice sheets are quite uncertain. The IPCC summarizes the situation (SMB is the Surface Mass Balance, measuring the net change in ice due to atmospheric processes):

> . . . for periods prior to 1970, significant discrepancies between climate models and observations arise from the inability of climate models to reproduce some observed regional changes in glacier and GIS [Greenland Ice Sheet] SMB around the southern tip of Greenland. It is not clear whether this bias in climate models is due to the internal variability of the climate system or deficiencies in climate models. For this reason, there is still medium confidence in the ability of climate models to simulate past and future changes in glaciers mass loss and Greenland SMB.[15]

Nevertheless, the report offers projections of Global Mean Sea Level rise under the various emissions scenarios discussed in Chapter 3. Under RCP2.6 (the most emissions-lite scenario, which has global emissions vanishing in the latter half of this century), the IPCC projects levels will rise 0.43 m (with a two-thirds chance of the value being between 0.29 and 0.59 m) over the twenty-first century, while for the extreme,

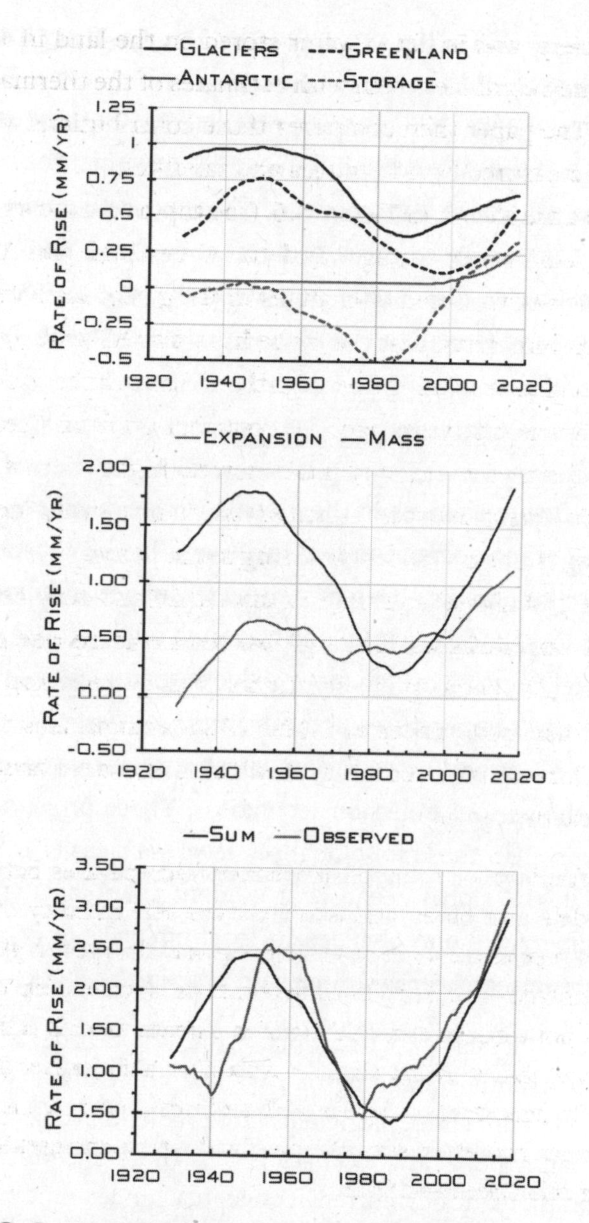

FIGURE 8.6 **Contributions to the rate of global sea level rise from 1929 to 2018.** The upper panel shows contributions to changes in the mass of water in the ocean from four sources. The middle panel shows the sum of those changes, as well as the change that results from expansion of ocean waters. The lowest panel shows the sum of all contributions compared with the observed rates of rise. All trends are calculated as thirty-year trailing averages. Uncertainties are small enough so that the variations are highly significant.

emissions-heavy RCP8.5, the projected rise is 0.84 m (two-thirds likely value between 0.61 and 1.10 m).[16] These projections correspond to average rates of rise of 4.3 and 8.4 mm/yr, respectively. Both rates are significantly larger than the current 3 mm/yr, though less than the maximal rate shown in Figure 8.2 of 12 mm/yr 9,000 years ago at the peak of ice-sheet melting. More importantly, these projections depend upon the dubious climate models we've discussed—and their shaky grasp of the changes in the Antarctic and Greenland ice sheets that account for the bulk of sea level rise.

Whatever the future changes in the global average, local sea level is what matters for planning adaptation measures, and in the coastal places that matter most to humans, it's been measured far longer and more accurately than global values. Local sea level rise is related to global rise, but it differs because of local factors like land motion due to tectonics or subsidence and changes in ocean temperatures and currents.[17] Nevertheless, the climate models offer projections, however imperfect, for various cities around the globe under the various emissions scenarios. These projections are even more uncertain than those for global sea level—it's easier to project average changes in the global ocean heat content than specific spatial variations in temperature and changes in local effects like currents. As the World Climate Research Programme said in 2017:

> Despite considerable progress during the last decade, major gaps remain in our understanding of past and contemporary sea level change and their causes, particularly for prediction/projection of sea level rise on regional and local scales . . . These uncertainties arise from limitations in our current conceptual understanding of relevant physical processes, deficiencies in our observing and monitoring systems, and inaccuracies in statistical and numerical modeling approaches to simulate or forecast sea level.[18]

The need for such caveats becomes clear when you compare the projections with the historical data. As an example, Figure 8.7 shows the monthly mean sea level anomaly (corrected for seasonal variation) at the tip of Manhattan, which has been measured by The Battery tide gauge, referenced at the beginning of this chapter, for more than 160 years. The solid line shows the average behavior, while the two arrows show the rate of rise between 2020 and 2100 projected in the AR5 under two different RCP scenarios.

The long-term rate of rise is 2.87 ± 0.09 mm per year, not very different from the 3 mm per year rate of rise seen in the GMSL over the past few decades. The AR5 gives projections of New York City sea level rise from 2000 to 2100 that range from about 550 mm (22 inches) to 800 mm (31 inches) as the scenarios range from RCP2.6 to RCP8.5; uncertainties in each projection are about ± 300 mm (12 inches).

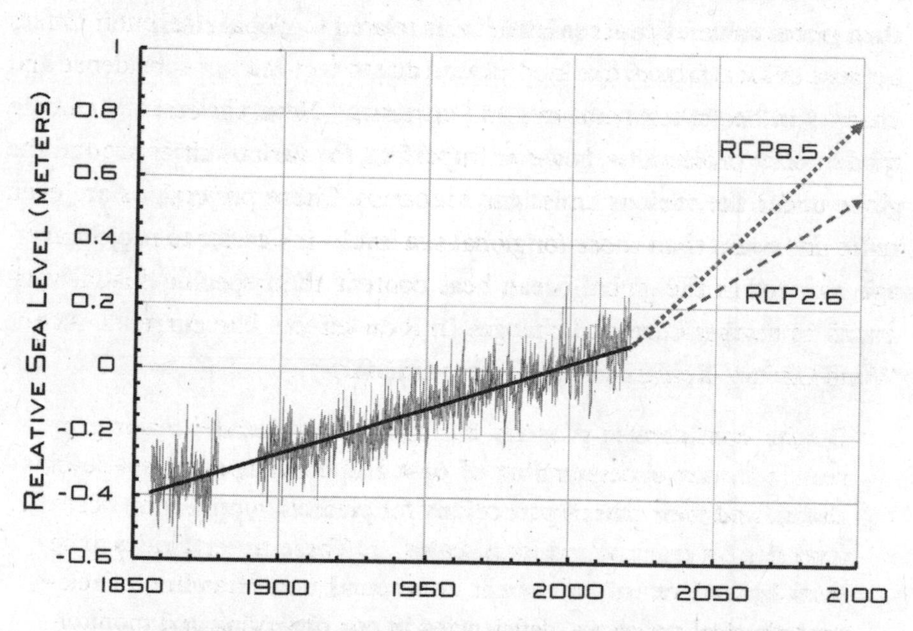

FIGURE 8.7 **Monthly mean sea level anomaly (after correction for the seasonal cycle) as measured since 1856 by the tide gauge at The Battery at the southern tip of Manhattan.** The straight line shows the trend; the arrows show the average rise from 2020 to 2100 projected in AR5 under two different scenarios.[19]

But, as I mentioned, graphs of the sea level itself, rather than the rate at which it's rising, can be deceptive and, in any event, we've got to take the long-term view. Figure 8.8, which shows the rate of rise for The Battery sea level over the past century, should calm any sense of alarm.

Each year's point reflects the trend during the thirty years preceding it, making this an excellent way to evaluate the big picture of the trend in sea level rise. You can see that the rate has varied a lot over the past century, from lows of less than 2 mm (0.08 inches) per year in the thirty years before 1930 and 1990 to highs of almost 5 mm (0.2 inches) per year in the three decades before 1955 and 2015, even as it's averaged about 3 mm per year over the whole century. The records of other tide gauges along the US northeast coast also show this sixty-year cycle, which is in sync with the Atlantic Multidecadal Oscillation (AMO) discussed in Chapter 4.[20] So it's reasonable to expect that the rate will decline again during the next few decades. The *average* rates projected by the IPCC,

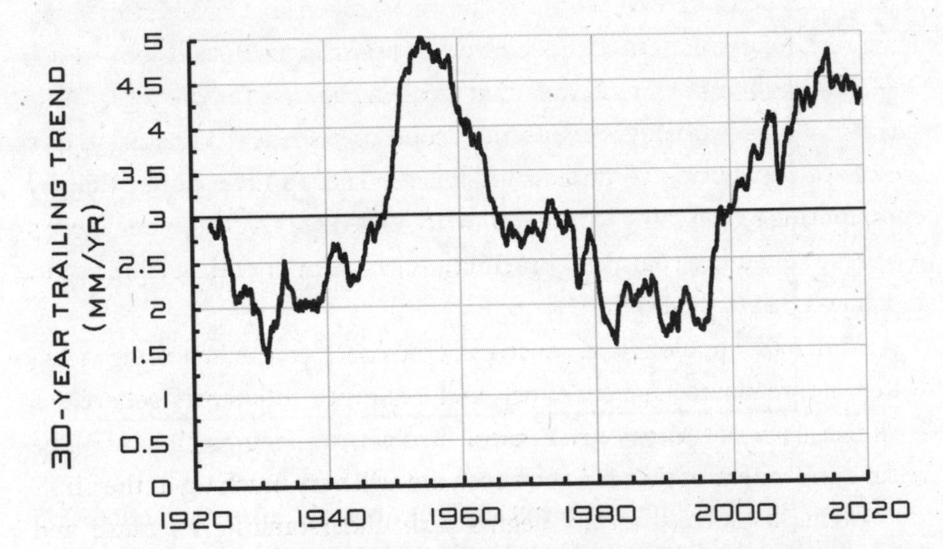

FIGURE 8.8 **Thirty-year trailing trends in the sea level at The Battery from 1923 to 2020.** Uncertainties (1σ) are about 0.35 mm/yr. The horizontal line indicates the average rate of 3.02 mm (0.12 inches) per year.

on the other hand—even the low RCP2.6 rate of 5.5 mm/yr—would be so unusually high that they would fall outside the scale of this chart. The next few decades will tell.

The media is as confused about the state of local sea level rise as the models are. For example, a recent article about sea level rise on Oahu, Hawaii, had dramatic pictures of flooding over the past few years and showed maps of which areas of Honolulu would be inundated if the sea level rose 30, 60, or 100 cm above the current level.[21] But it failed to mention that the NOAA tide gauge record for Honolulu shows an average rate of 1.5 mm (0.06 inches) of rise per year since 1905, meaning that, absent some *very* dramatic acceleration, it would take two hundred years to achieve even the lowest mapped rise of 30 cm (one foot). Omitting that kind of context is, unfortunately, even more common in media coverage than it is in the assessment reports.

In summary, we don't know how much of the rise in global sea levels is due to human-caused warming and how much is a product of long-term natural cycles. The CSSR and other assessment discussions of sea level rise omit important details that weaken the case for the rate of rise in recent decades being outside the scope of historical variability, and hence for attribution to human influences. There's little doubt that by contributing to warming we have contributed to sea level rise, but there's also scant evidence that this contribution has been or will be significant, much less disastrous.

Humanity's tendency to construct cities near coasts has made rising waters a threat since ancient times, and insurance companies believe sea level rise is one of the major risks associated with a changing climate.[22] The nature and extent of that risk, however, are still very much up in the air.

Solving a problem ideally begins with understanding its cause and what actions we can take to affect it. The message put forth that human-caused warming is the sole source of rising sea levels gives the impression that reducing emissions is a solution—alas, because ice melting lags behind warming (and also because of the persistence of CO_2 discussed in Chapter 3), even if we were the culprit and ceased all emissions tomorrow, global sea level would continue to rise. What's more, as we've seen, local sea level changes and their effects are far more complicated still, involving ocean currents, erosion, weather patterns, and land use and composition. Clear and unbiased communication of these nuances is essential. If sea level does become a serious threat in coming decades, there's no doubt that we'll be better prepared having devoted resources to further research and adaptation—unlikely to be a priority if we insist we already know all the answers.

SEA LEVEL RISE—2024 UPDATE

As I explained earlier in this chapter, the past variability in sea level rise is important in assessing recent trends. Unfortunately, AR6 hides that variability. The summary for policymakers of its Working Group I report reads:

> Global mean sea level increased by 0.20 [0.15 to 0.25] m between 1901 and 2018. The average rate of sea level rise was 1.3 [0.6 to 2.1] mm yr^{-1} between 1901 and 1971, increasing to 1.9 [0.8 to 2.9] mm yr^{-1} between 1971 and 2006, and further increasing to 3.7 [3.2 to 4.2] mm yr^{-1} between 2006 and 2018 (*high confidence*).[23]

This is a particularly devious obscuration of past variability, since the seventy-year period (1901–1971) contains as much variability over shorter periods as the more recent shorter periods cited (e.g., the earlier twentieth-century peak in the rate shown in Figures 8.5 and 8.6). And given the variability shown in those figures, it's hard to draw any conclusion from a twelve-year period.

In February 2022, NOAA issued its projections of sea level rise for various sites along the US coast.[24] They claim that by 2050, the sea will have risen one foot at The Battery in Manhattan. This corresponds to a slightly more rapid rise than the RCP8.5 extrapolation in Figure 8.7. It would put the 2050 point on Figure 8.8 at about 10 mm/yr, more than twice the current rate and about three times the average rate. In that historical context,

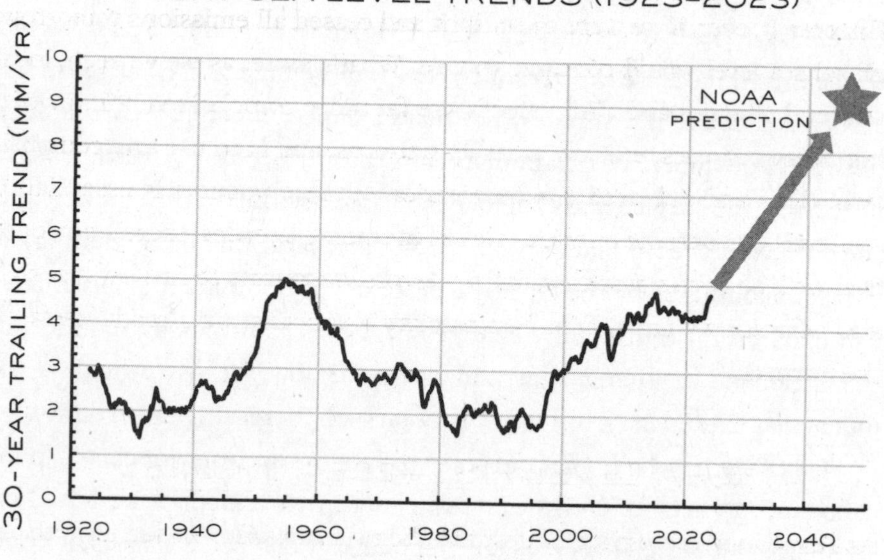

FIGURE 8.9 **Rate of sea level rise at the Battery in Manhattan.** Shown is the observed thirty-year trailing trend, together with the allegedly "locked in" NOAA prediction for 2050.[25]

NOAA's projection is remarkable—as shown in Figure 8.9, it seems a straight-line extension of the trend from 1990 to 2010, ignoring any past variation. But even more remarkable is that they say this rise is "locked in"—it will happen no matter what future emissions are. We should know in a decade or so whether that remarkable prediction has legs.

GREENLAND'S SHRINKING ICE—2024 UPDATE

The annual loss of ice from Greenland's ice sheet is a perennial source of alarm. A *Guardian* headline in December 2019 worried "Greenland's ice sheet melting seven times faster than in the 1990s"[26] while a NASA press release issued three months later warned "Greenland, Antarctica Melting Six Times Faster Than in the 1990s."[27] It's then not surprising that this book's claim that "Greenland's ice sheet isn't shrinking any more rapidly than it was eighty years ago" attracted some criticism.

The Greenland ice sheet extends over 1.7 million km^2 to cover much of Greenland in ice 2 to 3 km thick. If all that ice were to melt and flow into the sea, it would cause the oceans to rise by more than 7 m. Changes in the amount of ice are measured by the mass balance, which is the difference between how much ice accumulates on the surface due to atmospheric processes (the surface mass balance mentioned earlier in the chapter) and the outflow at the ice sheet's base (glacier discharge and basal melting). A dataset maintained by the Geological Survey of Denmark and Greenland (Mankoff et al.) tallies daily changes in these quantities by monitoring the Greenland weather, height, and glacier outflow.[28] It also provides estimates that extend back to 1840, although with increasing uncertainty.

Politifact, a website that purports to rate the accuracy of claims, published a fact-check of *Unsettled*'s statement in November 2021, based on an analysis of decadal averages of the Mankoff dataset.[29] That is, it compared average mass loss during the decades ending in 2020, 2010, and so on. Since the annual mass loss during the 1930s averaged 207 Gt/yr (with uncertainty between 60 and 354 Gt) while it was 248 Gt (with uncertainty between 153 and 343) during the 2010s, Politifact concluded that my statement was "mostly false."

The Politifact analysis erred in using arbitrary ten-year intervals. Nature doesn't care about the human construct of decades; using different ten-year averaging intervals, say those ending in 2015, 2005, and so on, can lead to a different conclusion.

That arbitrariness can be eliminated by calculating the ten-year trailing average of the annual mass loss for *every* year, not just once per decade. The results, shown in Figure 8.10, make it evident that the recent annual mass loss is no larger than in the 1930s and has been declining for the past decade. As the IPCC AR6 WGI report notes with high confidence in Section 9.4.1.1, "the mass loss varies strongly due to the large interannual variability of the SMB [Surface Mass Balance]."

Since Politifact refused to publish a retraction, even when confronted with the data, I published a *Wall Street Journal* Op-Ed containing the graph in February 2022.[30] I noted that the ups and downs over the decades, even

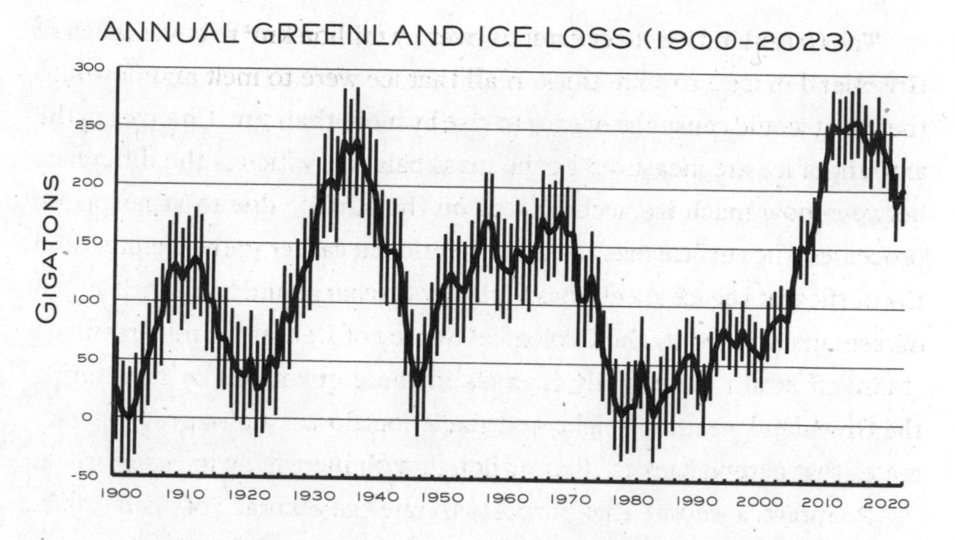

FIGURE 8.10 **Annual ice loss of the Greenland ice sheet** (10-year trailing average, 1900–2023). Data from Mankoff et al.

as warming human influences on the climate were growing steadily, suggested that natural cycles in the North Atlantic were playing an important, if not dominant, role in determining ice loss.[31]

I was disappointed by the reaction of the experts, who accused me of "cherry picking" by focusing on the decline over the past decade.[32] They said my argument was

> based on an incorrect interpretation of the plotted data . . . Mr. Koonin claims that "the annual loss of ice has been decreasing in the past decade even as the globe continues to warm." While that is factually correct, it is an invalid interpretation, considering only the last decade and excluding previous periods. This is often referred to as "cherry picking."[33]

But because I know well that any discussion of recent changes must be put in historical context, my Op-Ed had in fact clearly displayed and discussed more than a century of data. In contrast, you can't find annual mass balance values over multiple decades in either the annual reports issued by researchers or in the popular media.[34] And, of course, the *Guardian* article and NASA press release make no mention of past variability. As I noted in the Op-Ed:

> Much climate reporting today highlights short-term changes when they fit the narrative of a broken climate but then ignores or plays down changes when they don't, often dismissing them as "just weather."

9 APOCALYPSES THAT AIN'T

The media, and hence popular and political opinion, attribute all manner of impending societal catastrophes to human influences on the climate, including death and destruction, disease, agricultural collapse, and economic ruin. Luckily for us, the historical data doesn't support such claims, and projections of future impacts (exemplified by the familiar "could be as bad as . . .") stem from implausibly extreme scenarios fed into models that, as we've seen, are clearly not up to the task. But to understand just how badly The Science is misrepresented when it comes to such impacts, you really need to see the details of a few examples. This chapter offers three vignettes of headline "apocalypses that ain't." One is "climate-related deaths," a menace based on speculation, strained assumptions, and incorrect use of data. The second is a future agricultural "disaster" that is belied by the evidence and requires acrobatic distortion to even detect. And the third is purportedly "enormous" economic costs—which turn out, even based on the data presented, to be minimal, if not too small to measure.

I first met Michael Greenstone during my time in the Obama administration when he was leading an interagency effort to determine the economic impact of greenhouse gas emissions; our paths have crossed a few times since. Greenstone is now the director of the Energy Policy Institute at the University of Chicago, and I've found him an astute and careful energy economist. In 2019, Michael testified to Congress about some of the findings of his ongoing research into the economic effects of local and global changes in climate:

> . . . the increase in the global mortality rate due to climate change–induced temperature changes in 2100 is larger than the current mortality rate due to all infectious diseases . . . we estimate the full mortality risk due to climate change to be an additional 85 deaths per 100,000 in 2100 . . .[1]

Let's break down those numbers. In 2018, all infectious diseases globally accounted for about 75 deaths per 100,000—about one tenth of total deaths from any cause, which have an annual rate of 770 per 100,000. Since the global population is about eight billion, that amounts to six million deaths each year from infectious diseases.[2] So if future temperature changes were to cause at least as many additional deaths, that would indeed be a very big deal. (To set a scale for these numbers, US COVID-related deaths in 2020 amounted to 100 per 100,000. Globally, the rate was 23 per 100,000 or one-third the rate from all infectious diseases.)

Evaluating Greenstone's alarming assertion means asking some questions. In particular: What are the current numbers of climate-related deaths? And how have climate-related deaths trended over the past century? More basically, you might be wondering: What's a "climate-related death" in the first place?

Well, people don't die from climate. Climate changes slowly and societies largely adapt to that (or migrate). But people do die from climate-related weather events—droughts and floods, storms, temperature extremes, and wildfires. We've already seen that it is far from certain that changes in climate have already increased these phenomena, but let's start by looking at the record of weather-related deaths over the past century.

The Centre for Research on the Epidemiology of Disasters (CRED) within the Université catholique de Louvain maintains an Emergency Events database covering over 22,000 mass disasters globally, starting from 1900.[3] Data on deaths due to natural disasters are easily downloaded from that site and can be separated into those that are weather related (droughts, floods, storms, wildfires, and extreme temperatures) and others that are not (earthquakes, tsunamis, and volcanoes). While there are large year-to-year fluctuations in those counts because of particular events, and deaths in the early years have been underreported, some sense of the trends can be had by looking at averages over each decade, as shown in Figure 9.1.

One takeaway from this graph is that weather-related death rates fell dramatically during the past one hundred years even as the globe warmed 1.2°C (2.2°F); they're about 80 times less frequent today than they were a century ago. That's largely due to better tracking of storms, better flood control, better medical care, and improved resilience as countries have developed. A recent UN report confirms the trend over the past two decades.[4] A second point is that in the most recent decade, extreme temperatures caused 0.16 deaths per 100,000 each year, about five hundred times smaller than Greenstone's projection for 2100.

So how did Greenstone come up with his alarming prediction? In further testimony,[5] Greenstone provided some details of those findings from a paper that had just been published.[6] The analysis uses historical records to understand how temperatures influence deaths, and then combines that with climate model projections of temperatures to estimate deaths in 2100. This is done over 24,378 different geographical regions, accounting for differences in their current climates, incomes, and age distributions.

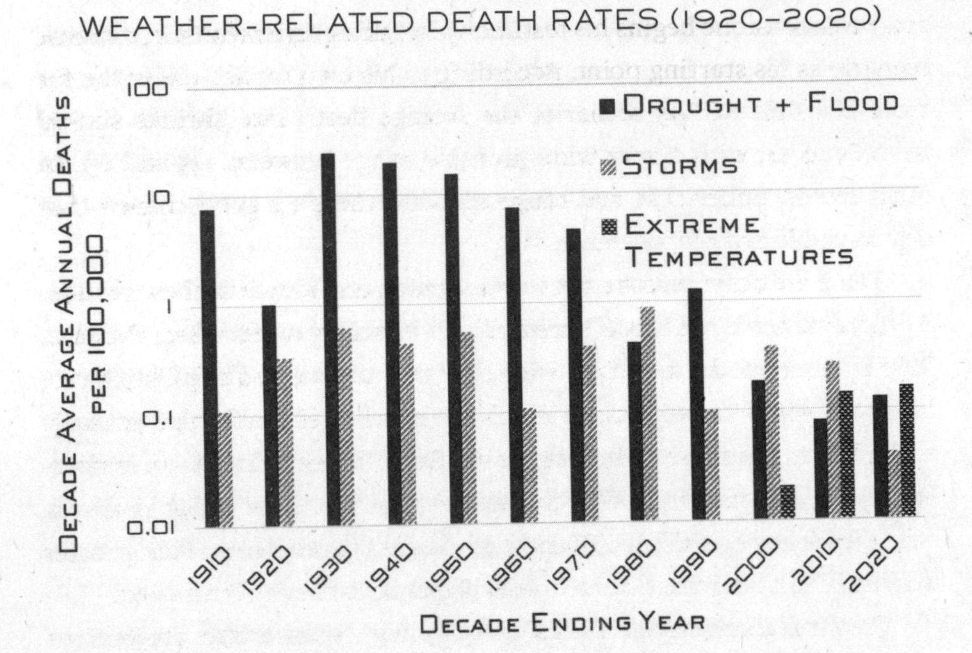

FIGURE 9.1 **Decadal averages of annual death rates from weather-related catastrophes over the past century.** The scale is logarithmic and decades are labeled by the year in which they ended. Deaths from wildfires in any decade are too small to be visible on this chart.[7]

The researchers also claim to have at least partially accounted for future economic development and adaptation to higher temperatures.

As the research paper (but not the coverage of Greenstone's testimony) acknowledges, the analysis is fraught with assumptions and uncertainties. In fact, the authors note that "our full set of estimates reveals a remarkable degree of uncertainty," at least some of which is "fundamentally unresolvable." Even apart from the economic and demographic dimensions, as we've seen in Chapter 4, there are great uncertainties in both the temperature changes the models project and in the emissions scenarios used to drive them. It turns out that the 85 deaths per 100,000 in 2100 quoted with great confidence in Greenstone's 2019 testimony is actually the average value under the implausibly high emissions scenario RCP8.5, with an 80 percent probability of the value lying between –21 and

201. So Greenstone begins his testimony by taking an unrealistic, extreme scenario as his starting point. According to his own model, under the far more plausible RCP4.5 scenario, the average death rate shrinks sixfold from 85 to 14, with a very wide probable range between −45 and 63. In other words, under that mid-range scenario there's a good chance that deaths would actually decrease.

There are other reasons not to place much confidence in these results. A crucial measure of a model's credibility is its ability to reproduce the past. That is, if the model were run using only the data we had as of, say, 1980, how well would its projections match the number of deaths that actually took place in the subsequent forty years? Such "hindcasting" is an important test of the climate models themselves. But that basic check is absent from Greenstone's analysis. Without it, the results require an even greater helping of salt than the climate model projections they're built upon.

Creating alarming headlines through highly uncertain projections of the future is one thing, but promoting the specter of climate-related deaths by distorting existing data is quite another. A 2019 article in *Foreign Affairs* by the Director-General of the World Health Organization, Tedros Ghebreyesus, was entitled "Climate Change Is Already Killing Us."[8] Yet the text doesn't deliver on the catchy title. Astoundingly, the article conflates deaths due to ambient and household air pollution (which cause an estimated 100 per 100,000 premature deaths each year, or about one-eighth of total deaths from all causes) with deaths due to human-induced climate change. The World Health Organization itself has said that indoor air pollution in poor countries—the result of cooking with wood and animal and crop waste—is the most serious environmental problem in the world, affecting up to three billion people.[9] This is not the result of climate change. It's the result of poverty. That pollution does indeed affect the climate (as we've seen, the aerosols are actually a cooling influence), but pollution deaths aren't caused by a changing climate; it's the pollution itself that kills. Such brazen misinformation by the WHO's leadership is particularly upsetting for its potential to diminish confidence in the organization's vital public health mission.

In August 2019, the *New York Times* ran the headline "Climate Change Threatens the World's Food Supply, United Nations Warns."[10] The article announced the release of the Summary for Policymakers of the IPCC's Special Report on Climate Change and Land (SRCCL).[11] The *Times*'s description of the report's findings followed the standard template for such coverage:

- it's already bad ("climate change is already hurting the availability of food because of decreased yields"),
- it's going to get a lot worse (". . . if emissions of greenhouse gases continue to rise, so will food costs"),
- but we can take prompt and drastic action to avert the worst (". . . there is still time to address the threats by making the food system more efficient").

By now, I trust that reading this litany of woe naturally raises the familiar questions about historical context and what the data shows. What have agricultural yields been doing in recent decades—have they already been impacted, and if so, how? What exactly do projections of future catastrophe say, and how solid are they?

Answering these questions requires careful reading of the IPCC Special Report itself. The SRCCL's Key Finding A.1.4 says:

> Data available since 1961 shows the per capita supply of vegetable oils and meat has more than doubled and the supply of food calories per capita has increased by about one third (*high confidence*).

That's supported by data in the report's Technical Summary, which shows that global production of both crop and animal calories has gone up dramatically since 1960, and that beginning in 1965, more than enough food calories have been produced every year to satisfy humanity's nutritional requirements. Indeed, annual deaths due to famine have averaged about two to four per 100,000 since 1980; the rate was ten to twenty times larger in the first half of the twentieth century.[12] That's not to say hunger is no longer an issue—poverty and problems with food distribution are among the factors leaving about 10 percent of the global population still undernourished, and the same Key Finding mentioned above notes that about a quarter of the food produced is lost or wasted.

But we do have the ability to feed all of humanity, and that ability stems largely from improvements in crop yields, such as those shown in Figure 9.2. In the fifty years from 1961 to 2011, global yields of wheat, rice, and maize (corn) have each more than doubled, and US corn yields have more than tripled.[13]

Crop yields depend upon several factors—plant genetics, nutrients available in the soil, agricultural practices, and the weather manifestations of climate (temperature, insolation, and rainfall). But you might be surprised to learn that the increasing concentration of carbon dioxide has been a significant factor in yield improvements, as it boosts the rate of photosynthesis and alters plant physiology to use water more efficiently.[14, 15] Increasing CO_2 in the atmosphere has also fertilized the natural world.[16] As the SRCCL notes in its Key Finding A.2.3, during the past four decades the Leaf Area Index (the fractional area covered by leaves) observed by

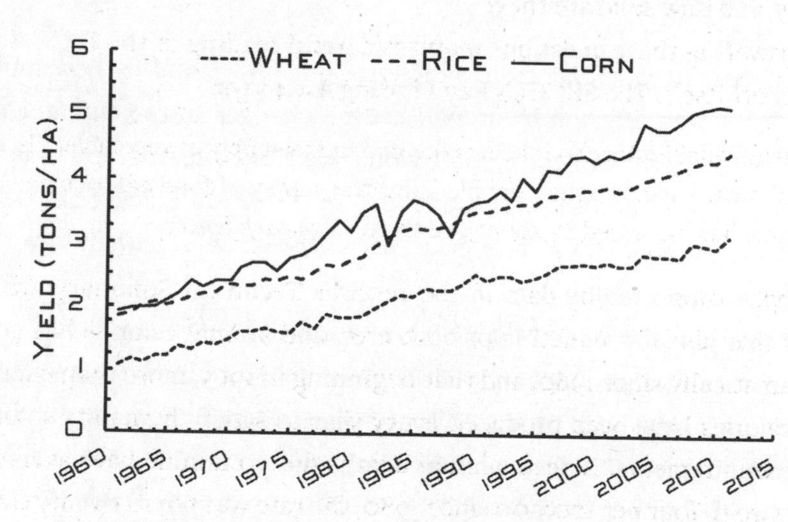

FIGURE 9.2 **Trends in the global yields of wheat, rice, and corn—the top three crops grown in the world.**[17]

satellites has increased markedly ("greened") over 25–50 percent of the vegetated areas of globe, while it has decreased ("browned") over less than 4 percent of the globe.

Despite the yield improvements over the past decades, the SRCCL makes the following claim:

> ... climate change between 1981 and 2010 has decreased global mean yields of maize, wheat, and soybeans by 4.1, 1.8 and 4.5%, respectively, relative to preindustrial climate, even when CO_2 fertilisation and agronomic adjustments are considered. Uncertainties (90% probability interval) in the yield impacts are −8.5 to +0.5% for maize, −7.5 to +4.3% for wheat, and −8.4 to −0.5% for soybeans. For rice, no significant impacts were detected. This study suggests that climate change has modulated recent yields on the global scale and led to production losses, and that adaptations to date have not been sufficient to offset the negative impacts of climate change, particularly at lower latitudes.[18]

In other words, although the actual wheat yield went up by about 100 percent from 1981 to 2010, it would have gone up even more (104 percent) if there hadn't been any human-caused changes in the climate. Similarly, the maize yield would have gone up by 77 percent, instead of by 70 percent.

Unfortunately, it's far from simple to judge how, and by how much, yields have been affected by human-caused changes in the climate. You need to know both what the climate would have been absent human influences and how agriculture was affected by those differences. In other words, we'd need to do a counterfactual analysis, one that can never be tested against observations.

Aside from the acknowledged limitations of the methodology and the climate and crop models used to make those estimates above,[19] the impacts they came up with are pretty small—comparable to the precision with which yields are measured in the first place (data from the UN's Food and Agriculture Organization[20] have a sampling precision of no better than 3 percent, among other uncertainties).

As for catastrophes to come, the IPCC's report contains plenty of qualitative warnings of future food problems, many based upon the dubious climate projections discussed in Chapter 4. Among them are a concern that crops grown under elevated CO_2 and significantly higher temperatures will have about 10 percent lower nutritional value, which the report notes could be mitigated by changes in crop genetics. But saying "yields will be impacted" is pretty meaningless, if not misleading, unless one also says by how much.

Alas, it's tough to find any quantitative projections of future yields in the SRCCL. However, the report's Key Finding A.5.4 has this to offer for a "middle of the road" future scenario in which development follows historical trends:

> . . . global crop and economic models project a median increase of 7.6% (range of 1–23%) in cereal prices in 2050 due to climate change (RCP6.0), leading to higher food prices and increased risk of food insecurity and hunger (*medium confidence*).

That's a remarkable statement, not just for what it says, but for what it's about—prices, not yields. Prices for food commodities are set in global markets by the balance of two large numbers, supply and demand, and small changes in one or the other can induce large price swings. Apparently the SRCCL's reasoning is that changes in climate will decrease yields, and thus supply, raising prices. But the profound difficulties in modeling supply and demand are compounded by the uncertainties in climate projections. And there are many factors beyond climate that affect supply.

But let's put all that aside and take the Key Finding at face value. The median projected price increase is 7.6 percent by 2050, or an average of about one-quarter of one percent per year. We'll go a step further, however, and take the highest projection offered by any of the models—23 percent over the next three decades, or an average of about three-quarters of a percent per year. What would be the impact of that sort of price increase?

Figure 9.3 shows a fifty-year history of corn and wheat prices. You can see that inflation-adjusted prices grew by a factor of two during the 1970s, and have been on a downward trend since, suggesting there's a lot more than a changing climate influencing prices. And there are substantial ups and downs over few-year periods that are much larger than any projected climate-related increases over the next three decades. In other words, even if they came to pass, climate impacts would hardly be apparent.

To sum up, agricultural yields, and the food supply overall, have surged during the past century even as the globe has warmed; 2020 saw record high grain production.[21] The IPCC assesses that whatever climate changes occurred between 1981 and 2010 had minimal impact on that strong growth. Projected price impacts of future human-induced climate changes through 2050 are not only uncertain, but also much smaller than past variations and so should hardly be noticeable amid ordinary market dynamics. In short, the science says that crop failures because of "climate" are yet another apocalypse that ain't.

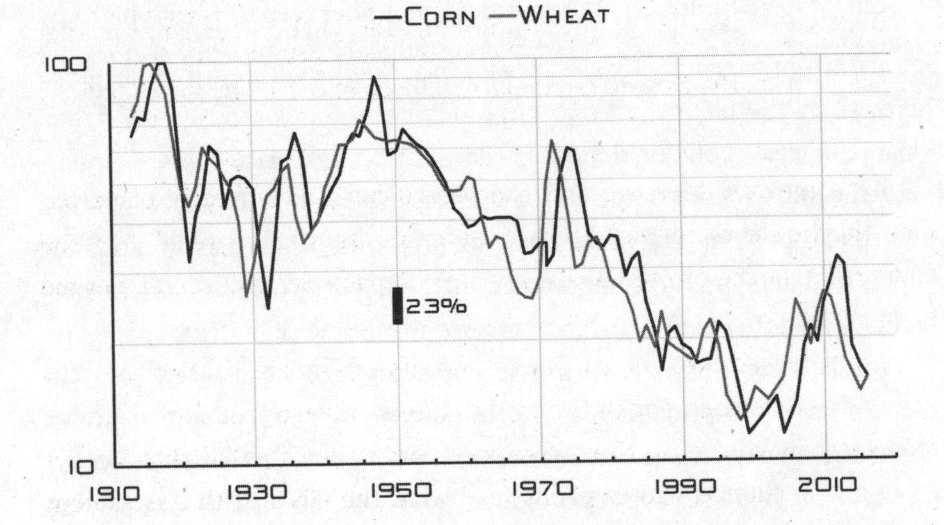

FIGURE 9.3 **Inflation-adjusted corn and wheat prices from 1913 to 2016.** The plot is on a logarithmic scale. Each price is relative to its maximum around 1920. The vertical bar shows that the maximum projected increase in 2050, 23 percent, is small compared with the historical variation in prices.[22]

In 2018, on the day after Thanksgiving (Black Friday), the second volume of the Fourth National Climate Assessment (NCA2018) was released. It deals with the projected impacts of human-induced climate change, and it immediately generated the now familiar headlines warning of impending economic disaster, among them:

- "Climate change will wallop the US economy" (NBC News[23])
- "Climate report warns of grim economic consequences" (Fox News[24])
- "Climate change could cost US billions" (*Financial Times*[25])
- "US climate report warns of damaged environment and shrinking economy" (*New York Times*[26])

Indeed, Key Message 2 of the report's Chapter 29 reads:

> In the absence of more significant global mitigation efforts, climate change is projected to impose substantial damages on the US economy, human health, and the environment. Under scenarios with high emissions and limited or no adaptation, annual losses in some sectors are estimated to grow to hundreds of billions of dollars by the end of the century.

Both the Key Message and the heated headlines greatly dismayed me—they're clearly intended to be frightening. Yet I had studied the issue and knew that the projected net economic impacts were minimal. Let me explain.

I first looked into the economic impacts of climate change the year before, in 2017, when one of the world's largest investment organizations requested my advice on climate science. Since they'd asked that I cover economic impacts, I had carefully read what the UN's Fifth Assessment Report (AR5) had to say on the matter.

Projections of the economic impacts of a changing climate are highly uncertain. Of course, we already know there are great uncertainties in how the climate will change because of inadequate climate models and uncertainty in future emissions. And climate uncertainties are larger at the regional level than they are at the global level. For example, for the first five or six years of the recent California drought, many climate scientists said that human influences on the climate increased the risk of drought.[27] Yet it took only about a year after the drought broke dramatically in 2016 for papers to appear claiming that a warming world would also mean a wetter California.[28] Perhaps this is just the process of scientific understanding being refined. Less charitably, I get the distinct sense that the science is unsettled enough that *any* unusual weather can be "attributed" to human influences.

In addition, climate is only one of many factors influencing economic development and well-being. Economic policies, trade, technology, and governance are also important, and these are different in different countries and can change in unpredictable ways. Economic measures are highly regional, and their future uncertainties are compounded by the uncertainty of regional climate predictions. It is particularly difficult to predict how, and how much, a rising temperature would damage a society economically in the face of so many unknowns—among them the role that might be played by adaptation measures like the raising of sea walls or shifts in what crops are cultivated that minimize, or sometimes even exploit, the impact of climate changes.

Despite those challenges, the AR5's Working Group II—whose part of the assessment is devoted to the ecological and societal impacts of the changes in climate outlined by Working Group I—does say something about how world economic activity would be affected by a warming globe. Figure 9.4 plots some twenty published estimates showing that the (by now familiar) projected global temperature rise of up to 3°C by 2100 would negatively impact the global economy by—wait for it—3 percent or less.

For my talk to the investors, I provided some important context that was missing from the UN report. An impact of 3 percent in 2100—some eighty years from now—translates to a decrease in the annual growth rate by an average of 3 percent divided by 80, or about 0.04 percent per year. The IPCC scenarios (discussed in Chapter 3) assume an average global

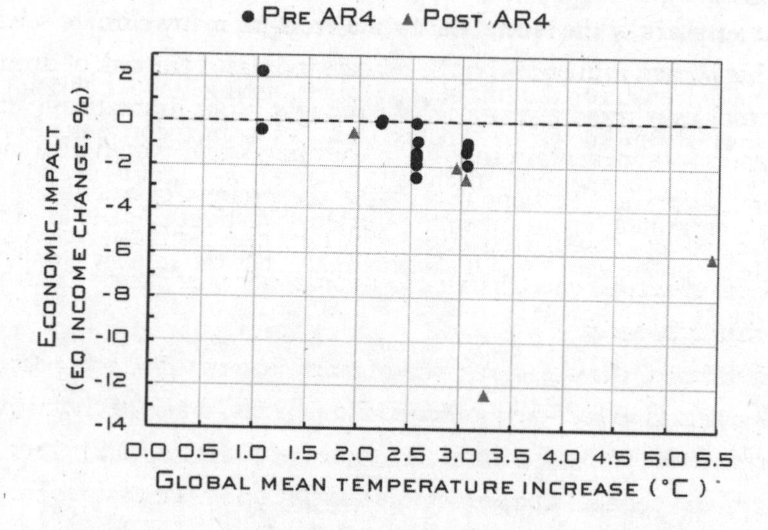

FIGURE 9.4 **Estimates of the net global economic impact in 2100 from rising global temperatures.**[29]

annual growth rate of about 2 percent through 2100; the climate impact would then be a 0.04 percent decrease in that 2 percent growth rate, for a resulting growth rate of 1.96 percent. In other words, the UN report says that the economic impact of human-induced climate change is negligible, at most a bump in the road. In fact, the first point in the Executive Summary to its Chapter 10 is:

> For most economic sectors, the impact of climate change will be small relative to the impacts of other drivers (*medium evidence, high agreement*). Changes in population, age, income, technology, relative prices, lifestyle, regulation, governance, and many other aspects of socioeconomic development will have an impact on the supply and demand of economic goods and services that is large relative to the impact of climate change.

A 2018 article written by one of the IPCC's coordinating lead authors reviewed a further four years of published papers and came to a similar conclusion:

> . . . the total economic impacts of climate change are negative, but modest on average, and that the severe impacts on less developed countries are caused primarily by poverty.[30]

The consensus on the minimal overall economic impact of rising temperatures is well known to experts, though it's an inconvenient one for those wishing to sound the alarm on climate. I was dumbfounded when I asked a prominent environmental policymaker about the UN assessment and the response was: "Yes, it's unfortunate that the impact numbers are so small."

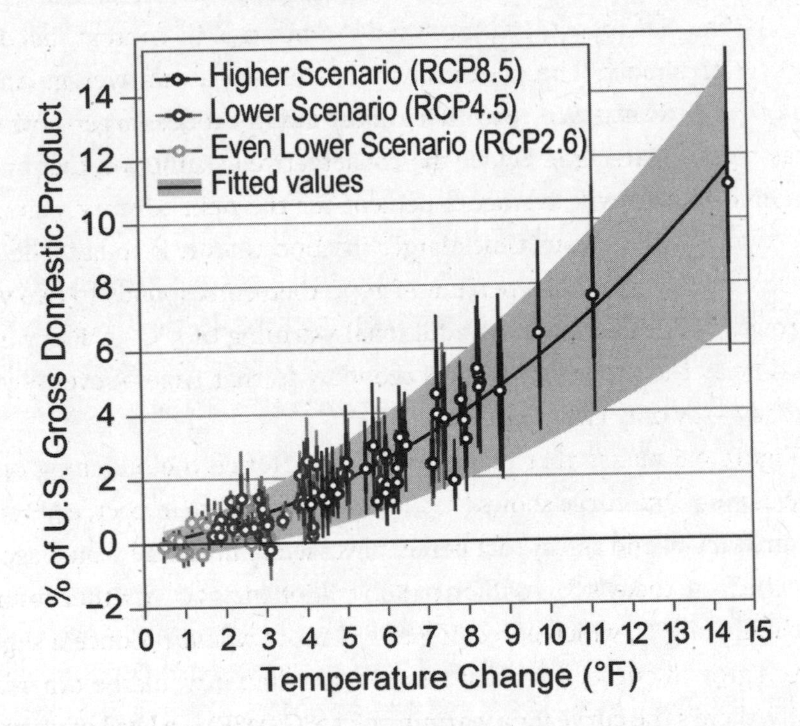

FIGURE 9.5 **Projected damages to the US economy at the end of the century.** The horizontal axis is the change in global average temperature (in degrees Fahrenheit) from the interval of 1980–2010 to that of 2080–2099. The dots show the median impact, and the lines and shading show the uncertainties. (Figure 29.3 of NCA2018.)

At any rate, this background left me primed to weigh in on the breathless coverage that accompanied the release of Volume II of NCA2018. The last figure in that report's last chapter (reproduced in Figure 9.5) is based on a 2017 paper published in *Science* magazine.[31] It shows that projected direct damages to the US economy at the end of the century grow with increasing global average temperature (shown as the anomaly relative to the 1980–2010 average). As in the IPCC projection for the world economy, the impacts on the US are small: a very large warming of 5°C (9°F) at the end of the century would diminish the US economy by 4 percent. (It's worth noting that this 5°C warming is relative to today's temperatures, which are up 1°C from preindustrial values, making this equal to 6°C of warming by the Paris Agreement accounting, which has set 1.5°C as a goal.)

Like the UN report, NCA2018 fails to put this in context, but I can do so quite simply: The US economy has grown at an average annual rate of 3.2 percent since 1930 (it's almost twenty times larger now than it was ninety years ago). Under the conservative assumption that annual economic growth will average 2 percent for the next seventy years, the US economy will be four times larger in 2090 than it is today. The purported climate impact of 4 percent in 2090 then corresponds to two years of growth. In other words, an additional warming of 5°C (9°F) by 2090 would delay the growth of the US economy to that time—seventy years from now—by only two years.

Figure 9.6 makes that point graphically. Notice the bunching of the three curves. One curve shows that, absent any climate impact, a US economy growing at the assumed 2 percent average annual rate would see the GDP rise from today's $20 trillion to $80 trillion in 2090. Another assumes a warming of 5°C, which according to NCA2018 would produce a slightly delayed growth curve, 4 percent less in 2090 than it would be otherwise. Finally, there's the curve for a warming of 7.2°C (13°F)—a level of warming well beyond what's projected under even the most extreme IPCC scenario. According to NCA2018, this would result in a 10 percent hit between now and 2090, which still amounts to only a five-year delay in growth seventy years from now.

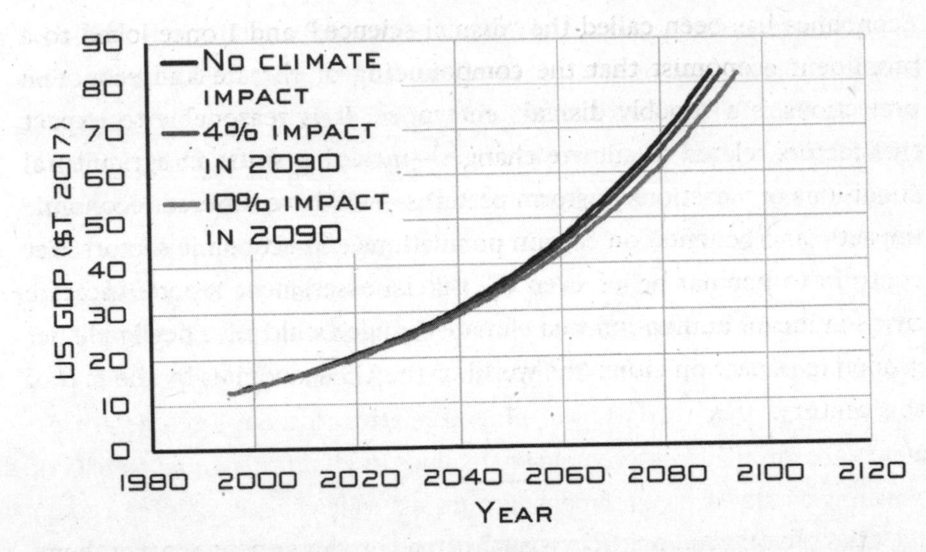

FIGURE 9.6 **Constant-dollar projections of the US GDP to 2090.** Shown are curves assuming a nominal 2 percent annual growth rate with no climate impact, as well as 4 percent and 10 percent climate-related impacts in 2090. The impacts are assumed to set in linearly over the time shown.

Within a few hours of the NCA2018's release on Black Friday, I had drafted a short Op-Ed saying more or less what I have said here, which the *Wall Street Journal* published online on Monday.[32] The next day, a prominent US energy economist sent an email thanking me for making the point—alas, that person could never express that thanks publicly. The next week, one of the authors of the original 2017 research paper from which the estimates used in the assessment report were drawn expressed dismay at the way their results were portrayed in the media.[33]

The climate science establishment, most notably the authors of NCA2018, reacted to my Op-Ed with silence. They did nothing to address the media's catastrophizing. Perhaps they were embarrassed by their own doom mongering. Or perhaps, like the policymaker I mentioned earlier who wished the impact numbers had been greater, it was precisely the coverage they'd been hoping for.

As you've no doubt noticed yourself, the notion of climate-related economic disaster remains alive and well in the media and political dialogue.

Economics has been called the "dismal science," and I once joked to a prominent economist that the compounding of climate and economic projections is a "doubly dismal" enterprise. It is reasonable to expect that factors related to climate changes—including shifts in agricultural conditions or variations in storm patterns—will have different economic impacts (and benefits) on certain populations and economic sectors. Yet contrary to popular belief, even the official assessment reports indicate that significant human-induced climate change would have negligible net economic impact on either the world or the US economies by the end of this century.

It's clear that media, politicians, and often the assessment reports themselves blatantly misrepresent what the science says about climate and catastrophes. Those failures indict the scientists who write and too-casually review the reports, the reporters who uncritically repeat them, the editors who allow that to happen, the activists and their organizations who fan the fires of alarm, and the experts whose public silence endorses the deception. The constant repetition of these and many other climate fallacies turns them into accepted "truths."

Over the course of this book, we've explored the chasm between what is presented as settled when it comes to climate and what the science actually tells us. So how did we get here? The next chapter will look more closely at the perfect storm of interests that leads to a fervent belief in a consensus that isn't.

IPCC'S DEATHS DECEPTION—2024 UPDATE

Unfortunately, the IPCC continues its misrepresentation of what the data says about climate-related fatalities. In discussing fatalities from extreme temperatures, the AR6 Synthesis Report (Section A.2.5) says only:

> In all regions increases in extreme heat events have resulted in human mortality and morbidity (*very high confidence*).

In fact, over the twenty years between 2000 and 2019, an average of about five million people per year died from extreme temperatures.[34] That's about 9.4 percent of all deaths. But you might be surprised to learn that some 90 percent of those deaths were from extreme cold rather than from extreme heat. And as the global average temperature increased by 0.5°C over those twenty years, fewer cold snaps caused a large decrease in cold-related deaths that more than outweighed the smaller increase in heat-related deaths. That is, the total number of deaths from extreme temperatures *decreased*. So the IPCC statement is completely factual but factually incomplete. It is difficult to believe this omission is an oversight instead of an attempt to persuade.

Heat-related deaths are a serious matter—most of the world lacks access to air-conditioning, and increasing urbanization means more people exposed to even greater heat. But a report on the effects of a changing climate, especially one aimed at informing the public and policy, should not cherry-pick which effects it describes.

ECONOMIC IMPACTS OF A CHANGING CLIMATE—2024 UPDATE

Several developments since publication have reinforced the discussion of the economic impacts of warming that I covered in the third section of this chapter.

The AR6 WGII report (Cross Working Group ECONOMIC) updated the global "damage function" shown in Figure 9.4.[35] There is now a greater number of estimates, many of which are in accord with "a few percent for a few degrees" conclusion I discussed in the book. But there is a greater spread among them, which caused AR6 to say:

> The wide range of estimates, and the lack of comparability between methodologies, does not allow for identification of a robust range of estimates with confidence (*high confidence*). Evaluating and reconciling differences in methodologies is a research priority for facilitating use of the different lines of evidence (*high confidence*). However, the

> existence of higher estimates than AR5 indicate that global aggregate economic impacts could be higher than previously estimated (*low confidence* due to the lack of comparability across methodologies and robustness of estimates).

Almost all that increase in uncertainty is due to the results of papers by Burke and collaborators in 2015[36] and 2018,[37] which differ from many of the others in their assumptions of how the changing climate impacts the economy. Some of the Burke results transcend the unlikely to become truly absurd.[38] For example, in 2100, Iceland becomes thirty times richer than today and Mongolia two hundred times richer (four times richer per person than the US!). Although the IPCC assessment now correctly disregards climate models that give absurd results, it hasn't yet reached that level of discernment for the economic models. But when it does, results revert to the "few percent for a few degrees."[39]

The subject of "tipping points" comes up frequently. While climate models give no indication that such catastrophes are in our future, a 2021 paper analyzed the global economic impact of eight different tipping points, ranging from outgassing of the permafrost to disintegration of the Greenland and West Antarctic ice sheets to slowing of the circulation in the Atlantic.[40] They find another 2 to 3 percent of impact at most, even when tipping points are considered in combination.

The US macroeconomic risks for "climate change" were also discussed in a white paper issued in March 2023 jointly by the Biden administration's Council of Economic Advisors and Office of Management and Budget; it's available on the White House website.[41] The report compiles and analyzes twelve independent peer-reviewed estimates of the impact of future warming on the US GDP, as shown in Figure 9.7.

The report findings are quite consistent with "a few percent for a few degrees": A global warming of 2.5°C above preindustrial (remember we're currently at 1.2 degrees, the stated Paris goal is 2.0° with an aspiration of 1.5°, and the UN projects about 2.7° by 2100) would make the US GDP about 1 percent smaller than it would be otherwise. In other words, if the US GDP grew at an average annual rate of 2 percent over the next seventy

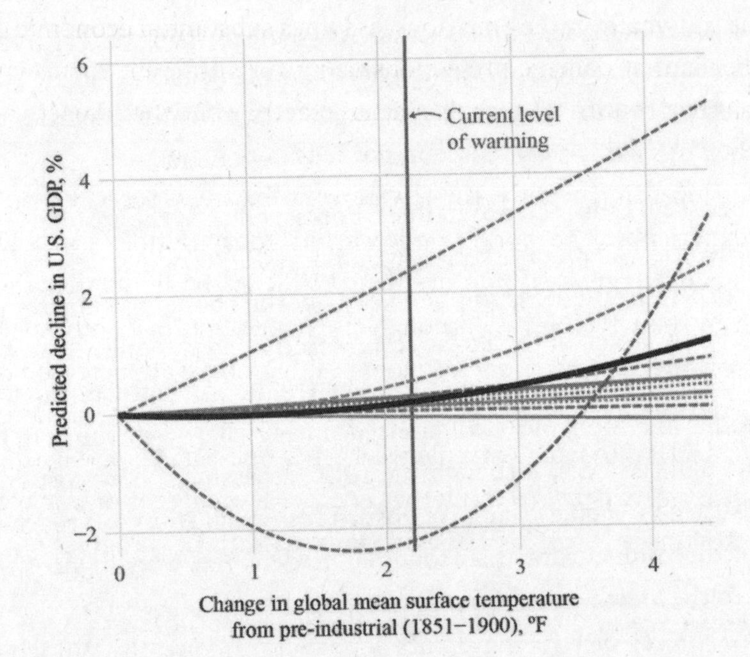

FIGURE 9.7 **Biden administration compilation of estimates of global warming's impact on the US GDP.** The thick black line is the aggregate of the dozen peer-reviewed estimates shown.[42]

years, it would be four times larger. A climate impact of 2 percent by the end of the century would reduce that growth to 3.92 times, a negligible difference.

The White House report also reinforces, perhaps inadvertently, how little climate change is projected to affect the US economy in the coming decades. Today's ratio of (privately held) debt to GDP is about 100 percent. Under a low global emissions scenario that achieves net zero by 2050, the ratio is projected to reach 111.2 percent by 2048, while it would be 112.6 percent under a high-emissions scenario deemed unlikely by the IPCC. Since many factors other than climate will be far more influential in determining the GDP and debt over the next twenty-five years, the small 1.4 percent difference between emissions scenarios is in the noise.

Why should we believe these projections when the climate models themselves are so deficient? One reason is that *every* IPCC scenario for the future, whatever the emissions, assumes substantial economic growth. Another is that damage projections using very different robust methods give similar results. It doesn't matter exactly what the impacts are, just that they're small—a few percent, not tens of percent.

But most importantly, the few percent for a few degrees is in accord with experience. The world is 2°F warmer today than in 1900. Even so, we've seen the greatest improvement EVER in the human condition, as shown in the table below. Although the population has quintupled since 1900, longevity, literacy, and economic activity have all increased dramatically, and the fraction of the globe in extreme poverty plummeted. And today's death rate from extreme weather is one-fiftieth of what it was in 1900. It beggars belief to think that another few degrees of warming over the next hundred years would significantly derail that progress.

Indicator	Change	Unit	"1900"	"Today"
Global temperature[43]	**1.3 warmer**	degrees C	-0.5 (1905)	+0.8 (2022)
Population[44]	**5X larger**	billions	1.65 (1900)	8.0 (2022)
Life expectancy[45]	**130% longer**	years	32 (1900)	72.6 (2019)
Literacy fraction[46]	**4X larger**	percent	21.4 (1900)	86.25 (2016)
GDP per capita[47]	**6.8X larger**	$2011	2,241 (1920)	15,212 (2018)
Extreme poverty[48]	**>7X smaller**	percent (<$1/day)	70 (1900)	<10 (2015)
Weather death rate[49]	**50X smaller**	per million	241 (1920)	5 (2008)

One might argue that it's only in recent decades that the climate has been broken. Yet Figure 9.8 shows that economic loss rates due to extreme weather have actually declined slightly over the past thirty years, averaging about 0.2 percent of global GDP. A wealthier world is a more resilient world.

You might not be very convinced by the argument that the world has prospered despite 1.3°C (2.3°F) of warming since 1900, thinking that at some point, like the proverbial frog in a warming pot, any additional warming will become catastrophic. People in 1900 might certainly have thought that the population increase alone would have been "too much"

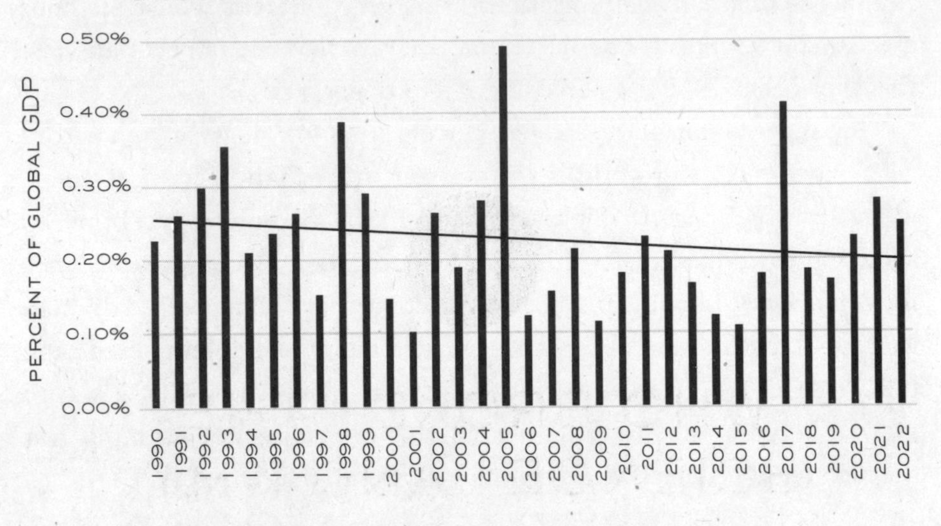

FIGURE 9.8 **Global weather losses as percent of global GDP (1990–2022).**[50]

and, in fact, the "Malthusian catastrophe" had then already been a concern for a century.[51] Instead, the human condition improved spectacularly. So, while there are no guarantees, it seems reasonable to expect that we will be in a better position to cope with a changing climate.

WHO BROKE "THE SCIENCE" AND WHY

If crucial parts of the science really are unsettled, as we've seen over the past chapters, why is the narrative of The Science so different? Can it really be that the multiplicity of stakeholders in climate matters—scientists, scientific institutions, activists and NGOs, the media, politicians—are all contributing to misinformation in the service of persuasion? And why has The Science gained such prominence over science?

Observing this scene over the years, I've given a lot of thought to how the communication of climate science works. I'm no expert on human behavior, but I have seen this process up close, and my direct experiences, along with some universal truths about humans, suggest not some secret cabal, but rather a self-reinforcing alignment of perspectives and interests. Let's look at the most important players in turn.

THE MEDIA

When I moved to the UK in 2004, I naturally began reading the British newspapers. I was struck by how much more international coverage there was than in the US, no doubt because the smaller UK necessarily has more foreign interactions, as well as ties to the rest of Europe and historical relationships with Commonwealth countries once part of the British Empire. And of course, soccer—that is, football—got many more column inches. But what surprised me most wasn't only a matter of content, but tone. The British papers were often overtly partisan, not just in their editorials, but also in their reporting. Although I had read widely among US national newspapers, including the *New York Times*, the *Wall Street Journal*, and the *Washington Post*, it was a revelation to see the stark differences in what was covered, and how it was covered, among the UK's *Guardian*, the (London) *Times*, the *Telegraph*, and the *Financial Times*.

In the years since, US media outlets have developed more explicit and more differentiated points of view themselves, and those have likewise seeped from their editorial pieces into their reporting. Most notably, as the age of the internet advanced, headlines became more provocative to encourage clicks—even when the article itself didn't support the provocation. Today, the shift toward the alarming—and shareable—has traveled well beyond the headlines. That's especially true in climate and energy matters.

Whatever its noble intentions, news is ultimately a business, one that in this digital era increasingly depends upon eyeballs in the form of clicks and shares. Reporting on the scientific reality that there's been hardly any long-term change in extreme weather doesn't fit the ethos of *If it bleeds it leads*. On the other hand, there is always an extreme weather story somewhere in the world to support a sensational headline.

Changes in staffing also contribute to the media miscommunication of the science. Many newsrooms are shrinking, and serious in-depth reporting is becoming less common. Many people reporting on climate don't have a background in science. This is a particular problem because, as we've seen, the assessment reports themselves can be misleading, especially to non-experts. Science stories are almost always stories of nuance; they require time and research. Unfortunately, the pace of the news cycle has only become more frantic, and reporters and editors have less time than ever. The diversity and ubiquity of modern media have increased the demand for fresh "content" and the competition to be the first to post a story. And as with scientists, a professional code that calls for lack of bias doesn't mean none creeps in.

As I interact with journalists, I realize that, for some, "climate change" has become a cause and a mission—to save the world from destruction by humans—so that packing alarm into whatever the story is becomes the "right" thing to do, even an obligation. This has been compounded by the rise of a new job category: "climate reporters." Their mission is largely predetermined; if they don't have a narrative of doom to report, they won't get into the paper (whether digital or print) or on the air.

Here's an example. A recent front-page story in the *Washington Post* reported that the Biden administration's climate policy would aim "to rapidly shrink the nation's carbon emissions," explaining that "a warming planet has made the issue increasingly hard to ignore, as the litany of climate-related catastrophes has grown with each passing year."[1]

Of course, as you have already read, the data does not at all support that "climate-related catastrophes" are growing "with each passing year."

There's much factual reporting in the full-page story that follows about plans for the new administration. But without those initial alarm bells, would the story have made it to the front page?

In short, the general lack of knowledge of what the science actually says, the drama of extreme weather events and their heart-rending impact on people, and pressures within the industry all work against balanced coverage in the popular media.

POLITICIANS

Politicians win elections by arousing passion and commitment from voters—by motivating and persuading. This is not new. H. L. Mencken's 1918 book *In Defense of Women* noted:

> The whole aim of practical politics is to keep the populace alarmed (and hence clamorous to be led to safety) by menacing it with an endless series of hobgoblins, most of them imaginary.[2]

The threat of climate catastrophe—whether storms, droughts, rising seas, failed crops, or economic collapse—resonates with everyone. And this threat can be portrayed as both urgent (by invoking a recent deadly weather event, for instance) and yet distant enough so that a politician's dire predictions will be tested only decades after they've left office. Unfortunately, while climate science and associated energy issues are complicated, complexity and nuance don't lend themselves at all well to political messaging. So the science is jettisoned in favor of The Science, and "simplified" for use in the political arena, which allows the required actions to be portrayed simply as well—*just eliminate fossil fuels to save the planet.*

Of course, this isn't a climate-specific problem, and the electorate—which abhors a gray area—bears part of the blame. It's hard to rally the base with uncertainty. There would surely be less support for, say, promoting renewable energy sources if they were more realistically portrayed as a *possible way* to mitigate a *possible future problem* instead of an *essential solution* to an *imminent crisis.* And uncertainty can be a political weapon. Politicians on the right who deny even the basics that science *has* settled—that human influences have played a role in warming the globe—are not above exploiting climate science uncertainties, offering them as "proof" that the climate isn't changing after all.

Politicians on the left find it inconvenient to discuss scientific uncertainties or the magnitude of the challenge in reducing human influences. Instead, they declare the science settled and label anyone who questions that conclusion "a denier," lumping conscientious scientists advocating for less persuasion and more research in with those openly hostile to science itself.

Some politicians have gone far beyond name-calling, brazenly attempting to undermine the scientific process. Two billionaire politicians, Michael Bloomberg and Tom Steyer, whose goal was "making the climate threat feel real, immediate, and potentially devastating to the business world," conspired with some scientists and others to produce a series of reports mischaracterizing the extreme emissions scenario RCP8.5 as "business as usual" (that is, a world without further efforts to rein in emissions).[3, 4] The reports were accompanied by a sophisticated campaign to infuse that notion into scientific conferences and journals.[5] Those who seek to corrupt the scientific process in that way are playing the same game as the anti-science crowd they loudly decry. Fortunately, the deception is now being called out in leading scientific journals.[6, 7]

Finally, it is standard practice to suggest that many of the politicians on the right who promulgate the idea of a "climate change hoax" are influenced by ties to industries negatively affected by restrictive environmental regulation. Alas, as the alternative energy industry grows, there is financial incentive for politicians to hype climate catastrophe as well. Science should not be partisan, but climate science's intersection with energy policy and politics all but guaranteed that it would become so.

SCIENTIFIC INSTITUTIONS

Trust in scientific institutions underpins our ability—and the ability of the media and politicians as well—to trust what is presented to us as The Science. Yet when it comes to climate, those institutions frequently seem more concerned with making the science fit a narrative than with ensuring the narrative fits the science. We've already seen that the institutions that prepare the official assessment reports have a communication problem, often summarizing or describing the data in ways that are actively misleading. In the next chapter, we'll delve a bit further into how this happens; I won't belabor the point here.

Other scientific institutions, or their leaders, have also been overwilling to persuade rather than inform. The National Academies of Sciences,

Engineering, and Medicine (NASEM) is a private, nonprofit institution chartered by the US Congress in 1863 to advise the nation. To quote from their website:

> The National Academies of Sciences, Engineering, and Medicine are the nation's pre-eminent source of high-quality, objective advice on science, engineering, and health matters.[8]

The Academies provide that advice largely through written reports sponsored by federal agencies. Some two hundred reports are published each year, dealing with a great range of topics in science, engineering, medicine, and the societal issues associated with them.[9]

Academies reports undergo an extensive authoring and review process. I know that process well, having led two Academies studies and reviewed the reports of several others, along with for six years overseeing all the Academies' report activities in Engineering and the Physical Sciences (including several in Energy, but none in Climate Science). This process does indeed result in reports that are almost always objective and of the highest quality. Unfortunately, as we've seen, their reviews of the National Climate Assessments (they don't write the assessments themselves) in 2014 and 2017/18 didn't quite meet that standard.

On June 28, 2019, the presidents of the National Academies of Science, Engineering, and Medicine issued a statement affirming "the Scientific Evidence of Climate Change." The sole paragraph dealing with the science itself read:

> Scientists have known for some time, from multiple lines of evidence, that humans are changing Earth's climate, primarily through greenhouse gas emissions. The evidence on the impacts of climate change is also clear and growing. The atmosphere and the Earth's oceans are warming, the magnitude and frequency of certain extreme events are increasing, and sea level is rising along our coasts.[10]

Even given the need for brevity, this is a misleadingly incomplete and imprecise accounting of the state of climate science. It conflates human-caused warming with the changing climate in general, erroneously implying that human influences are solely responsible for these changes. It invokes "certain extreme events" while omitting the fact that most types (including those that pop most readily to mind when one reads the phrase "extreme events," like hurricanes) show no significant trend at all. And it states that "sea level is rising" in a way that not only suggests that this, too, is solely attributable to human-caused warming, but elides the fact that the rise is nothing new.

I'm quite sure that this personal statement issued by the presidents in a news release was not reviewed by the usual Academies procedures; if it had been, its deficiencies would have been corrected. The statement therefore carries the weight of the Academies' name without being subject to its customary rigor. Ironically, the statement goes on to say the Academies "need to more clearly communicate what we know." Which in this case they didn't.

When communication of climate science is corrupted like this, it undermines the confidence people have in what the scientific establishment says about other crucial societal issues (COVID-19 being the outstanding recent example). As Philip Handler, a *prior* president of the National Academy of Sciences, wrote in the 1980 editorial I mentioned in the Introduction:

> It is time to return to the ethics and norms of science so that the political process may go on with greater confidence. The public may wonder why we do not already know that which appears vital to decision—but science will retain its place in public esteem only if we steadfastly admit the magnitude of our uncertainties and then assert the need for further research. And we shall lose that place if we dissemble or if we argue as if all necessary information and understanding were in hand. Scientists best serve public policy by living within the ethics of science, not those of politics.[11]

SCIENTISTS

This book's introduction described Stephen Schneider's false choice between being effective and being honest. But there are other factors that encourage climate researchers' monolithic portrayal of the science as settled, however vigorous their internal debates might be. Feynman closes his Cargo Cult speech by wishing the Caltech graduates

> the good luck to be somewhere where you are free to maintain the kind of integrity I have described, and where you do not feel forced by a need to maintain your position in the organization, or financial support, or so on, to lose your integrity.

I know from experience that such institutional pressures are real; whether you're working for the government, a corporation, or an NGO, there is a message to be adhered to. For academics, there is pressure to generate press and to secure funding through grants. There's also the matter of promotion and tenure. And there is peer pressure: more than a few climate contrarians have suffered public opprobrium and diminished career prospects for publicizing data that doesn't support the "broken climate" meme.

Carl Wunsch, a prominent oceanographer from MIT who has long urged scientists to be realistic in their portrayal of the science,[12] has written about the pressures on climate scientists to produce splashy results:

> The central problem of climate science is to ask what you do and say when your data are, by almost any standard, inadequate? If I spend three years analyzing my data, and the only defensible inference is that "the data are inadequate to answer the question," how do you publish? How do you get your grant renewed? A common answer is to distort the calculation of the uncertainty, or ignore it all together, and proclaim an exciting story that the *New York Times* will pick up.
>
> A lot of this is somewhat like what goes on in the medical business: Small, poorly controlled studies are used to proclaim the efficacy of some new drug or treatment. How many such stories have been withdrawn years later when enough adequate data became available?[13]

Scientists *not* involved with climate research are also to be faulted. While they're in a unique position to evaluate climate science's claims, they're prone to a phenomenon I call "climate simple." The phrase "blood simple," first used by Dashiell Hammett in his 1929 novel *Red Harvest*, describes the deranged mindset of people after a prolonged immersion in violent situations; "climate simple" is an analogous ailment, in which otherwise rigorous and analytical scientists abandon their critical faculties when discussing climate and energy issues. For example, the diagnosis was climate simple when one of my senior scientific colleagues asked me to stop "the distraction" of pointing out inconvenient sections of an IPCC report. This was an eyes-shut-fingers-in-the-ears position I've never heard in any other scientific discussion.

What causes climate simple? Perhaps it is a lack of knowledge of the subject, or fear of speaking out, particularly against scientific peers. Or perhaps it is simple conviction born more of faith in the proclaimed consensus than of the evidence presented.

Leo Tolstoy's 1894 philosophical treatise *The Kingdom of God Is Within You* contains the following thought:

> The most difficult subjects can be explained to the most slow-witted man if he has not formed any idea of them already; but the simplest thing cannot be made clear to the most intelligent man if he is firmly persuaded that he knows already, without a shadow of doubt, what is laid before him.[14]

Whatever its cause, climate simple is a problem. Major changes in society are being advocated and trillions will be spent, all based on the findings of climate science. That science should be open to intense scrutiny and questioning, and scientists should approach it with their usual critical objectivity. And they shouldn't have to be afraid when they do.

ACTIVISTS AND NONGOVERNMENTAL ORGANIZATIONS (NGOS)

My inbox fills with fundraising appeals from such organizations as 350.org, the Union of Concerned Scientists, and the Natural Resources Defense Council. If you believe there is a "climate emergency," have built an organization on that premise, and rely upon your donors' continuing commitment to the cause, projecting urgency is crucial. Hence statements like "The climate crisis is immense—we must be daring and courageous in response" (from the 350.org website[15]) or "Climate change is one of the most devastating problems that humanity has ever faced—and the clock is running out" (from the UCS website[16]). It's hardly in your best interest to tell your donors that the climate shows no sign of being broken or that projections of future disasters rely on models of dubious validity.

The media tend to accord NGOs an authoritative stance. But these are also interest groups, with their own climate and energy agendas. And they are powerful political actors, who mobilize supporters, raise money, run campaigns, and wield political power. For many, the "climate crisis" is their entire *raison d'etre*. They also have to worry about being outflanked by more militant groups.

I have no problem with activism, and the efforts of NGOs have made the world better in countless ways. But distorting science to further a cause is inexcusable, particularly with the complicity of those scientists who serve on their advisory boards.

THE PUBLIC

Fear of extreme weather events is understandable, and concerns about changes in climate are as old as humanity. Short-term weather events (storms, floods, droughts) have stressed and challenged societies, while changes extending over decades induced mass migrations or even destroyed entire civilizations. For example, repeated crop failures devastated communities in the southwest United States during the twenty-five-year-long Great Drought about 750 years ago.[17]

The notion that our behavior might be causing such calamities is also as old as humanity—as is the hope that we might avoid the worst of climate disasters by changing our behavior. Leviticus 26:3–4 promises regular rain (very important in the Middle East) and its ensuing benefits in return for doing the right thing:

> If ye walk in my statutes, and keep my commandments, and do them; Then I will give you rain in due season, and the land shall yield her increase, and the trees of the field shall yield their fruit.

We like to think public attitudes toward climate today are more discerning, but they still mostly involve unquestioning acceptance of wisdom handed down from on high. As around the world, most citizens in America are not scientists, and the educational system does not deliver much in the way of scientific literacy to the wider public. Most people do not have the ability to examine the science themselves, and they have neither time nor the inclination to do so. Many increasingly get their information from social media, where it is far too easy to promote misinformation or disinformation. And in my experience, people tend to believe—and trust—their chosen media in areas outside their expertise.

Michael Crichton, the bestselling author of *The Andromeda Strain* and *Jurassic Park*, lived near Caltech and was a prominent member of Pasadena's extended intellectual community until his death in 2008. Crichton, who was a physician before he became a writer, was an outspoken advocate for scientific integrity, and he looked askance at the public presentation of climate science (his 2004 novel *State of Fear* deals with that subject). Crichton's conversations with Caltech professor Murray Gell-Mann (the Nobel prize–winning physicist who was one of the first researchers to hypothesize quarks) led him to describe the "Gell-Mann Amnesia" effect:

> You open the newspaper to an article on some subject you know well. In Murray's case, physics. In mine, show business . . .
>
> In any case, you read with exasperation or amusement the multiple errors in a story, and then turn the page to national or international affairs, and read as if the rest of the newspaper was somehow more accurate about Palestine than the baloney you just read. You turn the page, and forget what you know.[18]

It certainly doesn't help that, at this point, even attempting a discussion of The Science is to enter a political minefield. When I tell people some of the things the assessment reports really say about climate, many immediately ask whether I was a Trump supporter. My reply is that I was not, and that, as a scientist, I have always supported truth.

As a scientist, I'm disappointed that so many individuals and organizations in the scientific community are demonstrably misrepresenting the science in an effort to persuade rather than inform. But you also should be concerned as a citizen. In a democracy, voters will ultimately decide how society responds to a changing climate. Major decisions made without full knowledge of what the science says (and doesn't say) or, even worse, on the basis of misinformation, are much less likely to lead to positive outcomes. COVID-19 offered a sobering illustration of this, and it's as true for climate and energy as it is for pandemics.

FIXING THE BROKEN SCIENCE

In early 2017, it had been three years since the American Physical Society workshop that opened my eyes to problems with "The Science." I had been tracking the misrepresentation of climate science by the media and politicians ever since, and I was freshly irked by the misleading presentation of hurricane data in the 2014 National Climate Assessment that I had stumbled upon six months earlier, as I described in Chapter 6. I was increasingly convinced that The Science needed a Red Team exercise, a concept I'd already been refining for a few years.

In such an exercise, a group of scientists (the "Red Team") would be charged with rigorously questioning one of the assessment reports, trying to identify and evaluate its weak spots. In essence, a qualified adversarial group would be asked "What's wrong with this argument?" And, of course, the "Blue Team" (presumably the report's authors) would have the opportunity to rebut the Red Team's findings. Red Team exercises are commonly used to inform high-consequence decisions such as testing national intelligence findings or validating complex engineering projects like aircraft or spacecraft; they're also common in cybersecurity. Red Teams catch errors or gaps, identify blind spots, and often help to avoid catastrophic failures. In essence, they're an important part of a prudent, belt-and-suspenders approach to decision-making. (Note that the use of "Red" and "Blue" is traditional in the military, where these exercises originated; it has nothing to do with US politics.)

A Red Team review of a climate assessment report could bolster confidence in the assessment, as well as demonstrate the robustness (or lack thereof) of its conclusions. It would both underscore the reliability of the science that stands up to its scrutiny and highlight for non-experts uncertainties or "inconvenient" points that had been obscured or downplayed. In short, it would improve and bolster The Science with science.

Of course, both the UN's IPCC and the US government claim that their respective assessment reports are authoritative because they're already subject to rigorous peer review before publication. So why call for yet another level of review? The most direct answer is that—as the previous chapters of this book have highlighted—these reports have some egregious failures. And an important reason for those failures is the way the reports are reviewed. Let me explain.

Science is a body of knowledge that grows by testing, one step building on the next. If each step is solid, researchers can get to some amazing places pretty quickly, like rapid vaccine development or modern information technology. To know that a researcher has produced a sound new piece of knowledge, other researchers scrutinize, and often challenge, results from experiments or observations, or formulate new models and theories. *Have the measurements been done properly? Were there adequate controls on the experiments? Are the results consistent with prior understanding? What are the reasons for an unexpected result?* Satisfactory answers to questions like those are the hurdle for accepting new results into the ever-growing body of scientific knowledge.

The peer review of scientific journals is one mechanism for scrutinizing and challenging new research results. In that process, individual independent experts analyze and criticize a draft paper describing the results; the authors' responses to those criticisms are adjudicated by a third-party referee, who will then recommend publication (or not) to the journal's editor or suggest how the paper should be revised.

I have participated in many such peer reviews over the course of my forty-five-year scientific career—sometimes as an author, sometimes as a reviewer, sometimes as a referee, and a few times as an editor. I can tell you from those experiences that peer review can improve a paper's presentation and will usually catch major errors, but it is far from perfect, and in no sense guarantees the correctness of what gets published.[1] Independent reproduction or refutation of the results by other researchers conducting their own studies is a much stronger guarantee of correctness. That eventually happens for all important findings.

But an assessment report is not a research article—in fact, it's a very different sort of document with a very different purpose. Journal papers are focused presentations written by experts for experts. In contrast, assessment authors must judge the validity and importance of many diverse research papers, and then synthesize them into a set of high-level statements meant to inform non-experts. So an assessment report's "story" really matters, as does the language used to tell it—especially for something as important as climate.

The processes for drafting and reviewing the climate science assessment reports do not promote objectivity. Government officials from scientific and environmental agencies (who might themselves have a point of view) nominate or choose the authors, who are not subject to conflict of interest constraints. That is, an author might work for a fossil fuel company or for an NGO promoting "climate action." This increases the chances of persuasion being favored over information.

A large group of volunteer expert reviewers (including, for the National Climate Assessment, a group convened by the National Academies) reviews the draft. But unlike the peer review of research papers, disagreements among reviewers and lead authors are not resolved by an independent referee; the lead author can choose to reject a criticism simply by saying "We disagree." The final versions of assessments are then subject to government approval (through an interagency process for the US government and often-contentious meetings of experts and politicians for the IPCC). And—a very key point—the IPCC's "Summaries for Policymakers" are heavily influenced, if not written, by governments that have interests in promoting particular policies. In short, there are many opportunities to corrupt the objectivity of the process and product.

I presented the Red Team idea in early February 2017 at the Fourth Santa Fe Conference on Global and Regional Climate Change, a forum that traditionally embraced a diversity of viewpoints. At the end of my pitch, I asked the few hundred people in the room for a show of hands and was surprised at the favorable reaction—most of the audience of experts thought that such an exercise could be useful, if executed well. Perhaps those researchers "in the trenches" were more uncomfortable than I'd realized with the way their science was being portrayed to non-experts. In any event, their endorsement emboldened me to take the idea to a wider audience.

The inaugural March for Science was to take place on April 22 (Earth Day), 2017, with rallies and marches in six hundred cities around the world. Since one of the march's goals was to call for evidence-based policy in the public's best interest, I thought it would be a good moment to make an important point about climate science and how it's communicated to non-experts. The moment seemed especially opportune since a major US government assessment (the first part of NCA2018, the Climate Science Special Report or CSSR) was scheduled to be released in the fall.

Two days before the March for Science, the *Wall Street Journal* published an opinion piece in which I advocated for a Red Team review of climate science assessments.[2] I used the misleadingly alarming description of sea level rise to illustrate the need for such a review, and outlined how it could be carried out.

My opinion piece drew almost 750 online comments from readers, the great majority of them supportive. Some in the Trump administration also took notice, and given the administration's reluctance to publicly accept even the basics of climate understanding, their interest in a climate science Red Team engendered some very strong objections to the proposal. Most prominent were pieces published in late July 2017 by John Holdren (the Obama administration's science adviser who had been the sponsor of the CSSR),[3] and one published the following week by Eric Davidson (president of the American Geophysical Union) and Marcia McNutt (president of the National Academy of Sciences).[4] Their essential point was that a Red Team exercise was superfluous since climate research, and the assessment reports, were already peer reviewed. As Davidson and McNutt put it,

> . . . if the idea is to have the red team poke holes in the mainstream scientific community's (the blue team) consensus on climate change, it discounts that such challenges have already been applied thousands of times while that consensus was gradually developed.

Holdren's language was more pointed:

> Some proponents may believe, naively, that such a rag-tag process could unearth flaws in mainstream climate science that the rigorous, decades-long scrutiny of the global climate-science community, through multiple layers of formal and informal expert peer review, has somehow missed.

It's telling that neither article addressed the NCA2014 misrepresentation of sea level data that I had highlighted nor explained how it had survived the "decades-long scrutiny" of "multiple layers of formal and informal expert peer review." That's especially disappointing since we scientists are trained to focus on specifics. Instead the opinion pieces offered only vague and anodyne assurances of the rigor with which the reports are written and reviewed. Of course, as I've already noted, while the *research* contained in them might indeed be subject to the type of peer review the public expects of scientific findings, the reports' summaries and conclusions are not, and the sea level example was only one of many report errors and misrepresentations, some of which I've described in this book's earlier chapters.

As administration interest in a Red Team review continued through mid-2019, there were further objections from non-scientist politicians who have been misled into believing that the science is settled. On March 7, 2019, Senator Schumer (together with Senators Carper, Reed, Van Hollen, Whitehouse, Markey, Schatz, Smith, Blumenthal, Shaheen, Booker, Stabenow, Klobuchar, Hassan, Merkley, and Feinstein) introduced Senate bill S.729

> ... to prohibit the use of funds to Federal agencies to establish a panel, task force, advisory committee, or other effort to challenge the scientific consensus on climate change, and for other purposes.[5]

Although the bill never went anywhere, and it certainly wasn't the first time Congress has attempted to stop an administration from doing something, I confess to being shocked—an "effort to challenge the scientific consensus" could easily include many climate science research studies, and enshrining a certain scientific viewpoint as an inviolable consensus is hardly the role of government (at least in a democracy). I can't imagine Congress trying to do such a thing for any other important field of research, say COVID-19 therapies. And as a student of history, I found the bill uncomfortably reminiscent of a 1546 decree by the Council of Trent that attempted to suppress challenges to Church doctrine.[6] Some human behaviors are timeless, even as we suppose modern times to be so enlightened.

I still believe that a Red Team review is an important tool that should be called upon to fix the assessment report process. But the way those assessments are communicated through the media also needs to be improved. In February 2018, after the National Academies established a Climate Communications Initiative,[7] I organized thirty-four other Academies members to author and send a letter urging that the initiative remain true to the Academies' stated stance of informing rather than persuading and of avoiding advocacy. We also urged that the Academies adopt a set of principles that would help ensure their communications on climate were transparent, complete, and unbiased.

The Academies' presidents, all of whom I know personally, responded in a polite email to me on February 21, 2018:

> We thank you and the other signatories to this letter for taking the time to provide this thoughtful input. We agree that the new Climate Communications Initiative must avoid advocacy, adopt guidelines for CCI material, and establish a review mechanism. We will share your letter with the Advisory Committee members, so that it may inform their deliberations, and let them know that they may reach out to you or other signatories for clarification or further input as they undertake their task.

I never heard from the Advisory Committee. But perhaps the public and decision makers would find better use for such principles, which can help them to approach climate "news" more critically. You might think it's almost impossible for a non-expert to know what (or who) to believe when it comes to climate science. But even if you're not willing or able to invest too much time in ferreting out the facts, it's still pretty easy to watch for a few red flags that should trigger skepticism. Here's what to look for:

Anyone referring to a scientist with the pejoratives "denier" or "alarmist" is engaging in politics or propaganda. And using the term "climate change" without distinguishing between natural and human causes signals a (perhaps deliberate) sloppiness in thinking. Many an article that purports to be about how humans have broken the climate (or how we must reduce our emissions to "fix" it) is nevertheless filled with examples of climate trends that are not attributable to (or fixable by) humans.

Any appeal to the alleged "97 percent consensus" among scientists is another red flag. The study that produced that number has been convincingly debunked.[8] And in any event, nobody has ever specified exactly what those 97 percent of scientists are supposed to be agreed upon. That the climate is changing? Sure, count me in! That humans are influencing the climate? Absolutely, I'm there! That we're already seeing disastrous weather impacts and face an even more catastrophic future? Not at all obvious (for reasons I hope you understand, having read this far).

Confusing weather and climate is another danger sign. One year's bad weather does not make for a changing climate; climate is determined over decades. And a headline might say "Most active storm season in thirty years!" . . . but if it happened before, when human influences were much weaker, natural variability must play a major role.

Omitting numbers is also a red flag. Hearing that "sea level is rising" sounds alarming, but much less so when you're told it's been going up at less than 30 cm (one foot) per century for the past 150 years. When numbers are included, omission of uncertainty estimates is another thing to watch out for in non-expert discussions of climate science, as has been recognized by at least one prominent journalist.[9]

Yet another common tactic is quoting alarming quantities without context. A headline that reads "Oceans are warming at the same rate as if five Hiroshima bombs were dropped in every second" does indeed sound scary, particularly as it invokes nuclear weapons.[10] But if you read further into that article, you learn that ocean temperatures are rising at only 0.04°C (0.07°F) per decade. And a basic science refresher would tell you that the earth absorbs sunlight (and radiates an equal amount of heat energy) equivalent to two thousand Hiroshima bombs each second. It's pretty easy to scare people in the service of persuasion if you don't give any numerical context.

Non-expert discussions of climate science also often confuse the climate that has been (observations) with the climate that could be (model projections under various scenarios). Chapters 3 and 4 have shown just how uncertain climate projections are, so watch out for the story that got onto the front page because it predicts a coming apocalypse "based on models." And wordings in press coverage such as "might be," "could possibly be," "as much as," and "cannot be ruled out" are more signals of our ignorance than they are prophesies of doom. At the very least, worst and best cases should be presented, though as I've discussed, readers should be particularly wary of worst-case scenarios described as "business as usual."

Anyone can (and should) read coverage of climate science with these red flags in mind. Checking on the consistency (or inconsistency) of coverage among various mass media outlets can help put a climate news story in context. Broadcast media is poorly suited for this, as their stories are brief and tuned to sound bites. (Beware particularly of weather presenters who have morphed into "climate and weather presenters"—reporting on thirty-year changes does not exactly qualify as "breaking news.") Print and online news media is better (with the exception of the headlines).

If you have time, checking up on what you've read in the media by looking at the primary research cited is a good next step. Summaries of original research papers are available from the journals in which they appear, and for a particularly important piece of research, the paper itself is sometimes available online without charge. There are also a few blogs that seriously and consistently cover recent climate science. On the consensus side, Real Climate (realclimate.org) is worth looking at, while Judith Curry's site Climate Etc. (judithcurry.com) hosts serious discussions from a non-consensus point of view.

But there's nothing like going directly to the data—which is, in the end, the arbiter of all science. Climate data is readily available online from the US government, for example, from the EPA (at www.epa.gov/climate-indicators) and NOAA (at www.noaa.gov/climate). So if you read a story about sea level rise, hurricanes, or average temperatures and want to dig a bit deeper, doing so requires only an internet connection and a sense of what questions might offer insight (hopefully something any reader of this book has by now).

UNSETTLED'S RECEPTION

When *Unsettled* was published in April of 2021, I was pleased that there were more than a few favorable reviews. Among those were Mark Mills[11] and Holman Jenkins[12] in the *Wall Street Journal* and Mark Thiessen[13] in the *Washington Post*. But the critical reviews have been disappointing, to say the least. Not so much because of the deficiencies they point out, which I could (and did) address. Rather, it was the quality of the discussion that I found most disheartening.

I was very careful in choosing *Unsettled*'s words and in trying to give an unbiased representation of the topics I covered. Yet I found critical reviewers putting words in my mouth, ignoring the caveats and nuance in my writing, or giving non-sequitur responses to the points I'd made. Incredibly, some prominent reviewers criticized not what I'd written but rather what a favorable reviewer said I'd written. The clichéd author's complaint that "they hadn't read the book" rang true (in fact, one prominent reviewer admitted as much to me). The critical reviews, alas, created markers that others could invoke to discredit *Unsettled*.

I won't catalogue the coverage the book received. But to illustrate the problems with the public scientific conversation around climate, I offer a few words on three of the more substantial critical reviews.

In May 2021, *Scientific American* published a highly critical review of *Unsettled* by Gary Yohe.[14] Yohe, a professor of economics and environmental studies at Wesleyan University, has been involved in preparing and reviewing numerous assessment reports for the IPCC and US government.

Yohe's main point is that I "falsely suggest that we don't understand the risks well enough to take action," accusing me of "making distracting, irrelevant, misguided, misleading and unqualified statements about supposed uncertainties." However, one of *Unsettled*'s goals was to accurately portray scientific uncertainties that are downplayed or ignored; how one weighs those uncertainties is quite a different, nonscientific matter. As I write in the introduction to Part II:

> Any discussion of how the world *should* respond to a changing climate is best informed by scientific certainties and uncertainties. But it's ultimately a discussion of values—one that weighs development, environment, and intergenerational and geographical equities in light of imperfect projections of future climates.

The specifics of Yohe's criticisms don't hold up to scrutiny. For example, he takes me to task for a statement about US temperatures on page 1:

> [H]eat waves in the US are now *no more common* than they were in 1900, and that the warmest temperatures in the US have not risen in the past fifty years.

Yohe writes on heat waves: "This is a questionable statement depending on the definition of 'heat wave,' and so it is really uninformative." And about the warmest US temperatures, he says, "According to what measure? Highest annual global averages? Absolutely not." But here are the figures from the CSSR report that support both parts of my statement:

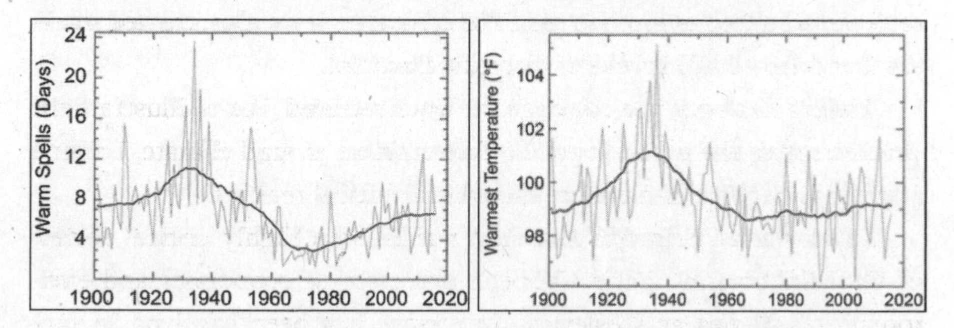

FIGURE 11.1 **Panels from CSSR Figure 6.4 (left) and Figure 6.3 (right) showing the frequency of US heat waves and warmest temperatures for the contiguous US.**[15] More recent data from the EPA shown in Figure 5.7 confirm the statement about heat waves.[16]

Yohe's criticism is particularly disappointing since he was a National Academies' reviewer of the CSSR report, which also cites a precise definition of heat waves.

Climate Feedback is a self-appointed fact-checking organization aligned with Facebook. Their 4,500 word critique wasn't based on *Unsettled* itself, but rather upon Mark Mills's favorable review of the book that had been published a week earlier.[17]

Facebook's fact-checkers included professors from Stanford, UCLA, MIT, and Penn State. Ironically, *Unsettled* favorably cites some of their work. But the oddest thing is that the "fact-check" doesn't contradict or challenge anything I wrote in *Unsettled*; rather, it provides contexts for Mills's statements. And their contexts, much too lengthy for Mills's nine-hundred-word review, are in all cases very nearly the same as what I'd written in *Unsettled*. Here's an example:

Two of three fact-checkers acknowledge that "tornado frequency and severity are also not trending up; nor are the number and severity of droughts" and go on to explain why that might not be true in the future, as I do in Chapter 6. But curiously, the other fact-checker disapproves of the statement because "it sets up a strawman" to disparage climate predictions. Highlighting that there are no climatologically significant trends in tornadoes, as Figure 6.8 shows is also true for many other severe weather phenomena, is a tonic to the widespread perception that humans have "broken the climate." Should we limit scientists to discussing only those few severe weather phenomena that *do* show a deleterious trend?

I posted a lengthy rebuttal to the Climate Feedback critique on my Medium page and published an abbreviated version as a *Wall Street Journal* Op-Ed.[18]

One month after *Unsettled*'s publication, *Scientific American* followed Yohe's review with a critique by twelve coauthors.[19] The great bulk of the one thousand words from those dozen distinguished scientists was devoted to ad hominem attacks. For example, I was called "a crank who's only taken seriously by far-right disinformation peddlers hungry for anything they can use to score political points." Considering that Vaclav Smil, Bill Gates's (and my) favorite energy author said I had "the credentials, expertise, and experience to ask the right questions and to give realistic answers," their name-calling is as demonstrably false as the rest of their critique.

The three specific criticisms they offered of scientific points I am alleged to have made were based, as was the Facebook "fact-check," upon another favorable review of the book,[20] not upon what I'd actually written. For example, they said I had portrayed sea level rise as steady over time, when the entirety of Chapter 8 is devoted to variations over the past century. *Scientific American* refused to publish a detailed rebuttal; I authored a brief Op-Ed in the *WSJ* and posted the rebuttal on my Medium page.[21]

My first public debate happened to be with one of the authors, Andrew Dessler from Texas A&M University. When I surprised him toward the end of that debate by asking whether he really believed I was a "crank," Dessler incredibly denied that he had any role in the article, saying he didn't know anything about it.

It has amazed me how much opprobrium I've attracted simply by making plain to nonexperts what was in the UN assessments, the peer-reviewed literature, and the official data. I'd like to think the response to this updated edition will be different, but I suspect otherwise.

PART II

THE RESPONSE

Scientists aren't fortune-tellers. There's no crystal ball that tells us how to (or even whether we'll need to) keep our planet safe from any given natural or human-caused climate problem that might arise. What we do have is our data, as imperfect as it is, and our ability to apply critical thinking and problem-solving skills to use that data to identify, or even anticipate, problems and come up with solutions.

Lots of people have lots of different ideas about what those solutions might be. You've probably heard at least a few. At one extreme, we could undertake a "moon shot" to completely eliminate human greenhouse gas emissions within the next several decades, as is advocated by many governments, the UN, and virtually every NGO. At the other extreme, we could carry on with business as usual, taking the position that the climate is pretty insensitive to human influences and that we'll be able to adapt to any changes that do occur.

There are many things we *could* do to reduce human influences on the climate (though they wouldn't necessarily stop the climate from changing). That *could* discussion is mostly about science and technology, since we'd need to know how the climate would change absent human influences and whether what we did would make a significant beneficial difference.

The *could* question is very different from the question of "What *should* we do?" Any discussion of how the world *should* respond to a changing climate is best informed by scientific certainties and uncertainties. But it's ultimately a discussion of values—one that weighs development, environment, and intergenerational and geographical equities in light of imperfect projections of future climates. And the *could* and *should* questions are different still from asking "What *will* we do?" Answering that involves assessing the realities of politics, economics, and technology development. Indeed, the simple truth is that there are many things the world *could* do and perhaps even *should* do—such as eliminating poverty—but which it *will* not do for various reasons. Importantly, making a judgment about *will* is not at all the same as stating an opinion about *should*.

I was asking myself those *could*, *should*, and *will* questions in 2004 when I left my job as professor and provost at Caltech to become BP's chief scientist and begin a serious study of energy technologies "beyond petroleum." I soon understood that I didn't have anything enlightening to say about the *should* question—my value judgments in such complex matters were not especially better than anyone else's, and I'm not a philosopher or ethicist. But within a year or so, by framing the issues clearly for non-experts and setting forth the advantages and drawbacks of various response strategies, I began contributing usefully to the *could* and *will* questions. That work came naturally to me, as it involved gathering, analyzing, and presenting data with scientific dispassion and was little different from what I had been doing in government advising.

I have been working on the *could* question for the past fifteen years now, and laying out the big picture in innumerable public presentations. The various assessment reports from the UN's IPCC urge that the world should (actually, *must*) reduce greenhouse gas emissions promptly to prevent the worst impacts of human-caused climate change. And those same reports urge that "mitigation" of emissions (largely of energy-related carbon dioxide) should be accomplished by transitioning to "low-carbon" energy sources and agricultural practices and by using less energy and food (conservation). The general goal has become to get to "net zero carbon" by midcentury. While in principle there are no absolute barriers to such reductions, multiple scientific, technical, economic, and social factors combine to make it highly implausible that the world *will* make them. Fortunately, not only is it far from certain that a climate disaster is pending (as we've seen in the first section of this book), but we have other strategies for responding to changes in climate, in particular adaptation and geoengineering.

So here's my high-level view of the context for society's response:

- Keeping human influences on the climate below levels deemed prudent by the UN and many governments would require that global carbon dioxide emissions, which have been rising for decades, vanish sometime in the latter half of this century.
- Emissions reductions would have to take place in the face of strongly growing energy demand driven by demographics and

development, the dominance of fossil fuels, and the current drawbacks of low-emissions technologies.

- These barriers, combined with the uncertainty and vague nature of future climate impacts, mean that the most likely societal response will be to adapt to a changing climate, and that adaptation will very likely be effective.

Let me take you through the data and analyses that support my view.

THE CHIMERA OF CARBON-FREE

In October 2004, I was sitting in a large conference hall in Kyoto, Japan, as a participant in the first Science and Technology in Society Forum. Mr. Koji Omi, a former Japanese minister of science and technology, had convened this global gathering of scientists, technologists, business executives, policymakers, and media to discuss the roles that science and technology might play in addressing global problems; the changing climate was foremost among them. The STS forum has since become an annual event, which I've attended in most recent years; I'm also a director of its American Associates. Meetings like these are an excellent way to get a sense of what's happening around the world at the intersection of science, technology, and policy.

My jet-lagged thoughts drifted away from the plenary presentations to the problem of how to talk to non-experts about reducing carbon dioxide's influence on the climate. Virtually all of the policy discussions focused on mitigating *emissions*, but it's the concentration of the gas in the atmosphere that influences the climate. There seemed to be a widespread lack of understanding of how emissions impact that concentration. Unfortunately, as I discussed in Chapter 3, that simple bit of science greatly increases the challenge of reducing human influences.

The concentration of carbon dioxide in the atmosphere grows by roughly half of the amount emitted each year. If the current concentration is 415 ppm, emitting 37 billion tons of CO_2 (as we currently do each year) will increase that concentration by about 2 ppm. But the concentration is the result of cumulative emissions, and as we saw earlier, the CO_2 we've added to the atmosphere doesn't go away when we stop emitting it. Emissions accumulate in the atmosphere and remain there for centuries as they are slowly absorbed by plants and the oceans. Modest reductions in emissions will only delay, but not prevent, the rise in concentration. So just how plausible was it that global CO_2 emissions could be reduced enough to stabilize, never mind reduce, human influences—whatever their impact—within the coming decades? I returned to London from Kyoto determined to find out, and to present my findings simply for others.

About a year later, I understood the science and the societal challenges well enough to see that a straightforward synthesis of a handful of basic facts led directly to the conclusion that even stabilizing human influences was so difficult as to be essentially impossible. Saying that directly and publicly while I was working for BP (and later the Obama Department of Energy) would probably have gotten me fired. But I was able to reconcile my organizational responsibilities with my scientific integrity by simply organizing and presenting the data and letting others follow the arrows for themselves.

I finally published the essence of my 2005 analysis as a *New York Times* Op-Ed[1] in November 2015, about a month before the Paris meeting where the world's countries were to pledge to reduce their emissions through 2030. The main points are quite simple:

- According to the IPCC, just stabilizing human influences on the climate would require global annual per capita emissions of CO_2 to fall to less than one ton by 2075, a level comparable to today's emissions from such countries as Haiti, Yemen, and Malawi. For comparison, 2015 annual per capita emissions from the United States, Europe, and China were, respectively, about 17, 7, and 6 tons.
- Energy demand increases strongly and universally with rising economic activity and quality of life; global demand is expected to grow by about 50 percent through midcentury as most of the world's people improve their lot.
- Fossil fuels supply 80 percent of the world's energy today and remain the most reliable and convenient means of meeting growing energy demand.
- The energy-supply infrastructure of electric generating plants, transmission lines, refineries, and pipelines changes slowly for unavoidable structural reasons.
- Developed countries would certainly have to reduce their emissions, but even if those were to halve, and per capita emissions of the developing world grew only to those of today's lower-emitting developed countries, annual global emissions would still increase by midcentury.
- The tension between emissions reductions and economic development is complicated by uncertainties in how the climate will change under human and natural influences and how those changes will affect natural and human systems.

All else being equal, it might be a good thing to eliminate, or even just reduce, CO_2 emissions. But all else *isn't* equal, so decisions must balance the cost and efficacy of mitigation measures against the certainties and uncertainties in climate science. And where your interests come out in that balancing act will depend in part upon what country you're in, how wealthy you are, and how much you care about (or whether you're a member of) the forty percent of humanity that doesn't have access to adequate energy.

I had been out of the government for four years in December 2015 when politicians and activists from 194 countries came together in Paris and agreed to limit human influences on the climate sufficiently to keep the global temperature from rising more than 2°C (3.6°F). A fact sheet released by the Obama White House shortly after that 21st Conference of Parties (COP21) stated:

> The Agreement sets a goal of keeping warming well below 2 degrees Celsius and for the first time agrees to pursue efforts to limit the increase in temperatures to 1.5 degrees Celsius. It also acknowledges that in order to meet that target, countries should aim to peak greenhouse gas emissions as soon as possible.[2]

Take a moment to read that again, carefully. When you do, you'll probably realize that there are a few unstated assumptions.

One is that there's an agreed-upon temperature baseline from which to measure warming, and that it's defined within 0.5°C—otherwise, we couldn't distinguish between 1.5°C and 2.0°C of warming. As you can see by looking back at Figure 1.1, if the baseline were the temperature in 1910, the world has already warmed by about 1.3°C, whereas if the baseline were the average from 1951 to 1980 (the zero line of that graph), the warming today would be 0.9°C, or 0.4°C smaller. Adopting a reference further back in time, say during the Little Ice Age around 1650, would mean even more apparent warming has already occurred (even as the world has prospered). Or perhaps we should choose a baseline from the Medieval Warm Period around 1000 AD, which would mean the warming to date would be only about 0.4°C (see Figure 1.8). In fact, the temperature from which we're supposed to judge future warming is ambiguous, with the IPCC reports usually measuring relative to the latter half of the nineteenth century and the Paris Agreement advocating a baseline of "preindustrial" values.[3] Without clarity on this point, future politicians and policymakers could declare either victory or failure, as might be convenient at the time.

Another assumption in the statement is that greenhouse gas emissions alone determine warming, and that we know how the climate will respond to them to within 25 percent (the precision required for our projections to distinguish the difference between warming of 1.5°C and 2°C). In fact, as discussed in Chapter 4, the climate's response to greenhouse gases is so uncertain that if we thought our emissions had been reduced enough to keep us "safe" at a warming of 2°C, the actual temperature might rise anywhere between 1°C and 3°C. And the newest generation of models is even more uncertain, as we've also seen in Chapter 4.

A third assumption is that warming of 1.5°C or 2.0°C will be net detrimental. In fact, many analyses suggest that a warming of less than 2°C is likely to have a small net positive economic impact, thanks to improved agricultural conditions and reduced heating costs in the temperate northern latitudes.[4] And, as we've seen in Chapter 9, warming of 2 to 5°C (3.6–9°F) is projected to have little net economic impact over time.

The Yale economist William Nordhaus first articulated the notion of a "guardrail" of about 2°C warming in the 1970s, work for which he would later be awarded the Nobel Prize in Economics.[5] But it is Hans Joachim Schellnhuber, a physicist turned climate scientist, who is known as "the father of the 2 degrees limit" for his unceasing promotion of that idea, especially across Europe.[6] Some dozen years before 1.5°C became fashionable, I had a chat with Schellnhuber during which I asked him, "Why 2°C and not 1.5 or 2.5 or whatever?" His response was something like "2°C is about right, and it's an easy number for politicians to remember." Evidently, politicians' memories have improved during the last decade.

Each of the assumptions in the Obama White House statement is dubious, if not just plain wrong, at least according to the science presented in the assessment reports and discussed in Part I of this book. But even if they're all correct, there is still the bottom-line assumption that the world *could* reduce emissions enough to keep the warming below 2°C. Unfortunately, scientific, technological, and societal realities make that the shakiest assumption of the lot.

In fact, according to the UN's IPCC, if the goal is to limit warming to 2°C, global carbon dioxide emissions must *vanish* by 2075; if the goal is a rise of no more than 1.5°C, this date becomes 2050, just thirty years from now.[7] In other words, to achieve the stated Paris goals, the world must almost completely forswear fossil fuels within the next thirty to fifty years. (Carbon capture schemes that remove carbon dioxide from the atmosphere can create negative emissions, so fossil fuel use wouldn't have to vanish entirely. I'll discuss those schemes and their feasibility in Chapter 14.)

But how realistic is it to believe that net global emissions can be eliminated within thirty to fifty years? Fossil fuels aren't burned "just because." They provide the energy that's essential to developed and developing societies. And the world will need much more energy in the coming decades, in part because of demographics. Today's global population of just under eight billion will grow to over nine billion by midcentury, with virtually all of that growth occurring outside the developed world.[8]

That last detail is important, because the economic betterment of most of humanity in the coming decades will drive energy demand even more strongly than population growth. Figure 12.1 shows the trajectories of annual per capita energy consumption for the four decades between 1980 and 2017 plotted against annual per capita GDP for some representative developed and developing countries and for the world as a whole.

People in every developing country (including China, India, Mexico, and Brazil) consume more energy as their economies grow—they build infrastructure, ramp up industrial activities, demand more food, electricity, transportation, and so on. People in developed countries show a high but slow-growing energy demand, with differences among them depending upon the nature of their economic activity, their infrastructure, and their climate (because of heating and cooling needs). Large energy-producing countries like Canada, the US, and Australia show relatively outsized energy demand. Perhaps the most salient point to be gleaned here is that

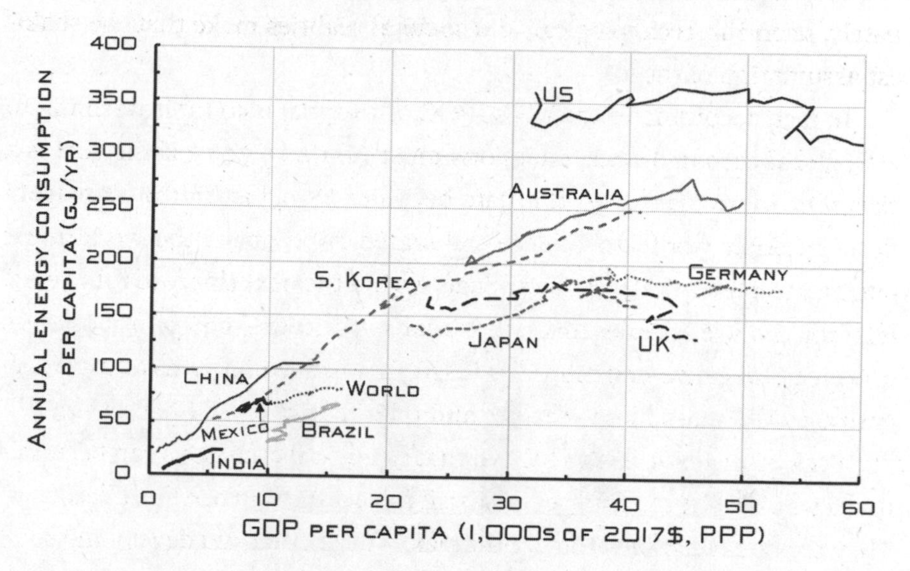

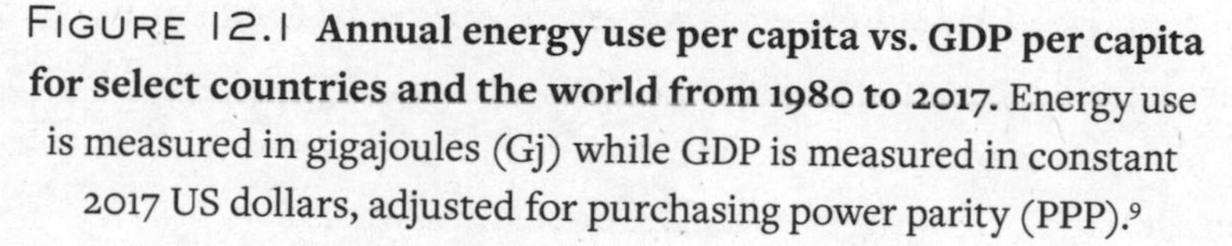

FIGURE 12.1 **Annual energy use per capita vs. GDP per capita for select countries and the world from 1980 to 2017.** Energy use is measured in gigajoules (Gj) while GDP is measured in constant 2017 US dollars, adjusted for purchasing power parity (PPP).[9]

there are only about 1.3 billion people in the developed countries of the Organization for Economic Cooperation and Development (OECD) that make up the upper portion of this chart, with some 6.5 billion people in the lower portion, increasing their energy use as they become better off.

Combined, the drivers of population and development are expected to grow energy demand by about 50 percent through 2050. Figure 12.2 shows projections by the US government's Energy Information Agency. The upper panel shows that growth will be strongest in Asia, with smaller and slower growth in the rest of the world. The lower panel shows that fossil fuels provide about 80 percent of the world's energy today (as they have for many prior decades), and that this dominance is projected to persist through midcentury under today's policies, although a strong growth in renewable sources like wind and solar will decrease the share of the world's energy provided by fossil fuels to about 70 percent.

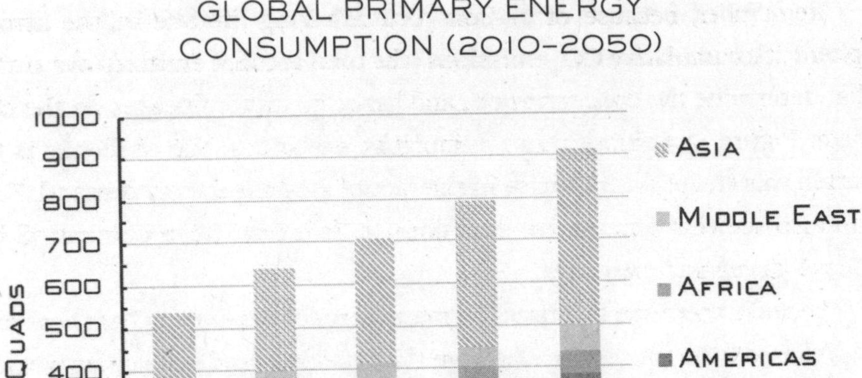

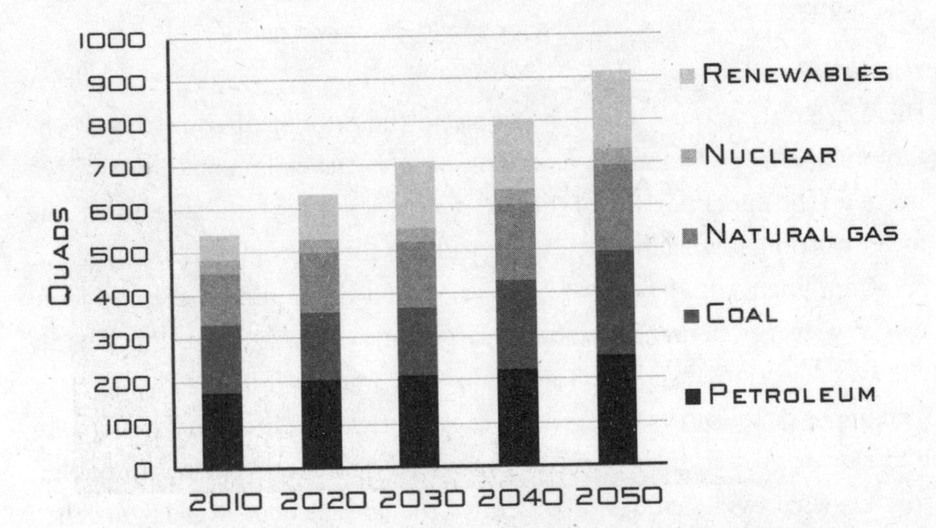

FIGURE 12.2 **Projections to 2050 of energy consumption by global region (upper panel).** Projections to 2050 of global energy sources (lower panel). Energy is measured in quads (quadrillion BTUs, or about 10^{18} joules); the US uses about 100 quads each year. Values for 2010 are historical and 2020 values are projections pre-COVID.[10]

Remember, because of carbon dioxide's long lifetime in the atmosphere, it's cumulative CO_2 emissions (the total amount emitted over time) that determine the concentration, and hence human influences on the climate. Figure 12.3 shows those cumulative emissions; the challenge is to flatten this steeply rising curve in the face of growing energy demand. You can also see that cumulative emissions to date have been dominated by today's developed countries.

Because there are five times as many people "developing" as there are "developed," the *total* emissions from the developed and developing worlds are now just about equal. But the very different growth rates of these two worlds have some sobering implications. First, over the course of this century, cumulative emissions (that is, the total amount emitted) from the developing world will be larger than those from the developed world. And

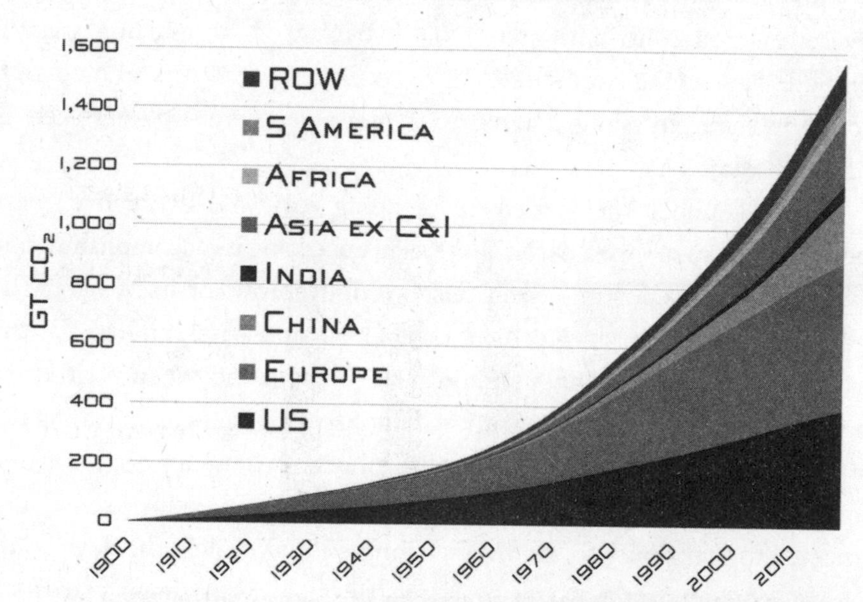

FIGURE 12.3 **Cumulative emissions of carbon dioxide by region from 1900 through 2018.** ROW is the Rest of the World beyond the regions listed. Emissions from fossil fuel use and cement production are included, but the much smaller emissions from land use changes are not.[11]

under current trends, every 10 percent reduction that the developed world makes in its emissions (a reduction it has barely managed in fifteen years) will offset less than four years of growth in the developing world. Finally, to give you a sense of the growth in emissions expected as the developing world improves itself: if nothing changed except that India's per capita emissions grew to be equal to those today of, say, Japan—one of the lowest emitting of the developed countries—global emissions would increase by *more than 25 percent*. These fundamental trends of demography and development show that eliminating global energy-related CO_2 emissions within some fifty years would require an enormous transformation of the world's energy system.

The world's governments expect to begin that transformation to a "net zero" world through the Paris Agreement. Developed countries pledged to cut their emissions by individually determined amounts (Intended Nationally Determined Contributions, or INDCs) by 2020, 2025, and finally 2030, while the developing countries pledged to make best efforts to moderate growth in their emissions and expand renewable energy sources such as wind and solar.

Progress under the agreement is being reviewed every five years (beginning in 2020), with each country self-reporting its accomplishments. The phrase "self-reporting" likely raised your eyebrows on its own, but the agreement also has no enforcement mechanism and is nonbinding—as the Trump administration demonstrated in November 2020, when it withdrew the US, and the Biden administration demonstrated again by initiating the process to rejoin as one of its first official acts in January 2021. Beyond their pledges of emissions reductions, the developed countries are also expected to contribute to the Green Climate Fund to help the developing countries moderate their own emissions through investments in carbon-lite energy projects. The fund's stated goal had been to grow to $100B per year by 2020, but governments had pledged a total of only $10.3B as of late 2019. A further $10B in pledges was secured by the end of 2020.

Although some view the Paris Agreement as a crucial step toward mobilizing national and international action to mitigate global emissions, it will hardly do anything directly to reduce human influences on the climate given the 100 percent reduction needed to stabilize the carbon dioxide concentration. The aggregate impact of the reductions pledged by all nations would reduce global emissions by less than 10 percent in 2030. These efforts are quite feeble compared to a requirement of zero emissions by 2075, let alone by 2050.

This was clear even in 2015, when the agreement was signed.[12] It is even more apparent now, as can be seen in Figure 12.4. The modest reductions

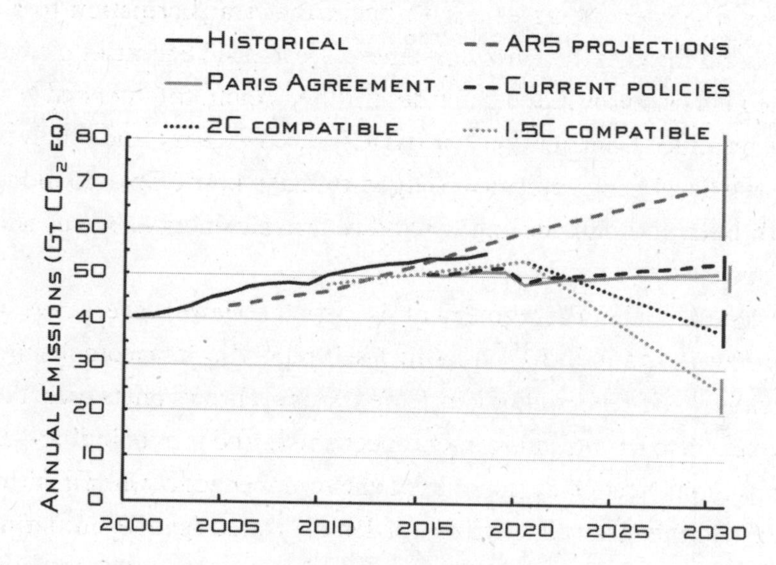

FIGURE 12.4 **Global annual greenhouse gas emissions from 2000 to 2030.** The historical record and IPCC AR5 projections from 2015 are shown, as are projections under both current policies and assuming all Paris targets and pledges are met. Future emissions paths thought to be compatible with global temperature increases of 1.5°C and 2°C are also shown. The vertical bars show the uncertainties in projections at 2030. Disruptions due to COVID-19 are evident in 2020.[13,14]

in 2030 emissions that would be achieved if all Paris pledges were met in 2030 are a long way from what would be needed to see emissions vanish.

Developments since 2015 have only strengthened the sense that the Paris Agreement is unlikely to even slow the growth of human influences on the climate, let alone reduce them. Here's the glum assessment of the UN's Emissions Gap Report of 2019, issued just before the Paris signatories convened that December at the COP25 conference in Madrid to review their progress:

> The summary findings are bleak. Countries collectively failed to stop the growth in global GHG emissions, meaning that deeper and faster cuts are now required . . . it is evident that incremental changes will not be enough and there is a need for rapid and transformational action.[15]

Figure 12.5 from the report shows the recent history of emissions from the top signatories of the Paris agreement.

Although the G20 countries (currently those with per capita GDP greater than or equal to Mexico's) are on track to meet their 2020 pledges,

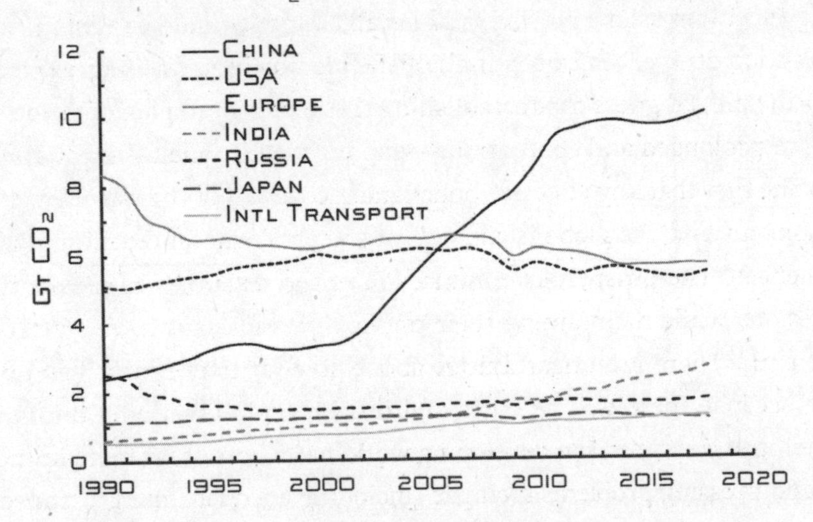

FIGURE 12.5 **Annual carbon dioxide gas emissions from selected countries and international transport, 1990–2018.**

they are projected to fall short in aggregate in 2030. In particular, as of the end of 2019 the US was on track to achieve its pledge of a 17 percent reduction in 2020 (relative to 2005), but will miss its 2030 goal without further policy changes. Japan was on track to meet its pledge of a 15 percent reduction by 2030, but it's building more coal-fired power plants following the Fukushima nuclear power disaster. The EU, which accounts for less than 10 percent of today's emissions, has pledged a 40 percent reduction by 2030 and has legislated a 100 percent net reduction (!) by 2050. In contrast, almost all high-emitting developing economies are expected to increase emissions significantly by 2030—China and India are building coal-fired power stations that will double and triple their emissions, respectively, while Russia (the world's fourth largest emitter) also proposes investments that will increase its emissions substantially.

In our globalized world, emissions restraints would have to be in place everywhere to be effective, or else carbon-intensive activities like heavy manufacturing will simply shift to regions without restraints. And we have yet to see serious international consideration of the emissions "embodied" when a low-emitting country imports goods manufactured in a high-emitting country. The EU has proposed a carbon-based border tax to level the emissions playing field for all countries and keep their industries competitive.[16] But even if all of the member countries agree to that in principle, I'd guess that establishing the details of implementation will lead to prolonged and contentious—and ultimately failed—negotiations.

The fact that any effective policy must cover all of the major emitting nations around the globe is the nub of the challenge in reducing human influences. The prosperous countries have the resources to reduce their emissions while maintaining their prosperity, and many have started on that path. There are uncertainties about how far they can or will go, in part because there will be costs and disruption in creating a minimally emitting society. Yet, the developing world has a host of far more immediate and pressing problems facing it (including adequate energy, transportation, housing, public health issues like clean water and sanitation, and education—not to mention recovery from COVID-19). The importance and urgency of these problems make emissions reductions a lower priority and, in fact, solving many of them would actually increase emissions. Unless emissions-lite technologies are developed to the point where they are essentially no more costly than emitting technologies, or efforts like the Green Climate Fund become much more substantial, it's natural to ask "Who will pay the developing world not to emit?" I have been posing that simple question to many people for more than fifteen years and have yet to hear a convincing answer.

An optimist would say that mitigation efforts must start somewhere, and that by raising awareness of the problem, reviewing country-by-country emissions every five years, and securing pledges of reductions, however insufficient, the Paris Agreement is the first step on a long and challenging journey. But considering the voluntary nature of even these modest steps and the record so far, there is ample reason to doubt that the Paris 2030 goal can be achieved. A similarly realistic view of the longer term is that the world is very unlikely to zero out its net emissions by 2075, let alone by 2050, and so society will largely respond by adapting. I'm in that latter camp, and I have more than a few informed camp mates.

THE PATH AND PRICE FOR NET ZERO—2024 UPDATE

"Net zero by 2050" has been a rallying cry of those seeking to reduce human influences on the climate. But exactly how might the world eliminate greenhouse gas emissions over the next few decades? One answer was released by the International Energy Agency (IEA) in May 2021 when it published *Net Zero by 2050: A Roadmap for the Global Energy Sector.*[17] While this is largely a *could* report (i.e., "What would it take?"), it is more commonly taken as a *should* report, justifying an immediate ban on investment in any new fossil fuel supply. It's therefore important to understand the plausibility of the report's assumptions about the evolution of the global energy system.

The fundamental assumption of the IEA's *Net Zero by 2050* report is that "the superiority of alternatives to hydrocarbons—principally wind and solar (nuclear barely gets a look in)—will cause demand for coal, oil, and natural gas to wither away."[18]

Here is a critique of some of *Roadmap*'s key assumptions for 2050 (as offered by Scott Tinker,[19] one of the world's leading energy thinkers):

- *No new oil and gas fields, and no new coal mines or mine extensions. Coal demand declines by 98 percent, oil consumption declines by 75 percent, and natural gas by 55 percent.* There are no hints of that in current trends. In fact, coal demand in Asia continues to expand significantly, with 2022 setting a record high for consumption. And gas consumption resumed its upward trend after a dip in 2021 that was due in part to the Ukraine invasion and milder temperatures in Europe.[20]
- *Global energy use declines even as population and the economy continue to grow.* As evident from Figure 12.1, this is contrary to all experience and hard to imagine even in the most aggressive assumptions about energy efficiency (refer to the rebound phenomenon discussed in Chapter 13).
- *Two-thirds of total energy supply in 2050 will come from wind, solar, bioenergy, geothermal, and hydro. Solar photovoltaic capacity increases 2,000 percent and wind increases 1,100 percent.* There are environmental, human rights, and national security issues associated with the mining, manufacturing, and waste disposal that such an expansion implies.
- *Per capita CO_2 emissions in developed economies (now about ten tons annually), and in emerging and developing economies (about four tons), decline to zero.* "In reality, per capita CO_2 emissions in emerging and developing nations, which have the largest and fastest growing populations and economies, will continue to rise long before they fall."[21]

And, incredibly, all of this would create a global system supplying about 8 percent *less* energy than today!

As for how much that would cost, it's very difficult to estimate, because it depends primarily upon the interplay of technology development, regulation, and economics. Regulatory requirements stimulate technology development, which in turn reduces the cost of meeting those requirements. Nevertheless, IEA estimates a cost of $4 trillion per year over the next thirty years globally, or about 4 percent of today's GDP.

A McKinsey report offers a comparable estimate of the spending required for a global net-zero transition:

> [E]quivalent to about 7.5 percent of GDP from 2021 to 2050. The required spending would be front-loaded, rising from about 6.8 percent of GDP today to about 9 percent of GDP between 2026 and 2030 before falling. In dollar terms, the increase in annual spending is about $3.5 trillion per year, or 60 percent, more than is being spent today, all of which would be spent in the future on low-emissions assets. This incremental spending would be worth about 2.8 percent of global GDP between 2020 and 2050. The increase is approximately equivalent, in 2020, to half of global corporate profits, one-quarter of total tax revenue, 15 percent of gross fixed capital formation, and 7 percent of household spending.[22]

Another estimate for the US only summarizes its findings in this way:

> The cost to 2050 will comfortably exceed $12 trillion for electrification projects, and $35 trillion for improving the energy efficiency of buildings. . . . The scale of this project suggests that a war footing and a command economy will be essential, as major cuts to other favoured forms of expenditure, such as health, education and defence, will be needed.[23]

A reasonable conclusion from these studies is that net zero by 2050 would cost, on average, between 3 and 7 percent of the economy, whether globally or for the US. Spending would be more in the early decades, less after 2040. But whatever the details, that's an enormous cost, even if the technologies were available. A transparent and informed discussion of the tradeoffs involved is long overdue.

COULD THE US CATCH THE CHIMERA?

I once asked a well-to-do audience in the US if they really understood what it would mean to eliminate their "carbon footprint" next year. That is, to zero out the emissions associated with their personal behaviors. Air travel, large homes (and surely second homes), and meat would all be *verboten*. There wasn't much enthusiasm for any of that, although some were interested in "meatless meat." More than a few thought that vague, unspecified "technology" and "policies" could let their children and grandchildren have a "carbon-neutral" existence without too much pain.

That kind of transformation in energy use would have to happen in every country around the globe, even as most of the world's people need *more* energy to have even the minimal quality of life we take for granted in the developed world. Countries vary greatly—in their degree of development, the nature of their economy, their endowment of energy resources, their climate, and their existing energy system. It's hard to generalize about paths to zero-emissions futures other than to say that technology, economics, policy, and behavior would all have to play a role. What might work for Switzerland will not work for Sri Lanka. But let's focus on just one country, and explore what it would take for the United States to get to zero greenhouse gas emissions.

The obvious place to start is by looking at which activities in the US are emitting greenhouse gases now. Figure 13.1 shows that transportation, electricity generation, and industrial activities account for the great majority of emissions and, importantly, neither the major sources of emissions nor their amounts have changed all that much over the past thirty years. While the top three sectors are huge challenges in their own right, the figure also shows that to "get to zero," agriculture, residential, and commercial emissions would also have to be dealt with, either by direct reductions or by offsetting activities like growing more trees that remove CO_2 directly from the atmosphere.

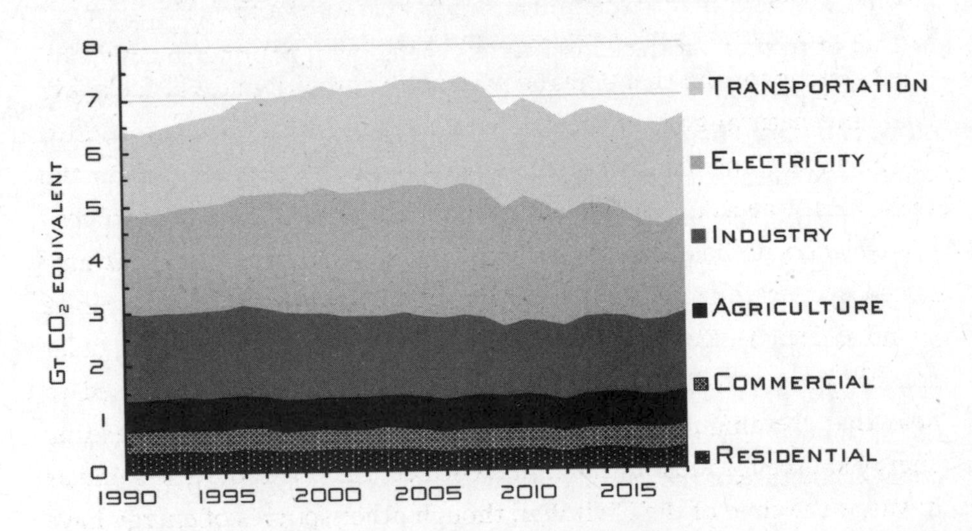

FIGURE 13.1 **US greenhouse gas emissions by economic sector, 1990–2018.**[1]

Still, the fact that the total in 2018 was about what it was in 1990 could be construed as progress toward reducing emissions, given that the country's population grew 31 percent during that time and its real GDP doubled. Total US energy-related emissions fell 16 percent from their peak in 2005, mostly because of changes in electricity generation led by natural gas replacing coal. The off-shoring of energy-intensive manufacturing also played a role, though it bears noting that this simply shifts emissions to other countries. In the end, the problem with this "progress" is that the average reduction of 1 percent a year from 2005 is pretty meaningless both because of the scientific realities of how concentration relates to emissions and because the reductions that matter are global—while US emissions declined in the period after 2005, global emissions still increased by about one-third (visible back in Figure 3.2).

The economic slowdown caused by the COVID-19 pandemic demonstrates just how challenging it will be to reduce emissions rapidly. Global CO_2 emissions during the first half of 2020 were down only 8.8 percent compared to the same period in 2019, with 40 percent of that reduction coming from surface transport and 22 percent from the electricity sector. And emissions rebounded promptly in many countries as restrictions eased.[2]

Enduring structural changes in the energy system take decades. Figure 13.2 shows that energy demand grew strongly as the US developed after 1950, and that demand was met by growth in fossil fuels (oil, coal, and natural gas). Nuclear power, introduced in the 1970s, added to the supply but didn't displace any of the other sources. But in the most recent decades, as demand growth has slowed, coal used to generate electricity has been displaced by the inexpensive natural gas produced by fracking and, to a lesser extent, by the growth of renewables (wind and solar). Even as new sources of energy have come online, older sources have not disappeared. For instance, you may be surprised to hear that the amount of energy provided by wood (by far the leading energy source for most of the nineteenth century) is today the same as it was at the time of the Civil War, though other sources of energy have grown enormously since then.

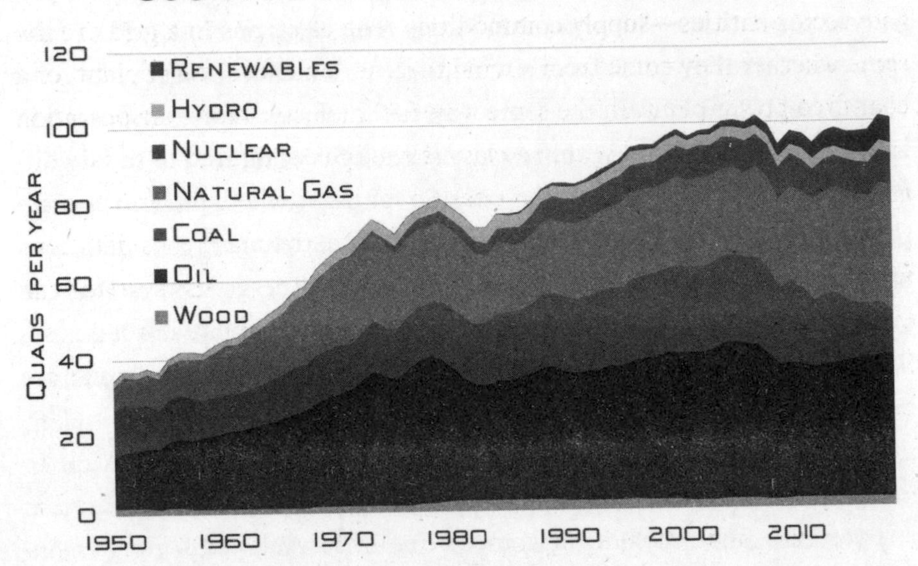

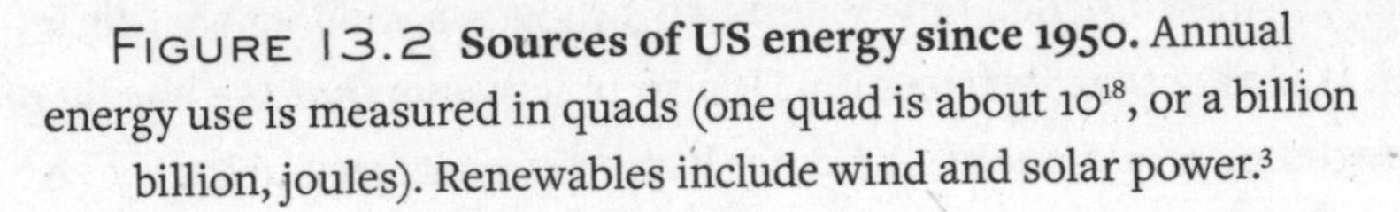
FIGURE 13.2 **Sources of US energy since 1950.** Annual energy use is measured in quads (one quad is about 10^{18}, or a billion billion, joules). Renewables include wind and solar power.[3]

Energy systems change slowly for good reasons.[4] An important one is that delivery of energy must be highly reliable—when fuel supplies are disrupted or there's an electrical blackout, society comes to a halt and chaos ensues. Indeed, the expectation in the US is that the regional electrical grids (not the more fragile local systems that distribute power to users) shouldn't be disrupted more than one day over a decade. That kind of high reliability comes only after decades of developing and optimizing hardware and operating procedures. Given that, it's natural (indeed, essential) to take a very conservative approach to making changes, including introducing new technologies.

Another factor hindering change is that energy supply facilities such as power plants or refineries require large up-front investments and last for decades (over which those investments are gradually paid off). They must also be compatible with other parts of the infrastructure—for example, fuels, fueling infrastructure, and vehicles must all work together.

Moreover, those energy facilities—which in the US are virtually all private sector entities—supply commodities. The electrons in a grid are the same whether they come from a wind turbine, a nuclear power plant, or a coal-fired power plant. In the same way, fuel molecules for transportation have to be standardized within a class (although configured to satisfy different regional air quality regulations). So for those who build and operate the energy supply system, cost and reliability are the main considerations after meeting safety and regulatory requirements. Energy-use systems can evolve somewhat more quickly than those for energy supply, although cars in the US remain on the road for as many as fifteen years and buildings are refreshed only after many decades.

Energy systems also change slowly because energy is important to everyone—and to everything we do. The energy infrastructure that enables heat, light, mobility, and much else in a developed society is evident everywhere, from power lines to filling stations to the outlet into which you plug your refrigerator. It's so ubiquitous that we hardly give it a thought as we go about our daily lives. Yet that very ubiquity not only creates the reliability requirement discussed above, it also generates direct interests from many different players: industry, consumers, governments, and NGOs. Those interests are often not aligned, which makes changes difficult to agree upon in principle, much less in the details. For example, there are often yearslong disputes over the routing of pipelines or the siting of power plants.

Thus, while it is often invoked as an example, a "Manhattan Project" isn't a very apt or useful way of thinking about energy change. The real Manhattan Project in the early 1940s produced a few specific atomic "gadgets" for a single customer (the US military); it did not aspire to transform a large system already embedded throughout society. It also didn't have to compete with an existing capability, while we already have perfectly serviceable ways of providing electricity and fuels. The Manhattan Project was carried out in secret, so public opinion and acceptance wasn't an issue (although it turned out that spies let the Soviet Union in on the secret). And finally, it had an almost unconstrained budget (it was a national wartime priority), while those aiming to produce alternatives to our current energy system have to worry not only about their costs, but also about consumer energy prices. These same categories of difference (except for secrecy) also apply to the commonly invoked "moon shot" analogy to the US space program of the 1960s.[5] So the phrase "energy revolution" is self-contradictory. Rather, bringing about energy change at scale—that is, reducing emissions enough to make a difference—would be a process of slow transition, more like orthodontia than tooth extraction.

For greenhouse gas emissions to decrease enough (and at a sufficiently rapid pace) to stabilize human influences on the climate in the foreseeable future, there would have to be dramatic changes in policies—the rules under which the energy system is created and operated. One possibility is outright regulation: *Coal-fired power plants shall cease operation within a decade* or *New gasoline-powered cars cannot be sold after 2035*. Alternatively, the government could induce lower emissions by imposing a financial penalty for every ton of greenhouse gas emitted into the atmosphere. Such a "carbon price" could be created by a tax or by a market in emissions rights, in which emitters could buy and sell government-issued permits giving them the right to emit CO_2. Whatever the mechanism, it's clear that even an accelerated energy transformation in the US would take decades.

To be effective, greenhouse gas policies (aka "climate policies") must have a number of features:

Consistency: The political timescale is a few years, the business timescale is a quarter, and the news cycle is a day—or hours. But CO_2 persists in the atmosphere for centuries, and the energy infrastructure lasts for decades; emissions policies would have to be consistent over decades as well. For someone to invest an extra billion dollars in an emissions-lite power plant lasting fifty years, they will need a reasonable expectation that those emissions reductions would still be valued decades later. (That's more than a few Congresses and presidential terms!)

The US (like almost all countries) is not well equipped to handle the long timescales inherent in reducing greenhouse gas emissions. Thus, we see politicians outbidding one another by proposing ever greater reductions thirty years in the future through programs that won't even begin until after they've left office. The haggling over emissions reductions that we've seen among EU countries and between EU industries and the governments suggests that implementing those programs won't be simple. Emissions policies would have to be sheltered from the political winds, much as the US attempts to isolate its monetary policy and interest rates by delegating decisions to the Federal Reserve.

James Madison understood the importance of predictable policies when he wrote Federalist Paper No. 62 more than two hundred years ago.[6] Madison was explaining why the US government needed a Senate that was more deliberate and stable than the House of Representatives; the following reason, among the more than a half dozen he listed, is highly relevant to the challenge of energy policy today:

> In another point of view, great injury results from an unstable government. The want of confidence in the public councils damps every useful undertaking, the success and profit of which may depend on a continuance of existing arrangements. What prudent merchant will hazard his fortunes in any new branch of commerce when he knows not but that his plans may be rendered unlawful before they can be executed? What farmer or manufacturer will lay himself out for the encouragement given to any particular cultivation or establishment, when he can have no assurance that his preparatory labors and advances will not render him a victim to an inconstant government?

Significance: Greenhouse gas emissions would have to be reduced significantly to even begin stabilizing human influences on the climate; the usual societal response of implementing a partial solution to a problem and declaring victory might buy us a delay, but not a reprieve. That means something or somebody would to have to change significantly—and that means political pushback.

As you can see from Figure 13.1, the three major sectors (electricity, transportation, and industry) all produce comparable emissions. But they'd be affected very differently by an economy-wide carbon price. For example, coal fueled about one-quarter of US electricity in 2019, and each metric ton of that coal was sold for about $39.[7] A carbon price of $40 for each ton of CO_2 emitted would effectively double that cost to power plant operators and so be a strong inducement for them to forswear coal. In contrast, that same carbon price would increase the effective price of crude oil by only about 40 percent above $60 per barrel. And if that cost were passed through to the pump, gasoline would increase by only some $0.35 per gallon. Since that's small compared to how much pump prices have varied historically, consumers wouldn't have much incentive to move away from gasoline. So reductions in emissions from power (and, as it turns out, heat) are much easier to encourage than reductions from transportation, fundamentally because oil packs a lot more energy per carbon atom than does coal.

Focus: Emissions-reduction policies would be most effective if they were focused—on reducing emissions. However, given the political need to secure broad support, it's inevitable that emissions policies get diluted by other, quite distinct issues, such as trade protectionism, energy security, or the promotion of particular technologies. For example, the US imposes tariffs[8] that significantly increase the cost of solar panels, while the EU has imposed[9] substantial import duties on energy-efficient light bulbs. And many state-level standards for electricity generation require a certain percentage to be renewable, rather than simply *low-carbon*. That disfavors nuclear power, one of the most significant emissions-lite technologies. When actions to reduce emissions are weakened by conflating them with other goals, it belies claims that we're facing an imminent existential threat.

Systems thinking: Energy is produced and delivered by systems, and so focusing efforts on changing just one piece of the puzzle might be not just ineffective, but even counterproductive. Having a high fraction of electricity generation that depends upon rapidly changing weather (wind turbines and solar panels) will threaten the grid's reliability unless there's reliable backup that can be called upon at short notice. Similarly, vehicles are useful only if there's a system to produce and deliver their fuel.

In 2011, I led the Department of Energy's Quadrennial Technology Review to develop strategies for government support of emerging clean energy technologies. In one town hall meeting, I faced advocates for four different vehicle technologies—internal combustion engines powered by biofuels, compressed natural gas, hydrogen-powered fuel cells, and battery-powered plug-ins. Each of them believed that their technology was the optimal vision for the future, and that all the government had to do was support the development of the appropriate fueling infrastructure. When I reminded them that the country could probably deploy no more than two new fueling technologies at scale, a squabble ensued. There are several reasons I believe that electricity will fuel the passenger vehicles of the future, but one of them is that the existing electrical grid is a good start on the fueling infrastructure. If a widespread transition to plug-in electric cars does come about, systems thinking will be even more important as the electrical and transportation systems would have to work together to accommodate charging millions of vehicles.

Technical practicality: Many policymakers believe that we only need to get the regulations and economic incentives right to see new technologies developed and deployed. But scientists and engineers know that there are powerful physical constraints that any technology must respect. For example, no policy can circumvent the fundamental limits on energy efficiency imposed by the Second Law of Thermodynamics—we cannot "create" energy, only convert it from one form to another, and that process of conversion will always "cost" some amount of energy itself.

To be effective, policies need to be informed by technical knowledge. It is possible to make plausible judgments about how rapidly technologies will (or will not) evolve over a decade or so. We must assess technologies against their ability to scale, their economics, and their potential to be improved. The deployment of ineffective "feel good" technologies is triply bad—aside from the obvious problem of being ineffective, it also lessens urgency by creating the *illusion* of doing something; perhaps worst of all, it diverts potential resources from more urgent needs. An apt example involves our old friend cooking oil: used vegetable oil can be processed into "carbon neutral" biodiesel. While that might solve the problem of what to do with the spent oil, we can't fry enough potatoes to make a significant difference in emissions. Processing *all* of the two hundred million metric tons of vegetable oil the world uses each year would satisfy just *one day* of global diesel demand. It's true that "there are no silver bullets," but some bullets have a bigger caliber than others.

Government has an important role to play in supporting both basic research and research into technologies that aren't yet ready for the marketplace—things like advanced solar, fission, fusion, and the next generation of biofuels. But it's also important that its policies promote the development and deployment of a technology at the right time: in other words, when the technology will be effective, not simply when the government wants it to be. There have been many examples around the world of premature deployment of technologies leading to little impact at huge expense. A good one is the "hydrogen highway" that California inaugurated in 2004. After sixteen years, just over seven thousand of California's thirty-five million vehicles run on hydrogen. The cost of hydrogen-powered cars (today more than double that of gasoline-powered cars) has hindered wider adoption, as has a lack of hydrogen filling stations—there are only forty-four in California today, mainly in cities.[10] And perhaps most importantly, hydrogen is currently produced from natural gas, releasing carbon dioxide, and will be for at least the next few decades. So much for "zero emissions."

Promotion of conservation rather than efficiency: One often hears that we can reduce emissions by using energy more efficiently. It *is* a bad thing to waste energy. But efficiency is about how well we use energy, not about how much we use—*that* is conservation. And for the purposes of reducing emissions, conservation is what matters. Economists have known for 150 years about "rebound"—that is, that efficiency improvements often lead to less conservation than might be expected. For example, you might keep your lights on more if you know they're using less electricity. Or you might tend to drive an energy-efficient car more than a gas-guzzler. In my opinion, the only sure ways to promote conservation are through regulation or price increases. Either of those is a difficult act for a government to pull off.

What would the US energy system look like under an effective greenhouse gas policy? The answer would emerge from a complex interaction among technology development, economics, policy, and behavior. There are certainly ways to change how we produce and use energy that would significantly reduce US greenhouse gas emissions.[11] In fact, the country could have any sort of energy system it wants, as long as it doesn't violate the laws of physics. But because of the central role that energy plays in society, creating an emissions-free energy system will be broadly disruptive—both economically and behaviorally. The question is whether the country will choose to invest the financial and political capital needed to bring that transformation about. Given the barriers I've discussed, and the other more tangible and immediate demands on the nation's attention and resources, I think that's unlikely to happen anytime soon.

Even if such a transformation did come to pass, it would make very little direct difference, if any, to the climate: the US accounts for only some 13 percent of global greenhouse gas. Of course, there is the argument that the rest of the world would follow our lead. But how likely are they to do so when their energy needs are so pressing and the benefits of reductions so murky?

THE GRID'S TROUBLED TRIANGLE—2024 UPDATE

The almost universally agreed-upon strategy for reducing greenhouse gas emissions is to electrify most heating and transportation (especially passenger cars and light trucks) while transitioning to a zero-emissions

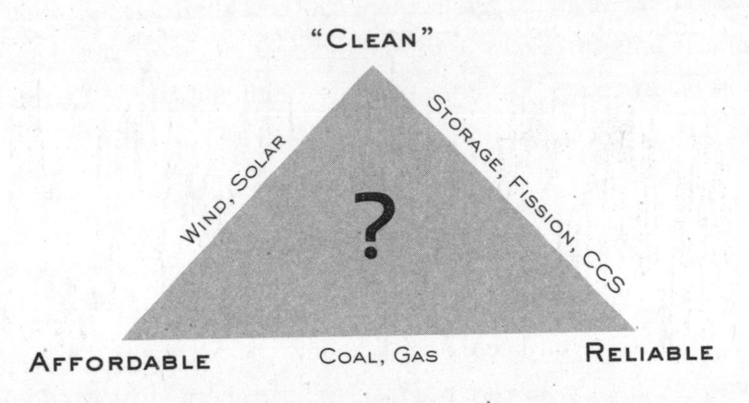

FIGURE 13.3 **The three currently incompatible desiderata of a modern electrical grid.**

electrical grid. Although building heat, industrial heat, and electrical vehicles (EVs) each pose their own challenges, a decarbonized grid is the centerpiece of the strategy.

The ideal electrical grid would be reliable, affordable, and "clean" in the sense of both local pollution and greenhouse emissions. Unfortunately, these three criteria are currently incompatible. Generation by coal and natural gas is both reliable and affordable, but not clean in the sense of greenhouse gas emissions. In contrast, the wind turbines and solar panels that receive so much attention and funding can indeed be the cheapest ways of producing electricity today. Unfortunately, they are unreliable—solar doesn't work at night or when it's cloudy, and the wind comes and goes hourly. So this grid needs a backup system for when weather-dependent renewables fail—reliable low-emitting technologies like natural gas with carbon capture, or nuclear, or some form of storage (like giant batteries or pumped hydropower). But these technologies are not affordable.

Reliable backup isn't too expensive when it has to fill in for less than a day—during a cloudy day (or at night) when solar panels don't work or

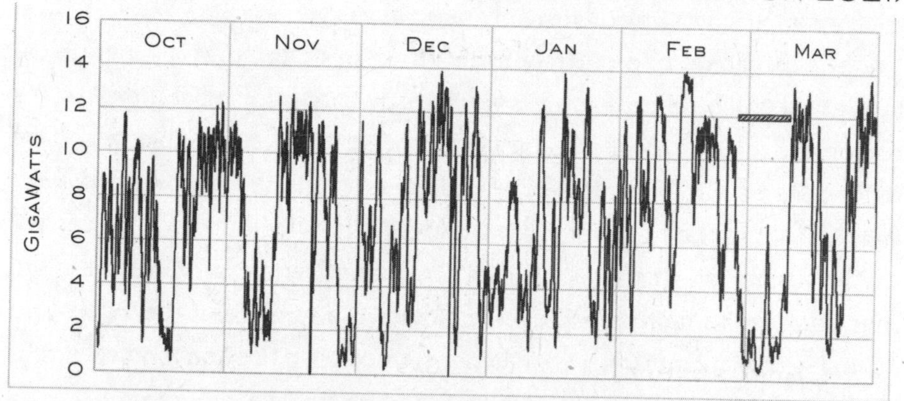

FIGURE 13.4 **Daily average generation by Britain's wind farms** (October 2020–March 2021). The 11-day *dunkelflaute* in late February/early March is highlighted.[12]

during a calm day when wind turbines don't turn. But there are occasions, up to several weeks long, when neither wind nor solar will generate much power at all. Those times are so important that the Germans coined a word for them: *Dunkelflaute*—a dark stillness. You can see a few *dunkelflauten* in Figure 13.4, which shows six months of daily average wind generation in the UK. There's an eleven-day period when wind generation averaged only 11 percent capacity—a once-in-twenty-year event. Grids in Texas, California, and elsewhere experience similar periods of still, dark conditions.

To ride through those long *dunkelflauten*, the backup system must be at least as capable as the wind and solar alone. But it will also be at least as expensive, since solar and wind are today the cheapest ways of (unreliably) generating electricity. In other words, the most expensive part of a renewables-heavy grid is reliability, and it becomes more and more expensive as you demand higher reliability. Studies that explore different mixes of renewable and backup technologies using decades of weather and demand data show that having sufficient backup to ensure high reliability (the federal standard is >99.99 percent) more than doubles the cost of electricity compared to a grid with only natural gas generation.[13]

It is incorrect, and entirely misleading, to assert that a renewables-heavy grid will be cheap—unless you're okay with poor reliability. And it's reasonable to ask: "If the backup system needs to be so capable, why have renewables at all?" Indeed, the UK grid coped with the early March *dunkelflaute* shown in the figure because fossil fuels provided up to 73 percent of the power during that time. In short, wind and solar can never be more than adjuncts to more reliable technologies. The failure of the Texas grid during winter storm Uri in 2021 demonstrated both the importance of reliability and the inability of wind turbines to provide it.

Solar and wind have other drawbacks. They need a lot more land—because sunlight and wind are much less concentrated than fossil or nuclear energy—to produce the same amount of electricity: Wind requires four times as much land as gas, seven times as much as coal, and thirty times as much as nuclear. And you need to cover that land with enormous structures. To produce the same amount of electricity, wind takes ten times as much concrete and steel as nuclear power. There's also the cost of building transmission lines to bring renewable electricity from remote areas to demand centers[14] and the sclerotic process for permitting any new infrastructure.[15]

There are several avenues by which we will resolve the grid's troubled triangle of reliable/affordable/clean. Most straightforward is to accept some increase in the cost of electricity. Less simple will be more sophisticated information technologies to allow for better grid management (for example, forecasting and coordinating when and where backup would be called upon). Longer term is research, development, and demonstrations aimed at bringing down the costs of storage and low-emission generation.

About the latter, it's constructive to see several small modular fission reactor designs moving from the development phase to demonstration. And longer term, there has been progress, and private-sector fervor, in producing practical energy from nuclear fusion. It was a major scientific achievement when the National Ignition Facility (NIF) achieved ignition in December 2022.[16] That is, it produced more fusion energy than the laser energy used to heat and compress a tiny target of hydrogen. Passing the ignition milestone was especially gratifying to me personally, as I had chaired the National Academies study in 1996 that recommended construction of the NIF while judging "the achievement of ignition appears likely but not guaranteed."[17] There are still several decades of work needed to know whether inertial (or other forms of) fusion will be practical.

CRITICAL MINERALS—2024 UPDATE

The first edition barely touched on a major challenge in the large-scale deployment of wind and solar energy. It's one that only recently burst into the public consciousness in a kind of "oops" moment as the world hurtled down the path of an all-renewables grid. Compared to conventional energy technologies, renewables use a lot more high-value materials like copper, molybdenum, and dysprosium, as shown in Figure 13.5.

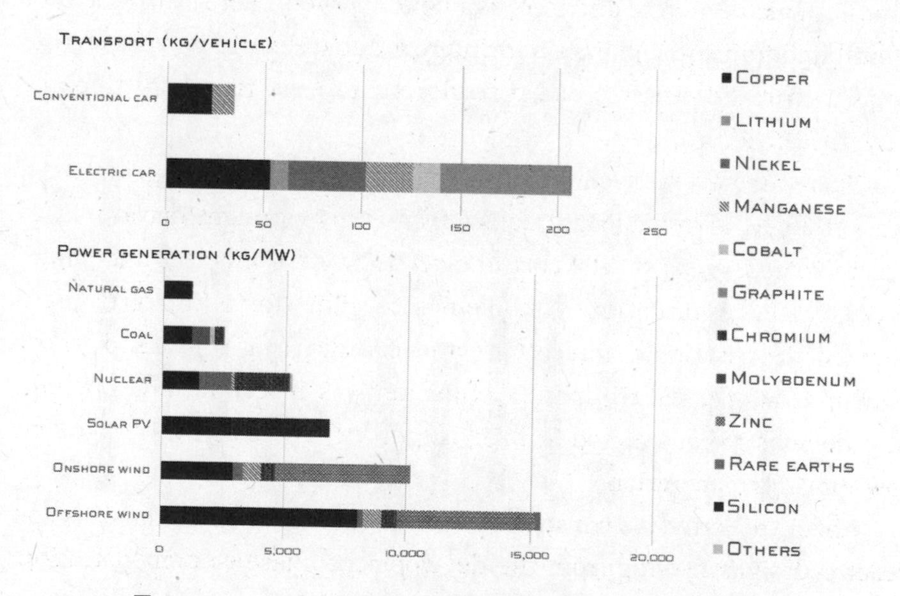

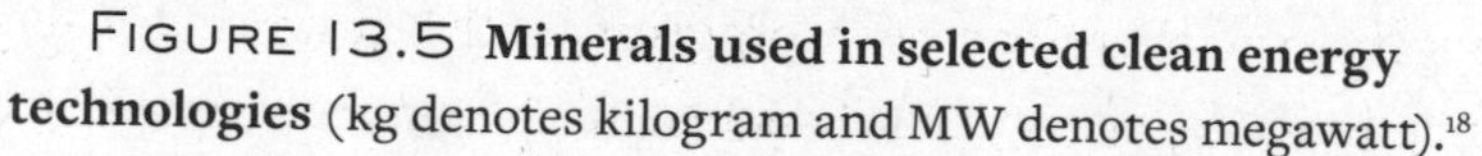

FIGURE 13.5 **Minerals used in selected clean energy technologies** (kg denotes kilogram and MW denotes megawatt).[18]

For example, an EV uses almost seven times as much high-value material as a conventional car, including 2.5 to 3 times as much copper. And onshore wind generation uses almost nine times as much as a natural gas turbine.

Unfortunately, those high-value materials and their processing are currently concentrated in inconvenient countries. The Democratic Republic of the Congo produces 75 percent of the world's cobalt, while China is a major player in extracting rare earths and graphite and in processing an array of critical minerals, as shown in Figure 13.6. And although China uses less than 40 percent of the world's solar panels, it makes 75 percent of all panels, 97 percent of the wafers, 85 percent of the cells, and 79 percent of the polysilicon used in them.

Chinese manufacturing costs are lower due in part to the scale of their activities and their mastery of supply chains. But cheap (coal-fired) electricity, loose environmental standards, and forced labor[19] are also important factors. In other words, clean energy is affordable because energy used to manufacture its components is not clean. The US government last year

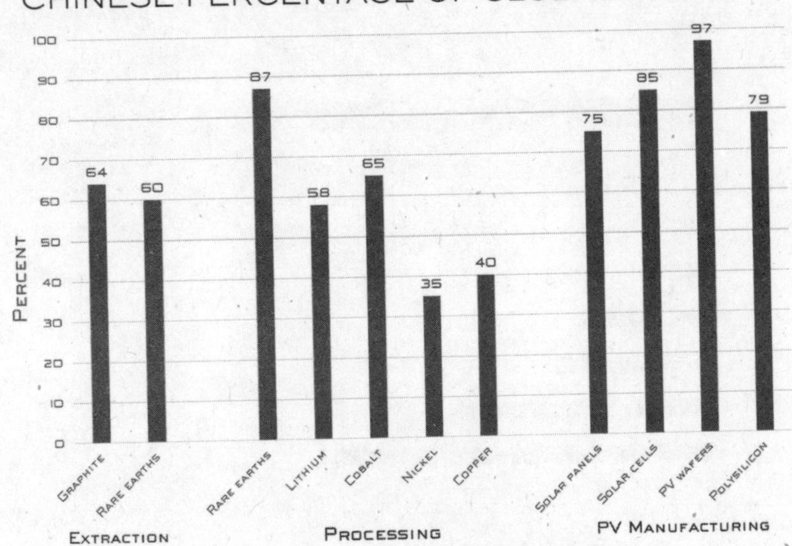

FIGURE 13.6 **Chinese percentage of global activity in various critical mineral supply chains.**[20]

imposed sanctions on some Chinese material for solar panels,[21] which has driven up costs. How much environmental degradation and forced labor in China would be okay to keep those panels cheap?

The Inflation Reduction Act begins an effort[22] to bring onshore or "friend shore" the supply chains for critical minerals. Some of the drawbacks of fossil fuels that disturb many people would still exist in a high-renewables world—there would still be international trade to lower commodities costs but also to threaten security of supply. And there will still be pollution from extracting and processing the enormous quantities of materials that renewables require. But since critical minerals are used to manufacture energy equipment, rather than providing the energy itself, disruption of one of those supply chains would not have the immediate impact associated with disruption of the supply of a fossil fuel.

In addition, renewables may not remain the cheapest form of generation. If wind, solar, EVs, and the like are deployed at the envisioned pace, mineral supplies will have a hard time keeping up. For example, by the middle of the next decade, copper demand is expected to double based upon current US and EU targets,[23] but the supply will be 20 to 25 percent short, because new mines will have lower quality ore than existing ones and each takes sixteen years to start up.

PLANS B

As the daunting challenges of effectively mitigating greenhouse gas emissions became clear to me, I grew interested in other, perhaps more feasible, strategies for responding to a changing climate. One such strategy was geoengineering—if humans were unintentionally exerting a net warming influence on the climate through greenhouse gas and aerosol emissions, could we do something intentional to counteract that influence? More colloquially, could we usefully "hack the planet"? The other strategy was simply to adapt to the changing climate, both by planning for future probabilities and responding to changes in the present. Together, geoengineering and adaptation constitute the "Plans B."

Most of those convinced the climate is in crisis avoid discussions of these strategies. Yet in the unlikely (in my opinion) event that human influences push the climate over some "tipping point," with deleterious changes happening very rapidly (see several bad Hollywood movies for dramatizations of such), the world would have no recourse other than to try to adapt—and to geoengineer. It's therefore important to know what options there are, and to understand their respective benefits, downsides, costs, unintended side effects, and so on. In fact, it's irresponsible for anyone who believes that there's an impending "climate catastrophe" *not* to support such research, especially since, as we've just seen, current mitigation efforts are highly unlikely to be successful at restraining human influences. We understand the importance of contingency planning in other areas of our lives—it's why we buy insurance, why we don't counsel students to apply to only one college, and so on. When it comes to climate, regarding the exploration of Plans B as "betraying" our efforts toward Plan A is dangerously wrongheaded.

The notion of geoengineering has a long history that began with weather modification.[1] The prospect of controlling weather has always been an enticing one, and as humanity learned more about how weather worked, appeals to physical processes supplanted appeals to deities. In the 1830s, James Pollard Espy, the US government's first official meteorologist—who was known as "The Storm King"—proposed inducing rain by setting "great fires" in the Appalachian forests to create clouds.[2] Needless to say, neither Congress nor the public had much enthusiasm for that idea.

Cloud-seeding experiments in the 1930s and 1940s were more serious and informed efforts to modify the weather. And in 1974, the Soviet climatologist Mikhail Budyko proposed that if warming ever became a pressing problem, a haze in the stratosphere could be created to cool the planet, as happens naturally after major volcanic eruptions. So when I began looking into geoengineering in the mid-2000s, I was hardly the first or the only scientist to think this was a strategy worth exploring, even if only to understand what the downsides might be.

But I soon discovered that any mention of geoengineering to governments or NGOs was met with tight-lipped silence, if not actual hostility. The focus was on reducing emissions, and any distraction from that goal, especially one that could allow the world to continue to use fossil fuels, was not to be contemplated.

But scientists are trained to explore all possible solutions to a problem, and an important part of science advising is to lay out the full spectrum of options, along with the advantages and drawbacks of each. So I persisted in learning about, and quietly discussing, geoengineering. I was eventually able to secure enough support to convene nine other scientists with a diversity of skills to have a serious look at the subject. The funding came from Novim—then a start-up foundation; today Novim continues to assemble teams of experts "to break down complex issues in such a way as to render them comprehensible to non-scientists." Our group met for a week in August 2008 to consider what an exploratory geoengineering research program would look like. Novim issued a formal report on the study in July 2009.[3]

Geoengineering was still *verboten* in most circles when I joined the US government in May 2009. John Holdren, the president's science adviser, had ignited a media firestorm in April of that year merely by mentioning the concept publicly, and he was forced to backtrack soon after.[4] Administration higher-ups subsequently discouraged my own efforts to fund an exploratory research program based on the Novim study. Again, the mandate was to keep the focus on reducing emissions.

But a decade later, as the challenges of reducing emissions have become evident to all, geoengineering can be discussed in polite company, even by governments. The UK Royal Society, commendably frank and forward-looking, broke the ice with a study[5] issued in September 2009, and the US National Academies issued reports on two separate approaches to "Climate Intervention"—the geoengineering strategies we're about to discuss—in 2015.[6,7]

There are at least two ways to counter warming of the planet. One is to make the earth a bit more reflective (increase its albedo), so that it absorbs a bit less energy from the sun. This strategy is termed Solar Radiation Management, or SRM, and would be appropriate whether the warming is natural or the result of human influences. Alternatively, we could pursue

Carbon Dioxide Removal (CDR), which is just what it sounds like: sucking some of the CO_2 back out of the atmosphere to directly offset human emissions. These two strategies are very different in terms of their practical challenges and potential impacts (positive and negative), but both merit discussion. Let's start with SRM.

Humans have been inadvertently increasing the earth's albedo for almost two centuries, as the burning of sulphur-laden coal produces tiny particles (aerosols) in the lower atmosphere that enhance the planet's reflectivity. One of my first calculations upon joining BP in 2004 had to do with that aerosol cooling. The company was embarking on a campaign to brand natural gas as "a bridge to a low-carbon future," as it produces only half as much carbon dioxide per unit of energy as coal. However, I quickly estimated, literally on the back of an envelope, that a sizable portion of that CO_2 reduction would be negated by the loss of aerosol cooling from the coal. BP management was not pleased when I pointed that out.

There are many ways we might further enhance the albedo, including brightening the land surface with "white roofs" on buildings, engineering crops to be more reflective, brightening the ocean with microbubbles on the surface, and putting up giant reflectors in space, to name a handful. However, creating aerosols in the stratosphere might be the most plausible way to make a significant global impact. The haze in the stratosphere that occurs naturally after major volcanic eruptions demonstrably cools the planet for a few years as the haze particles settle out. As we saw in Chapter 2, such cooling events are evident in the global temperature record.

It is well within the capabilities of current technology to create a stratospheric haze via any of a number of methods, including additives to jet fuel or artillery shells that disperse the gas hydrogen sulfide (H_2S, which smells like rotten eggs) at high altitude. This would not be a onetime exercise: the haze would have to be refreshed constantly, as the particles settle out over a year or two. The amount of sulfur that would have to be added to the stratosphere each year would be only about one-tenth of that humans currently emit at much lower altitudes, so direct health impacts would be minimal. And projected costs are low enough that a small nation or even a single wealthy individual could carry out the entire project themselves. As far as I know, the notion that "chemtrails" are evidence of covert geoengineering is entirely unfounded.

But Solar Radiation Management has several significant downsides. First, if the haze were not sustained, the global temperature would rebound rapidly when the cooling influence was lost (it would be like suddenly closing your parasol in the bright sunlight). Second, increasing the albedo doesn't exactly cancel out greenhouse warming—greenhouse gases warm all of the time over all of the globe, while albedo modification cools only when and where the sunshine it reflects is significant; it's ineffective at night and less effective during the winter, particularly at high latitudes. Climate models also suggest that SRM would cause small changes in precipitation and other aspects of the climate system, although these would be different from those caused by greenhouse gases alone, and would differ from place to place based on local conditions. In short, there are likely collateral effects that, at least for some, could be worse than the warming we're trying to counteract.

Thus even if it's technically and economically feasible, SRM would raise thorny societal questions that demand international cooperation. Who gets to determine whether it should be undertaken? There will no doubt be winners and losers in the resultant climate changes; if they're harmful for some, will there be compensation? Given the difficulty of pinpointing the causes of climate and weather phenomena (and our poor track record in doing so), how will those changes be attributed to SRM?

Then there are the ethical issues—and no doubt strong public opposition—attendant in deliberately messing with the climate in this way. What's more, because the costs are low enough that a single small nation, a subnational organization, or even a wealthy individual could "just do it," there is also the possibility of rogue SRM. How would the world respond in that case?

Nevertheless, Solar Radiation Management merits serious research, and in fact the US Congress has recently provided funds for exploratory work.[8] The first and largest part of that work should be a more careful and intense monitoring of the climate system, to establish a baseline against which to judge the effects of intervention. Intense observations of the effects of future volcanic eruptions would also be important. Fortunately, that monitoring would also lead to a better understanding of the climate system itself.

Beyond monitoring, we'd have to begin by asking *What field experiments would be permissible?* and *Who, and through what process, would give permission to do them?* The scientific and policy communities are just beginning to grapple with these questions. For now, since we've seen in previous chapters that extreme weather events show little sign of an imminent climate catastrophe, there's time to figure all that out.

In lieu of making the earth shinier, we could geoengineer our way out of some warming by directly removing CO_2 from the atmosphere. Carbon Dioxide Removal is a twin of mitigation—taking it out of the atmosphere instead of (or in addition to) putting less of it up there in the first place.

There are seeming advantages to CDR. It would make the issue of "whose CO_2 is it" less relevant and thus less contentious; assigning responsibility for emissions is one of the largest impediments to current international efforts to reduce them. It would also allow the continued use of fossil fuels as demand, economics, and technology might require (although some would consider that a drawback). Finally, since CDR works by directly "undoing" human influences, there would likely be little collateral impact to worry about—it would simply be restoring the CO_2 concentration to what it would have been.

It's not difficult to design a chemical plant that would capture CO_2 directly from the atmosphere. The capture technologies are similar to those used in the exhaust systems of power plants,[9] though there's the additional challenge of having to move a large amount of air through the system. So the real questions are about scale and cost.

The scale of carbon dioxide removal required to materially reduce human influences is daunting. The annual global consumption of energy materials is measured in gigatons. Every year the world as a whole uses about 4.5 Gt of oil and 8 Gt of coal, so removing even 10 Gt of CO_2 per year (about one-third of current emissions) would require a comparable infrastructure just to capture and handle the material. Needless to say, this wouldn't be cheap. Recent estimates are that it would cost upward of $100 to capture and compress one ton of CO_2, meaning a cost of at least one trillion dollars to remove 10 Gt of CO_2 per year.[10]

And then there is the issue of what to do with the CO_2 once it's removed from the atmosphere. The world today uses only 0.2 Gt of CO_2 each year—about 0.13 Gt to produce urea (fertilizer) and 0.08 Gt to enhance oil production (pumping CO_2 into oil fields through "injection wells" helps move the oil underground toward "production wells"). Those current uses of CO_2 are one hundred times too small to accommodate what would need to be taken out of the air. Unfortunately, it's difficult to imagine significant new uses. The largest global material flows after fuels are cement (less than 3 Gt each year) and plastics (about 0.5 Gt per year). And since CO_2 is the product of burning fuels to produce energy, it would take more energy (presumably "clean") to turn it back into a fuel. In short, the best thing to do with CO_2 removed from the atmosphere is to sequester it, either underground or in the ocean. Needless to say, this would be an enormous undertaking at the scale required.

Instead of using chemical plants to remove carbon dioxide from the atmosphere, another option might be to remove it using natural plants—that is, vegetation. About 200 Gt of carbon flow up and back between the earth's surface and the atmosphere each year as part of a seasonal cycle, more or less in equilibrium, as discussed in Chapter 3. By digging fossil fuels out of the ground, humans are adding about 8 Gt of carbon (in the form of 30 Gt of CO_2) to that cycle each year. About half of that excess is absorbed through photosynthesis. If we could induce more photosynthesis, more CO_2 could be removed. Hence the calls for planting some trillion trees (reforestation) to save the planet.[11] But while we might be planting trees now, forest growth takes decades; that's much too slow to be a response to the kind of "climate emergency" that would trigger SRM geoengineering, though it would remove CO_2 longer term. And we've yet to understand just how much CO_2 forests could absorb or what the ecological impacts of vast areas of new growth would be.

There has been a recent push for a massive US government research program to improve carbon dioxide removal technologies.[12] No doubt progress could be made. For example, it should be possible to genetically alter plants to better capture and store carbon (though this would surely be accompanied by environmental concerns about the widespread deployment of these altered plants).[13] Even so, I find it difficult to imagine that this could be achieved on the scale necessary to meaningfully reduce human influences on the climate. There is, however, money to be made if the cost to remove a ton of CO_2 from the atmosphere could be brought below the going price of carbon. As is often the case in the climate/energy business, it would then be possible to do well financially without actually doing much good.

Let's talk about our other Plan B: Adaptation.

I lived for thirty years in Southern California, where earthquakes are an unpleasant reality. As the US Geological Survey describes it:

> Each year the southern California area has about 10,000 earthquakes. Most of them are so small that they are not felt. Only several hundred are greater than magnitude 3.0, and only about 15-20 are greater than magnitude 4.0. If there is a large earthquake, however, the aftershock sequence will produce many more earthquakes of all magnitudes for many months.[14]

The small quakes are merely a nuisance, but the larger ones damage buildings and roads and can kill people. In the past century, California's earthquakes have killed hundreds of people and injured thousands more.

They are an entirely natural phenomenon; we can't stop them, and, as it happens, we can't predict them, except in a statistical sense.

Despite the earthquake hazard, my family and I didn't rush to move away from Pasadena—the weather, the community, and Caltech were powerful attractions. However, like millions of others in Southern California, we did take precautions. We bolted our house to its foundation and the shelves to the walls, we carried earthquake insurance, we stored a few days of food and water, we drilled our children in earthquake safety, and we formulated emergency contact and travel plans. Our family's steps complemented those of the larger society, where building codes, first-responder preparation, and, now, a warning system help minimize quake damage. In short, we all adapted, individually and as communities—shaping our infrastructure and behavior to thrive where we chose to live. People living in other places, with other natural hazards such as frequent floods and seasonal storms, have adapted in other, comparable ways.

Given the enormous challenges of effectively reducing human emissions, and the various concerns that make geoengineering likely to be deployed only *in extremis*, it seems all but certain that our efforts to reduce emissions will be complemented, if not overshadowed, by adaptation to a changing climate. To put this in the context of the questions I mentioned at the beginning of Part II: this is not a statement about what I think we *should* do; rather it is my judgment about what we *will* do.

Here's why I think adaptation *will* be our primary response:

- **Adaptation is agnostic.** Humans have been successfully adapting to changes in climate for millennia, and for most of that time, they did so without the foggiest notion of what (besides the vengeful gods) might be causing them. While the information we have now will help guide adaptation strategies, society can adapt to climate changes whether they are natural phenomena or the result of human influences.
- **Adaptation is proportional.** Modest initial measures can be bolstered as and if the climate changes more.

- **Adaptation is local.** Adaptation is naturally tailored to the different needs and priorities of different populations and locations. This also makes it more politically feasible. Spending for the "here and now" (e.g., flood control for a local river) is far more palatable than spending to counter a vague and uncertain threat thousands of miles and two generations away. Further, local adaptation does not require the global consensus, commitment, and coordination that have proved so far elusive in mitigation efforts.
- **Adaptation is autonomous.** It is what societies do, and have been doing, since humanity first formed them—the Dutch, for example, have been building and improving dikes for centuries to claim land from the North Sea. Adaptation will happen on its own, whether we plan for it or not.
- **Adaptation is effective.** Societies have thrived in environments ranging from the Arctic to the Tropics. Adapting to a changing climate always acts to reduce net impacts from what they would be otherwise—after all, we wouldn't change society to make things worse!

Despite the obvious importance of adaptation and its possible interplay with efforts to reduce emissions, today the two strategies are addressed separately—with much greater focus on mitigation. That imbalance might be due to the fact that adaptation is the "business as usual" response to ongoing natural climate change, but it might also be that we have no simple framework for thinking about adaptation. We have such a framework for emissions—the so-called "stabilization wedges,"[15] as illustrated in Figure 14.1. This approach catalogs a number of emission-reduction strategies and gives a sense of the scale each could achieve during this century, thus allowing the cost and efficacy of each strategy to be easily assessed and compared. This framework encourages a "systems perspective" on how various strategies should be prioritized and how they might interact with each other. It has led to specific policy proposals such as emissions pricing through charges or cap and trade, renewable electricity standards, and efficiency mandates.

While the wedges concept is by no means perfect, it does provide a useful way to think about emissions policy. As David Hawkins of the Natural Resources Defense Council put it:

> The wedges concept is sort of the iPod of climate policy analysis. It's an understandable, attractive package that people can fill with their own content.

There are, as yet, no analogous adaptation wedges laying out the methods, costs, efficacies, and policy levers of various adaptation measures. Rather, the discussion of adaptation is, at best, a dog's breakfast of anecdotes and possible strategies—more verbiage than content. Policy analysis of adaptation is relatively undeveloped. While numerous case studies have identified adaptation measures that would reduce adverse climate impacts, they notably do not address implementation issues, do not perform cost/benefit analyses of different adaptation strategies, and do not compare adaptation and mitigation efforts. And little attention is given to the

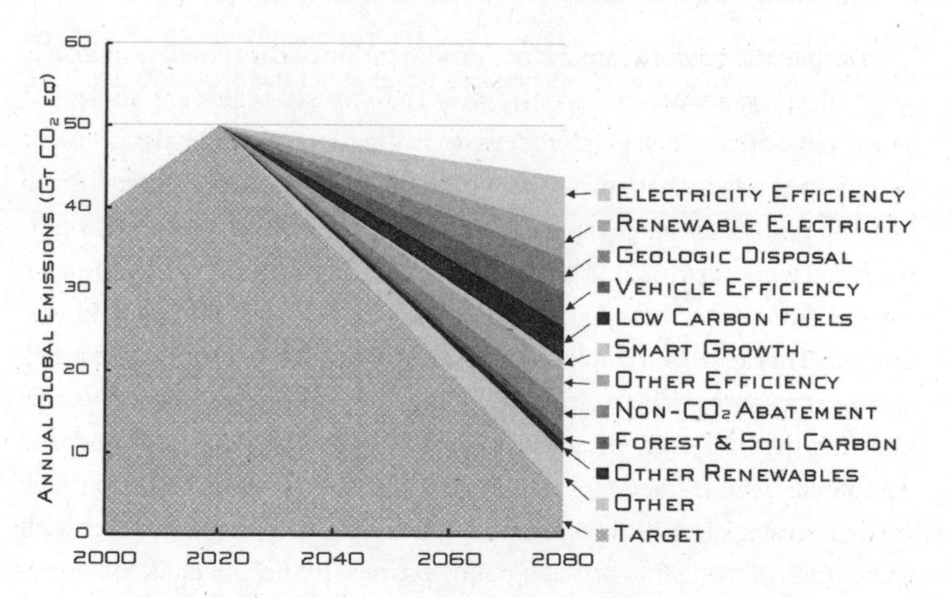

FIGURE 14.1 **An illustration of the stabilization wedges framework for emissions reductions.** Values are notional.[16]

specific bureaucratic, political, and fiscal changes that would be needed to move from analysis to deliberation to action.

Like most things, effective adaptation is easier in wealthier societies, which have the institutional and economic resources to change as circumstances require. Less developed countries are more fragile. Thus, the best way to enable adaptation globally is to encourage the economic development of less developed countries and strengthen their governance (such as the rule of law or the ability to formulate and execute national strategies). In that sense, the task of enabling adaptation becomes that of alleviating poverty, which would be a good thing for many reasons having nothing to do with the climate.

Of course, as alluded to above, any investments we make in adaptation measures today will be most effective at reducing the burden of future climate impacts if we understand what those impacts are likely to be—how the climate will change. Unfortunately, we've seen that there are still great uncertainties in what one might have to adapt to. Model projections of regional and local climates are currently nowhere near good enough to give guidance beyond vague statements like "sea level will continue to rise." At the very least, we should be prepared for the climate changes and extreme weather events that have happened before; in general, alas, we're not.

EASY ON THE ENERGY TRANSITION

2024 UPDATE

As I noted earlier, *Unsettled*'s publication gave me the opportunity to speak to many audiences about the realities of climate science and the energy system. Particularly useful in refining my own thinking has been a series of seven Oxford-style debates focused on the urgency of any energy transition. Such contests involve one side offering arguments in favor of a particular resolution, the other in opposition. The difference in audience polls taken before and after the arguments determines the winner—the side that swung the most votes.

I faced off against a different opponent in each debate. The first debate[1] was sponsored by the Soho Forum in Manhattan; the subsequent six, held at six different universities, were sponsored by the Steamboat Institute.[2]

Rather than focus on the more esoteric points of climate science, the debates were framed around the broader resolution:

Climate science compels us to make large and rapid reductions in greenhouse gas emissions.

Although my opponents and I often agreed on many things, I abandoned *Unsettled*'s descriptive stance to take the negative position in every debate. I did that in part because an intellectual tussle is a great way to engage broad audiences. But more importantly, in thinking through the arguments, I realized that many of the actions proposed (or already undertaken) to reduce greenhouse gas emissions will be damaging far beyond any conceivable impacts from human-caused climate change. My arguments must have been convincing because, at least according to the audience polls, I changed the most votes in six of the seven debates, with one resulting in a statistical tie.

Here's the essence of the arguments I used. You might find it useful to inform your own thinking, or in discussing climate and energy issues with others. [Data and references supporting these statements can be found throughout this book.]

Introduction: The proposition

Climate science compels us to make large and rapid reductions in greenhouse gas emissions

fails dramatically. The word "compels" makes the proposition unjustified, "us" makes it immoral, and "large and rapid" makes it a techno-economic fantasy. More succinctly, "We needn't do it," "We shouldn't do it," and "We can't do it." Let me take each of those three points in turn.

We needn't do it: Climate science itself doesn't "compel" anything. Rather, we must balance the scientific certainties and uncertainties of a changing climate against the world's growing demand for reliable and affordable energy and the cost and efficacy of whatever steps are to be taken to reduce emissions. In striking that balance, we should consider that humanity has prospered during the past century under comparable warming and the IPCC science does not project catastrophe.

The proposition is also unjustified because large and rapid reductions in emissions are disruptive if not damaging, perhaps even more so than the "climate threat" they're attempting to mitigate. William Nordhaus's Nobel Prize–winning work showed that there is an optimal pace to reduce emissions—moving too quickly causes turmoil and deploys immature technologies. His 2018 Nobel lecture stated that an economically optimal decarbonization could let the global temperature rise in 2100 exceed 6° C.[3] Of course, that's based on assumptions that can be, and have been, challenged, but Nordhaus's main takeaway is "don't panic"—if you're going to reduce emissions, take the time to do it gracefully.

In sum, the official scientific assessments, as well as common sense, do not at all justify large and rapid reductions in greenhouse gas emissions.

We shouldn't do it: The proposition is immoral because of the word "us." In fact, most of the world is asking, "Who do you mean by 'us'?"

People are better off as they use more energy. The 1.5 billion in the developed world enjoy abundant and affordable energy; we can't imagine life without it. But the globe's other 6.5 billion people are energy starved. The development of most of the world and some increase in population will drive a 50 percent increase in global energy demand by midcentury, with most of that growth in Asia.

Energy inequalities around the globe are astounding. Someone in the US consumes thirty times more energy than someone in Nigeria. And three billion of the world's eight billion people use less electricity every year than does the average US refrigerator.[4] Energy poverty also means cooking with wood and dung, and smoke in the kitchen kills some two million people each year. And while dining by candlelight is romantic, studying by candlelight is not.

Reliable and affordable energy is the overwhelming priority for developing nations. Minimizing local pollution comes second, while reducing greenhouse gas emissions comes in last—nice to have but by no means essential. Since fossil fuels are currently the most effective way for developing nations to get their energy, total carbon dioxide emissions are expected to increase in coming decades, even as the developed world's emissions decline slowly. As Anthony Downs put it: "The elite's environmental deterioration is often the common man's improved standard of living."[5]

So, again, when the proposition says "compels us," the response from the developing world is, "What do you mean 'us'?" The Indian prime minister has protested that the path for development is being closed to developing nations,[6] while the president of Niger has said that Africa is being punished by Western decisions and will fight to exploit the fossil fuels they have.[7]

It is immoral for the developed world to deny the developing nations the energy they need. And it is the height of eco-colonialism to restrain their development by mandating ineffective energy systems, especially since the developed world has neither the capacity nor the will to pay their "green premium" (the incremental costs of reducing emissions).

The proposition is also immoral because continued exaggerations like "science compels" induce a pernicious eco-anxiety. Some 60 percent of young people globally feel very worried about climate change, and more than 45 percent say it affects their daily lives.[8] Eco-anxiety also plays a role in making young people reluctant to have children.[9]

The prominent climate scientist Michael Mann correctly said that "climate doomerism can be harmful, because it robs us of agency, the agency we still have in determining our future."[10]

We can't do it: Finally, there is the fantasy of large and rapid reductions. Energy systems are recalcitrant for good reasons. They involve massive investments in assets that last decades, their parts need to work together (for example, cars, fuel, and the fueling infrastructure must all be compatible), and there are many stakeholders whose interests don't often align. It also takes time to refine the hardware and operating procedures that ensure that an energy system is highly reliable. So energy systems are best changed slowly and steadily over decades, more like orthodontia than the tooth extraction implied by "large and rapid."

Here's a sobering example of what happens when you try to move too fast. In April 2021, President Gotabaya Rajapaksa of Sri Lanka banned chemical fertilizers (which are made from natural gas) in that nation of twenty-two million, motivated by environmental considerations. There are some important lessons to be learned from what ensued. In particular, it took less than a year for that ill-considered action to crash the Sri Lankan economy, ultimately leading to starvation, riots, and a change in government.[11]

A *Foreign Policy* piece described that crisis in Sri Lanka as "a farrago of magical thinking, technocratic hubris, ideological delusion, self-dealing, and sheer shortsightedness."[12] We have glimmerings of the same sort of thing happening in Germany, the UK, and California, where hasty and ill-conceived greening of the energy system has degraded reliability and increased costs in an ineffective effort to avoid vague and uncertain climate problems several generations hence. It would indeed be a crisis if that were to happen across the US. We should be very careful, since precipitous emissions reductions are far more dangerous than climate change itself.

We have the time, in fact must take the time, to reduce emissions gracefully. Rapidly reducing carbon dioxide emissions wouldn't even reduce human influences on the climate anytime soon, because CO_2 accumulates in the atmosphere and persists for a century or more. And reducing emissions won't stop the climate from changing—it varies an awful lot on its own.

Summary: A dispassionate look at the science, as well as trends in demographics, development, and energy technology, shows that large and rapid reductions in greenhouse gas emissions are unjustified, immoral, and fantastical. Global net-zero emissions by 2050 is a fantasy, and net zero is quite unlikely even by 2100. But the consequences of missing that goal will not be catastrophic.

That doesn't mean the world, or we in the US, shouldn't do anything. Here's what I think we should do.

- We should sustain and improve climate science, for our knowledge of the earth's climate system is not what it should be. Paleoclimate studies tell us how and why climate has changed in the past; current observations with improved coverage, precision, and continuity tell us what the climate system is doing today; and models give a sense of what might happen in the future. There is a particular need for greater statistical rigor in the analyses and for more focused efforts to reduce large model uncertainties.

- We must improve climate communications to the public and decision makers. We need to stop foretelling climate apocalypses, even as we acknowledge human influences on the climate are growing, and we should be working to reduce them. The public must have an accurate view of both climate and energy that gets beyond slogans like "We are on a highway to climate hell with our foot still on the accelerator."[13] Non-experts are savvy enough to dismiss unsupported scare stories, and credibility is eroding. It is encouraging that the new head of the IPCC seems to understand this.[14]
- We must acknowledge that energy reliability and affordability take precedence over emissions reductions. A good start is President Biden's admission that oil and gas will be necessary in the US for at least a decade (actually, it will be far longer than that). Europe's energy crisis is self-inflected—fossil fuel investments and domestic production were abandoned in favor of unreliable import partners and unreliable wind and solar generation; incredibly, Germany also shuttered its nuclear plants. It was easy to see that these steps would lead to trouble, but mitigation was deemed more important than reliability and affordability.
- Governments must embark on thoughtful and graceful decarbonization programs that incorporate technology, economics, regulation, and behavior, and that estimate costs, timescales, and actual impacts on the climate.
- Research, development, and demonstration of emissions-lite technologies are essential, especially to reduce the so-called green premium. But programs that go beyond demonstration to meaningful deployment should not be the scattershot mandates and incentives currently in vogue. Small fission, grid storage and management, batteries, non-carbon chemical fuels, and carbon capture and storage would be on my list of the most promising early-stage areas today.

- Developed countries must acknowledge the inevitability, if not desirability, of meeting the developing world's energy needs. Most of the world today is energy starved, and fossil fuels are the most convenient and reliable way of meeting that demand; they provide about 80 percent of the world's energy now, as they have for the past many decades. Without costly backup systems, weather-dependent wind and solar generation cannot provide appropriate energy access for the energy poor. I have asked many advocates of rapid global decarbonization what they would do to meet the developing world's energy needs. I've yet to hear an answer that respects technical, economic, demographic, and political realities.
- There needs to be a greater focus on alternative strategies for dealing with a changing climate, most importantly adaptation. Adaptation is autonomous—it's what humans do, it's effective, it's proportional, and it's local and hence achievable. If nothing else, governments should work to facilitate adaptation.

Policymakers need to realize that large and rapid reductions in emissions are overkill—they risk far more damage to humanity than any conceivable impact from climate change itself. But there is a sensible path forward that will moderate human influences on the climate while responding to the growing demand for reliable and affordable energy. The policy challenge is to put aside the fervor, the emotion, the apocalyptic rhetoric, and understand what is settled and not settled in climate science. That would allow us to identify that sensible path and then begin to follow it.

CLOSING THOUGHTS

Writing this book has been an opportunity to collect and synthesize experiences over a fifteen-year journey in climate and energy. I began by believing that we were in a race to save the planet from climate catastrophe. Since then, I've evolved to become a public critic of how The Science of climate science is presented. And, throughout, I've been a student and strategist, constantly seeking to learn more and concerned with how the energy system could be transformed to meet evolving needs.

My journey has impressed me with the richness of the subjects. There aren't many (probably not *any*) other discussions that connect the composition of fifty-million-year-old microfossils with the regulation of the electrical grid. I'm also impressed by how much *is* understood about the climate, yet still a bit surprised by the vast gaps in our knowledge and what those gaps tell us about the complexity of the climate system. And I've been impressed with the importance with which the subjects are treated by most researchers in the field. The climate is changing, humans are playing a role, and yet our global energy needs are growing, too; we must be mindful of what that might mean for the future.

But I've been dismayed along the way as well. First by the willingness of some climate scientists—abetted by the media and politicians—to misrepresent what the science says, and then by the many other scientists who are silently complicit in those misrepresentations. The public deserves better. By demonstrably misinforming non-experts about what we know and don't know about the changing climate, they deny governments, industry, and individuals the right to make fully informed decisions about how to respond.

I've also been dismayed by how difficult it is for people, the media included, to understand what's actually written in the assessment reports. The newspaper opinion pieces I've written over the past six years have been a poor remedy for that—they're limited in length, in technical level, and in format (no graphs!). This book has allowed me to tell a richer, and I hope more informative, story.

I'm often asked "What are your takeaways?" or "What's the elevator speech?" My response is usually something like "Climate and energy are complex and nuanced subjects. Simplistic descriptions of 'the problem' or putative 'solutions' will not result in wise choices." If my fellow passenger is willing to ride the elevator further, we'll often have a longer discussion—which I always end with an urging not to take my word for any of it, but to look carefully at the data and assessments for themself.

My greatest hope is that decision makers, journalists, and the larger public will find some surprises as they read this book, and that they'll then turn to scientists and say something like "I've checked some of the things that guy Koonin says are in the assessment reports—and he's right. How come I haven't heard those things before? And what else am I not being told?" This could be the start of many awkward, but ultimately essential, conversations.

I have deliberately written this book in a descriptive manner rather than a prescriptive one: I've presented facts, the certainties and uncertainties in what they imply, and the options for choices to be made in response. That's the appropriate stance for a scientist to take when advising non-experts, whether those non-experts are other scientists, the public, or decision makers in government or industry, and whether the subject is climate and energy, nuclear terrorism, or the human genome project. But while responsible scientists are careful to keep the *should* questions distinct from the *could* or *will*, none of us can avoid having opinions. And I'm asked often enough "So what do *you* think we should do about the climate?" that I feel obligated, now that I've finished laying out the facts, to respond.

We can begin with sustained and improved observations of the climate system (the atmosphere, ocean, cryosphere, and biosphere). This is essential if we hope to understand what the climate is doing, how it is being affected by human and natural influences, and what it might do in the future. We've seen that climate changes resulting from human influences are small or subtle and happen over decades, so precision and persistence are essential, even in the face of institutional or funding vagaries.

We also need to better understand the tremendously complex climate models we've built. An awful lot of effort is being devoted to not very informative model simulations under varying emissions scenarios. It would be much better spent trying to understand why the climate models fail in describing the recent past and are so uncertain in their projections of the future. In short, there should be more thinking and less unproductive computing.[1]

We need to improve the science itself, and this begins with open and honest discussion that goes beyond slogans and polemics, and is free of accusations of skullduggery. Scientists should be welcoming of debate, challenges, and opportunities for clarification. Science starts with questions; it's hard to encourage new research if we insist they've all been answered. In fact, as I've shown in this book, there are still plenty of important, even crucial, questions about climate that are as yet unsettled. The truth is that real science is never entirely settled—that's how we make progress; it's what science is all about. Let's further our understanding, rather than repeating orthodoxy.

Climate science would also be improved by deliberate efforts to involve scientists from other fields in studying climate. The data is rich and accessible and the problems are scientifically interesting and societally important. The injection of working scientists from outside the field who have skills in statistics or simulation would complement the perspectives of those in the field currently.

We also need to get better at communicating climate science. Societal decisions that balance costs and drawbacks against risks and benefits must be made fully informed of the certainties and uncertainties in our scientific understanding. The public deserves complete, transparent, and unbiased assessment reports. A Red Team exercise like that I described in Chapter 11 would be a healthy addition to the climate science assessment process; it has proved its usefulness in other complex matters of national importance. A first Red Team review could involve the close public scrutiny of the UN's AR6 report or of the US government's National Climate Assessment. It could focus on the issues I've raised and the misrepresentations I've identified in this book. How do these upcoming reports deal with the failures of the most recent generation of models? Do they even mention, let alone highlight, that there have been no long-term trends in hurricanes or that net economic impacts of a 3°C warming (far above the Paris goal) are projected to be minimal? I'd think that kind of scrutiny of the assessments would be essential, particularly since the Biden administration is advocating some $2 trillion of spending on climate and energy.

At the same time, we need to reduce the hysteria in climate journalism. Journalists themselves need help to better understand the material they are presented with, and the public needs the tools to become more critical consumers of media coverage of climate (and many other topics, for that matter).

It also makes sense to pursue "easy" emissions reductions, most obviously stopping methane leaks. A few percent of the methane escapes from the production and distribution systems for natural gas; that's money lost and so there is a financial incentive to stem leaks (often more motivating to producers than climate concerns). The emission of more exotic greenhouse gases such as the CFCs and HCFs used as refrigerants and fire suppressors could also be reduced without much impact on society (unfortunately, the impact on human influences would be similarly slight).

Cost-effective efficiencies that lead to emissions reductions are also low-hanging fruit, particularly when there are side benefits. For example, advanced coal-fired generating plants that gasify the coal rather than burn it directly will also reduce local pollution and increase efficiency. And for vehicles, more gasoline-efficient engines, as well as a move toward hybrid and electric cars, can both reduce local chemical and noise pollution and enhance energy security by reducing dependence on the volatile global oil market.

A third "easy" step toward reducing emissions is further research and development in emissions-lite technologies. Cost and reliability are the primary factors by which new technologies will be judged feasible, and the focus should be on advances that overcome those stumbling blocks. Small modular fission reactors, improved solar technologies, and, in the longer term, nuclear fusion are all promising areas of research, as is how to economically store massive amounts of electricity on the grid. A win-win strategy is to develop and deploy more efficient, yet cost-effective, end-use technologies, from building ventilation systems to household appliances, as has happened with lighting technologies over the past few decades. Particularly promising here is the use of information-based approaches to transportation (such as suggesting more efficient routes for a trip or better monitoring and control of engine performance) and building operations (such as turning down the heating or cooling in unoccupied rooms).

We also need to have a frank conversation about the proper role of government in these efforts (how much R&D to support, how and how much to encourage deployment of a new technology). One of my jobs in the Department of Energy was to start that conversation among Congress, the executive branch, and the private sector; I hope that discussion is about to resume. At least in the US, the government's role in transforming the energy system has been a point of political contention for decades.

I'm less bullish on "forced and urgent" decarbonization, either through a price on carbon or by way of regulation. The impact of human influences on the climate is too uncertain (and very likely too small) compared to the daunting amount of change required to actually achieve the goal of eliminating net global emissions by, say, 2075. And for me, the many certain downsides of mitigation outweigh the uncertain benefits: the world's poor need growing amounts of reliable and affordable energy, and widespread renewables or fission are currently too expensive, unreliable, or both. I would wait until the science becomes more settled—that is, until the climate's response to human influences is better determined, or, failing that, until a values consensus emerges or zero-emissions technologies become more feasible—before embarking on a program to tax or regulate greenhouse gas emissions out of existence or to capture and store massive amounts of carbon dioxide from the atmosphere.

I believe the socio-technical obstacles to reducing CO_2 emissions make it likely that human influences on the climate will not be stabilized, let alone reduced, in this century. If the effects of those influences become more evident and more severe than they have been to date, of course, the balance of costs and benefits might shift, and society might well shift along with it. But I'd be surprised if this happened anytime soon.

Advocating that we make only low-risk changes until we have a better understanding of why the climate is changing, and how it might change in the future, is a stance some might call "waffling," but I'd prefer the terms "realistic" and "prudent." I can respect the opinions of others who might come to different conclusions, as I hope they would respect mine. Those differences can only be resolved if we realize that they're ultimately about values, not about the science.

Another prudent step would be to pursue adaptation strategies more vigorously. Adaptation can be effective. As mentioned in Chapter 14, humans today live in climates ranging from the Tropics to the Arctic and have adapted through many climate changes, including the relatively recent Little Ice Age about four hundred years ago. Effective adaptation would combine credible regional projections of climate change with a framework for assessing the costs and benefits of various adaptation strategies. As we've seen, we're a long way from having either of those. So the best strategy is to promote economic development and strong institutions in developing countries in order to improve their ability to adapt (and their ability to do many other positive things as well).

Finally, should there be significant deterioration of the global climate, from whatever cause, humanity would be well served to know whether deliberate intervention into the climate system (geoengineering) is a plausible strategy. A research program into geoengineering options like those discussed in the previous chapter is therefore prudent and, as I've noted, the intense monitoring of the earth system that would be a first step in that research program would, in any event, also improve our understanding of the climate system.

What I think we *should* do, in short, is to begin by restoring integrity to the way science informs society's decisions on climate and energy—we need to move from The Science back to science. And then take the steps most likely to result in positive outcomes for society, whatever the future might hold for our planet. As President Biden exhorted in his inaugural address, "We must reject the culture in which facts themselves are manipulated, or even manufactured."

ACKNOWLEDGMENTS

This book has been almost three years in the making. While I alone am responsible for its content, I'm grateful to the many people who helped along the way.

The list starts, of course, with my parents. From a very early age, my father imparted a sense of wonder and curiosity about the natural world and about "how things work," while my mother's energy and optimism have sustained me throughout my career.

My wife, Laurie, and children, Anna, Alyson, and Benjamin, supported my decision to write the book, as they have all else in my professional life. Their tolerance of the few-year metaphorical absence that followed made the writing so much easier, while their readings of successive drafts helped ground the presentation.

My agent, Keith Urbahn at Javelin, agreed that my perspective on climate and energy needed to reach a broader audience.

My publisher, Glenn Yeffeth at BenBella Books, had the courage to take on a book that will surely displease many, even though (or perhaps because) it's accurate, frank, and accessible.

My editor, Alexa Stevenson, asked insightful questions and deftly reshaped my prose and graphics. Her efforts made the book much better, and more readable, than it would have been otherwise.

Professors Will Happer, William van Wijngaarden, John Christy, and Demetris Koutsoyiannis provided figures or data that help me tell the science story.

Many people reviewed one or more drafts of *Unsettled*. While not all have agreed with everything I've written, their comments and insights were very valuable to me as the book matured. Among those readers were Henry Abarbanel, Kathleen Alexander, Jesse Ausubel, Wolfgang Bauer, Peter Blair, Kevin Chilton, John Christy, Marius Clore, Tamar Elkeles, Nancy Forbes, Phil Goode, Mike Gregg, Will Happer, Patrick Hogan, Barry Honig, Stephen Hsu, Eugene Illovsky, Eli Jacobs, Seymour Kaplan, Brian Koonin, Karlheinz Langanke, Dick Lindzen, Harriet Mattar, Andy May, Dan Meiron, Robert Powell, Roy Schwitters, Noah Trepanier, Craig Wiener, David Whelan, Jonathan Wurtele, and June Yu.

NOTES

PREFACE

1. IPCC, 2021: *Climate Change 2021: The Physical Science Basis. Contribution of Working Group I to the Sixth Assessment Report of the Intergovernmental Panel on Climate Change* [Masson-Delmotte, V., P. Zhai, A. Pirani, S.L. Connors, C. Péan, S. Berger, N. Caud, Y. Chen, L. Goldfarb, M.I. Gomis, M. Huang, K. Leitzell, E. Lonnoy, J.B.R. Matthews, T.K. Maycock, T. Waterfield, O. Yelekçi, R. Yu, and B. Zhou (eds.)]. Cambridge: Cambridge University Press. In press, https://doi.org/10.1017/9781009157896.
2. IPCC, 2022: *Climate Change 2022: Impacts, Adaptation, and Vulnerability. Contribution of Working Group II to the Sixth Assessment Report of the Intergovernmental Panel on Climate Change* [Pörtner, H.-O., D.C. Roberts, M. Tignor, E.S. Poloczanska, K. Mintenbeck, A. Alegría, M. Craig, S. Langsdorf, S. Löschke, V. Möller, A. Okem, B. Rama (eds.)]. Cambridge: Cambridge University Press. https://doi.org/10.1017/9781009325844.
3. IPCC, 2022: *Climate Change 2022: Mitigation of Climate Change. Contribution of Working Group III to the Sixth Assessment Report of the Intergovernmental Panel on Climate Change* [Shukla, P.R., J. Skea, R. Slade, A. Al Khourdajie, R. van Diemen, D. McCollum, M. Pathak, S. Some, P. Vyas, R. Fradera, M. Belkacemi, A. Hasija, G. Lisboa, S. Luz, J. Malley (eds.)]. Cambridge: Cambridge University Press. https://doi.org/10.1017/9781009157926.
4. IPCC, 2023: *Climate Change 2023: Synthesis Report. Contribution of Working Groups I, II and III to the Sixth Assessment Report of the Intergovernmental Panel on Climate Change* [Core Writing Team, H. Lee and J. Romero (eds.)]. Geneva, Switzerland: IPCC, pp. 35–115, https://doi.org/10.59327/IPCC/AR6-9789291691647.
5. IPCC, 2023. "Headline Statements." https://www.ipcc.ch/report/ar6/syr/resources/spm-headline-statements.

6. IPCC, 2023: Summary for Policymakers. In: *Climate Change 2023: Synthesis Report. Contribution of Working Groups I, II and III to the Sixth Assessment Report of the Intergovernmental Panel on Climate Change.*
7. Meinshausen, M., C.-F. Schleussner, K. Beyer, et al. "A perspective on the next generation of Earth system model scenarios: towards representative emission pathways (REPs)," *Geoscientific Model Development* [preprint], https://doi.org/10.5194/gmd-2023-176, in review, 2023.
8. USGCRP, 2023: *Fifth National Climate Assessment.* Crimmins, A.R., C.W. Avery, D.R. Easterling, K.E. Kunkel, B.C. Stewart, and T.K. Maycock, eds. U.S. Global Change Research Program, Washington, DC. https://doi.org/10.7930/NCA5.2023.
9. United Nations Environment Programme (2023). Executive summary. In *Emissions Gap Report 2023: Broken Record—Temperatures hit new highs, yet world fails to cut emissions (again).* Nairobi. https://doi.org/10.59117/20.500.11822/43922.
10. NOAA. "Greenhouse gases continued to increase rapidly in 2022." NOAA News Release, April 5, 2023. https://www.noaa.gov/news-release/greenhouse-gases-continued-to-increase-rapidly-in-2022.
11. United Nations Environment Programme (2022). *Emissions Gap Report 2022: The Closing Window—Climate crisis calls for rapid transformation of societies.* Nairobi. https://www.unep.org/emissions-gap-report-2022.
12. United Nations Environment Programme (2023). *Emissions Gap Report 2023.*
13. Statista. "Carbon dioxide emissions from energy consumption in the United States from 1975 to 2022." September 12, 2023. https://www.statista.com/statistics/183943/us-carbon-dioxide-emissions-from-1999/.
14. Statista. "Carbon dioxide emissions in the European Union from 1965 to 2022." September 13, 2023. https://www.statista.com/statistics/450017/co2-emissions-europe-eurasia/.
15. Pratt, Kevin, ed. "End of ICE Age? Enthusiasm Cools for Ban on Sale of New Internal Combustion Engine Cars." *Forbes* Advisor, April 19, 2023. https://www.forbes.com/uk/advisor/car-insurance/2030-internal-combustion-engine-ban/.
16. Fischer, Justin. "These Are the States Banning New Sales of Gas and Diesel Vehicles." CarEdge, April 15, 2023. https://caredge.com/guides/states-banning-ice-cars.
17. See https://www.energyinst.org/statistical-review/insights-by-source.
18. Horowitz, Julia. "'Drill, baby, drill' is back in Europe as gas crisis looms." CNN Business, August 1, 2022. https://www.cnn.com/2022/08/01/energy/gas-fields-europe-energy-crisis-russia/index.html.
19. The Associated Press. "After the Fukushima disaster, Japan swore to phase out nuclear power. But not anymore." NPR, December 22, 2022. https://www.npr.org/2022/12/22/1144990722/japan-nuclear-power-change-fukushima.
20. Sergeev, Angel. "General Motors Will Stick with Internal Combustion Engines Amid EV Expansion." Motor1.com, November 21, 2022. https://www.motor1.com/news/622700/general-motors-stick-combustion-engines/.

21. Dickie, Gloria. "Despite COP28 deal on fossil fuels, 1.5C goal likely out of reach." Reuters, December 14, 2023. https://www.reuters.com/sustainability/climate-energy/despite-cop28-deal-fossil-fuels-15c-goal-likely-out-reach-2023-12-14/.

INTRODUCTION

1. Climate Change Statement Review Subcommittee. "American Physical Society Climate Change Statement Review Workshop Framing Document." American Physical Society, December 20, 2013. https://www.aps.org/policy/statements/upload/climate-review-framing.pdf.
2. Koonin, Steven E. American Physical Society Climate Change Statement Review Workshop transcript. Brooklyn, NY: American Physical Society, 2014. https://www.aps.org/policy/statements/upload/climate-seminar-transcript.pdf.
3. Koonin, Steven E. "Climate Science Is Not Settled." *Wall Street Journal*, September 19, 2014. https://www.wsj.com/articles/climate-science-is-not-settled-1411143565.
4. Pierrehumbert, Raymond T. "Climate Science Is Settled *Enough*." *Slate*, October 1, 2014. https://slate.com/technology/2014/10/the-wall-street-journal-and-steve-koonin-the-new-face-of-climate-change-inaction.html.
5. Karanth, Sanjana. "Rep. John Lewis Delivers Emotional Speech on the 'Moral Obligation' to Impeach Trump." HuffPost, December 18, 2019. https://www.huffpost.com/entry/rep-john-lewis-emotional-speech-moral-obligation-impeach-trump_n_5dfaa5fee4b01834791ab306.
6. Feynman, Richard P. "Cargo Cult Science." Caltech 1974 commencement address, June 1974. http://calteches.library.caltech.edu/51/2/CargoCult.htm.
7. Schneider, Stephen H. "The Roles of Citizens, Journalists, and Scientists in Debunking Climate Change Myths." Mediarology, 2011. https://stephenschneider.stanford.edu/Mediarology/Mediarology.html.
8. Spencer, L., Jan Bollwerk, and Richard C. Morais. "The not so peaceful world of Greenpeace." *Forbes*, November 11, 1991. Full article at http://luna.pos.to/whale/gen_art_green.html.
9. Bell, Larry. "In Their Own Words: Climate Alarmists Debunk Their 'Science.'" *Forbes*, February 6, 2013. https://www.forbes.com/sites/larrybell/2013/02/05/in-their-own-words-climate-alarmists-debunk-their-science/.
10. Botkin, Daniel B. "Global Warming Delusions." *Wall Street Journal*, October 18, 2007. https://www.wsj.com/articles/SB119258265537661384.
11. Shabecoff, Philip. "U.N. Ecology Parley Opens Amid Gloom." *New York Times*, May 11, 1982. https://www.nytimes.com/1982/05/11/world/un-ecology-parley-opens-amid-gloom.html.
12. Onians, Charles. "Snowfalls Are Now Just a Thing of the Past." *Independent*, March 20, 2000. https://web.archive.org/web/20150912124604/http:/www.independent.co.uk/environment/snowfalls-are-now-just-a-thing-of-the-past-724017.html.

13. Harris, Paul, and Mark Townsend. "Pentagon Tells Bush: Climate Change Will Destroy Us." *Guardian*, February 22, 2004. https://www.theguardian.com/environment/2004/feb/22/usnews.theobserver.
14. Handler, Philip. "Public Doubts About Science." *Science*, June 6, 1980. https://science.sciencemag.org/content/208/4448/1093.
15. Census Bureau. "U.S. and World Population Clock." Population Clock, January 1, 2020. https://www.census.gov/popclock/.
16. Koonin, Steven E. "An Independent Perspective on the Human Genome Project." *Science*, January 2, 1998. https://science.sciencemag.org/content/279/5347/36.summary.
17. Glanz, James, with Andrew C. Revkin. "Some See Panic As Main Effect of Dirty Bombs." *New York Times*, March 7, 2002. https://www.nytimes.com/2002/03/07/us/a-nation-challenged-senate-hearings-some-see-panic-as-main-effect-of-dirty-bombs.html.
18. Resnikoff, Ned. "Kerry Compares Climate Change Deniers to 'Flat Earth Society.'" MSNBC, February 17, 2014. http://www.msnbc.com/msnbc/kerry-slams-climate-change-deniers.

PART I

1. Intergovernmental Panel on Climate Change (IPCC). "The Guidance Note for Lead Authors of the IPCC Fifth Assessment Report on Consistent Treatment of Uncertainties." https://www.ipcc.ch/site/assets/uploads/2017/08/AR5_Uncertainty_Guidance_Note.pdf.
2. IPCC, 2013: *Climate Change 2013: The Physical Science Basis. Contribution of Working Group I to the Fifth Assessment Report of the Intergovernmental Panel on Climate Change* [Stocker, T.F., D. Qin, G.-K. Plattner, M. Tignor, S.K. Allen, J. Boschung, A. Nauels, Y. Xia, V. Bex and P.M. Midgley (eds.)]. Cambridge University Press, Cambridge, United Kingdom and New York, NY, USA, 1535 pp. https://archive.ipcc.ch/report/ar5/wg1/. Figure 1.11.
3. Intergovernmental Panel on Climate Change (IPCC). "The Intergovernmental Panel on Climate Change." IPCC, January 1, 2000. https://www.ipcc.ch/.
4. US Global Change Research Program (USGCRP). "About USGCRP." GlobalChange.gov, January 1, 2000. https://www.globalchange.gov/about.
5. IPCC, 2007: *Climate Change 2007: The Physical Science Basis. Contribution of Working Group I to the Fourth Assessment Report of the Intergovernmental Panel on Climate Change* [Solomon, S., D. Qin, M. Manning, Z. Chen, M. Marquis, K.B. Averyt, M. Tignor and H.L. Miller (eds.)]. Cambridge University Press, Cambridge, United Kingdom and New York, NY, USA, 996 pp. https://www.ipcc.ch/assessment-report/ar4/.
6. IPCC, 2013: *Climate Change 2013: The Physical Science Basis. Contribution of Working Group I to the Fifth Assessment Report of the Intergovernmental Panel on Climate Change* [Stocker, T.F., D. Qin, G.-K. Plattner, M. Tignor, S.K. Allen, J. Boschung, A. Nauels, Y.

Xia, V. Bex and P.M. Midgley (eds.)]. Cambridge University Press, Cambridge, United Kingdom and New York, NY, USA, 1535 pp. https://archive.ipcc.ch/report/ar5/wg1/.

7. IPCC. "Managing the Risks of Extreme Events and Disasters to Advance Climate Change Adaptation (SREX)." IPCC, January 1, 2000. https://archive.ipcc.ch/report/srex/.
8. IPCC. "Special Report on the Ocean and Cryosphere in a Changing Climate." January 1, 2000. https://www.ipcc.ch/srocc/.
9. IPCC. "Climate Change and Land." Special Report on Climate Change and Land, January 1, 2000. https://www.ipcc.ch/srccl/.
10. USGCRP. "Assess the U.S. Climate." GlobalChange.gov, January 1, 2000. https://www.globalchange.gov/what-we-do/assessment.
11. USGCRP. *Climate Science Special Report: Fourth National Climate Assessment, Volume I*. US Global Change Research Program, Washington, DC, 2017. https://science2017.globalchange.gov/.
12. USGCRP. "Fourth National Climate Assessment, Volume II: Impacts, Risks, and Adaptation in the United States: Summary Findings." NCA4, January 1, 1970. https://nca2018.globalchange.gov/.
13. AAAS. "About." How We Respond, January 1, 2000. https://howwerespond.aaas.org/report/.

CHAPTER 1

1. Data accessed 11/27/20 from Berkeley Earth (http://berkeleyearth.org/archive/data/); UK MetOffice (https://www.metoffice.gov.uk/hadobs/monitoring/index.html); NOAA (https://www.ncei.noaa.gov/access/metadata/landing-page/bin/iso?id=gov.noaa.ncdc:C00934); and NASA (https://data.giss.nasa.gov/gistemp/).
2. Hansen, J. E., and S. Lebedeff. "Global Trends of Measured Surface Air Temperature." *Journal of Geophysical Research*, November 27, 1987. https://pubs.giss.nasa.gov/abs/ha00700d.html.
3. Data from Koutsoyiannis, D. "Hydrology and Change." *Hydrological Sciences Journal* 58, no. 6 (2013): 1177–1197. http://www.itia.ntua.gr/en/docinfo/1351/.
4. Mufson, Steven, Chris Mooney, Juliet Eilperin, and John Muyskens. "Extreme Climate Change in America." *Washington Post*, August 13, 2019. https://www.washingtonpost.com/graphics/2019/national/climate-environment/climate-change-america/.
5. Raw station data for West Point/NYC downloaded from berkeleyearth.lbl.gov.
6. NASA. "Data.GISS: GISS Surface Temperature Analysis (v4): Global Maps." NASA, January 1, 2000. https://data.giss.nasa.gov/gistemp/maps/index_v4.html.
7. The UN Framework Convention on Climate Change, Article 1 (2). https://unfccc.int/files/essential_background/background_publications_htmlpdf/application/pdf/conveng.pdf.
8. USGCRP. *Climate Science Special Report: Fourth National Climate Assessment, Volume I*. Figure 1.3.

9. "Current Status of Argo." Argo, January 1, 2000. https://argo.ucsd.edu/about/.
10. Rhein, Monika, and Stephen R. Rintoul, et al. "Observations: Ocean." IPCC, 2013. https://www.ipcc.ch/site/assets/uploads/2018/02/WG1AR5_Chapter03_FINAL.pdf.
11. Cheng, L., John Abraham, Jiang Zhu, Kevin E. Trenberth, John Fasullo, Tim Boyer, Ricardo Locarnini, et al. "Record-Setting Ocean Warmth Continued in 2019." *Advances in Atmospheric Sciences* 37 (2020): 137–142. https://link.springer.com/article/10.1007/s00376-020-9283-7.
12. Zanna, Laure, Samar Khatiwala, Jonathan M. Gregory, Jonathan Ison, and Patrick Heimbach. "Global reconstruction of historical ocean heat storage and transport." *Proceedings of the National Academy of Sciences*, January 2019. https://www.pnas.org/content/116/4/1126.
13. Gebbie, G., and P. Huybers. "The Little Ice Age and 20th-Century Deep Pacific Cooling." *Science*, January 4, 2019. https://science.sciencemag.org/content/363/6422/70.
14. NOAA. "What Are 'Proxy' Data?" National Climatic Data Center (NOAA), January 1, 2000. https://www.ncdc.noaa.gov/news/what-are-proxy-data.
15. IPCC AR5 WGI, Figure 5.7. https://www.ipcc.ch/site/assets/uploads/2018/02/WG1AR5_Chapter05_FINAL.pdf.
16. Voosen, Paul. "Record-shattering 2.7-million-year-old ice core reveals start of the ice ages." *Science*, August 15, 2017. https://www.sciencemag.org/news/2017/08/record-shattering-27-million-year-old-ice-core-reveals-start-ice-ages.
17. Fergus, Glen. Adapted from https://en.wikipedia.org/wiki/File:All_palaeotemps.png.

CHAPTER 2

1. Banks, P., J. M. Cornwall, F. Dyson, S. Koonin, C. Max, G. Macdonald, S. Ride, et al. "Small Satellites and RPAS in Global-Change Research Summary and Conclusions." JASON, The Mitre Corporation, January 1992. https://fas.org/irp/agency/dod/jason/smallsats.pdf.
2. National Aeronautics and Space Administration. "Measuring Earth's Albedo." NASA Earth Observatory, 2014. https://earthobservatory.nasa.gov/images/84499/measuring-earths-albedo.
3. Image courtesy of P. Goode et al., Big Bear Solar Observatory.
4. Flatte, S., S. Koonin, and G. MacDonald. "Global Change and the Dark of the Moon." JASON, The Mitre Corporation, 1992. https://fas.org/irp/agency/dod/jason/dark.pdf.
5. Palle, E., P. R. Goode, P. Montañés-Rodríguez, A. Shumko, B. Gonzalez-Merino, C. Martinez Lombilla, F. Jimenez-Ibarra, et al. "Earth's Albedo Variations 1998–2014 as Measured from Ground-Based Earthshine Observations." *Geophysical Research Letters*, May 5, 2016. https://agupubs.onlinelibrary.wiley.com/doi/full/10.1002/2016GL068025.
6. Turnbull, Margaret C., et al. *Astrophysical Journal* 644 (2006): 551. https://iopscience.iop.org/article/10.1086/503322/meta.

7. Clough, G. Wayne. "Antarctica!" *Smithsonian Magazine*, May 1, 2010. https://www.smithsonianmag.com/arts-culture/antarctica-13629809/.
8. Harde, Hermann. "Radiation Transfer Calculations and Assessment of Global Warming by CO_2." *International Journal of Atmospheric Sciences*, March 20, 2017. https://www.hindawi.com/journals/ijas/aip/9251034/.
9. Baum, Rudy M. "Future Calculations." Science History Institute, April 19, 2019. https://www.sciencehistory.org/distillations/future-calculations.
10. Simulations courtesy of W. A. van Wijngaarden and W. Happer, corresponding to a clear atmosphere with a typical mid-latitude temperature profile and a surface temperature of 288.7 K. The methodology is described in https://arxiv.org/pdf/2006.03098.pdf.
11. NASA. "Global Effects of Mount Pinatubo." NASA, 2020. https://earthobservatory.nasa.gov/images/1510/global-effects-of-mount-pinatubo.
12. Iverson, Nels A., et al. "The first physical evidence of subglacial volcanism under the West Antarctic Ice Sheet." *Nature*, September 13, 2017. https://www.nature.com/articles/s41598-017-11515-3.
13. CMIP5 forcings downloaded from https://data.giss.nasa.gov/modelforce/.

CHAPTER 3

1. Scripps Institution of Oceanography, UCSD. "Home: Scripps CO_2 Program." Scripps CO_2 Program, 2020. https://scrippsco2.ucsd.edu/.
2. Amos, Jonathan. "Scientists estimate Earth's total carbon store." BBC.com, October 1, 2019. https://www.bbc.com/news/science-environment-49899039.
3. Estimates from IPCC AR5, Chapter 6, Figure 6.1. https://www.ipcc.ch/site/assets/uploads/2018/02/WG1AR5_Chapter06_FINAL.pdf.
4. Olivier, J.G.J., and J.A.H.W. Peters. "Trends In Global Co_2 And Total Greenhouse Gas Emissions." PBL Netherlands Environmental Assessment Agency, The Hague, May 26, 2020. https://www.pbl.nl/sites/default/files/downloads/pbl-2020-trends-in-global-co2-and-total-greenhouse-gas-emissions-2019-report_4068.pdf.
5. Berner, Robert A., and Zavareth Kothavala. "Geocarb III: A Revised Model of Atmospheric CO_2 over Phanerozoic Time." *American Journal of Science* 301 (2001): 182–204. http://www.ajsonline.org/content/301/2/182.abstract.
6. NOAA. "Atmospheric CO_2 Growth Rates: Decadal Average Annual Growth Rates, Mauna Loa Observatory (MLO), 1960–2019." CO_2 Earth, November 2020. https://www.co2.earth/co2-acceleration.
7. IPCC Fifth Assessment Report (AR5), WGI (The Physical Science Basis), WGI Box 6.1 Figure 1.
8. National Research Council. *Radiative Forcing of Climate Change: Expanding the Concept and Addressing Uncertainties*. Washington, DC: The National Academies Press, 2005: Figure 3–4. https://www.nap.edu/read/11175/chapter/5#73.

9. US Department of Commerce, NOAA. "Global Monitoring Laboratory—Data Visualization." NOAA Earth System Research Laboratories, October 1, 2005. https://www.esrl.noaa.gov/gmd/dv/iadv/graph.php?code=MLO&program=ccgg&type=ts.
10. Motavalli, Jim. "Climate Change Mitigation's Best-Kept Secret." Climate Central, February 1, 2015. https://www.climatecentral.org/news/climate-change-mitigations-best-kept-secret-18613.
11. van Vuuren, D. P., J. Edmonds, M. Kainuma, et al. "The Representative Concentration Pathways: An Overview." *Climatic Change* 109, 5 (2011). https://doi.org/10.1007/s10584-011-0148-z.
12. O'Neill, B. C., E. Kriegler, K. Riahi, et al. "A New Scenario Framework for Climate Change Research: The Concept of Shared Socioeconomic Pathways." *Climatic Change* 122, 387–400 (2014). https://doi.org/10.1007/s10584-013-0905-2.
13. van Vuuren, D. P., et al. "The Representative Concentration Pathways: An Overview," Figure 11. https://doi.org/10.1007/s10584-011-0148-z.
14. Burgess, Matthew G., et al. "IPCC baseline scenarios have over-projected CO_2 emissions and economic growth." Environmental Research Letters, November 25, 2020. https://doi.org/10.1088/1748-9326/abcdd2.

CHAPTER 4

1. Koonin, S. E., T. A. Tombrello, and G. Fox. "A 'Hybrid' R-Matrix-Optical Model Parametrization of the $^{12}C(\alpha, \gamma)^{16}O$ Cross Section." Nuclear Physics A, available online October 26, 2002. https://www.sciencedirect.com/science/article/abs/pii/0375947474907155.
2. Koonin, Steven E. 1985. *Computational Physics*. Benjamin-Cummings. Koonin, Steven E., and Dawn C. Meredith. 2018. *Computational Physics: Fortran Version*. Boca Raton: CRC Press. https://www.amazon.com/Computational-Physics-Fortran-Steven-Koonin-ebook/dp/B07B9YXMZ8.
3. Abarbanel, H., P. Collela, A. Despain, S. Koonin, C. Leith, H. Levine, G. MacDonald, et al. "CHAMMP Review." fas.org. JASON, The Mitre Corporation, 1992. https://fas.org/irp/agency/dod/jason/chammp.pdf.
4. Buizza, R., and M. Leutbecher. "The Forecast Skill Horizon: ECMWF Technical Memoranda." European Centre for Medium-Range Weather Forecasts, June 2015. https://www.ecmwf.int/en/elibrary/8450-forecast-skill-horizon.
5. Adapted from Figure 5.2 of Kendal McGuffie and Ann Henderson-Sellars. *The Climate Modelling Primer, 4th Edition*. Hoboken, New Jersey: Wiley-Blackwell, 2014. https://www.wiley.com/en-us/The+Climate+Modelling+Primer%2C+4th+Edition-p-9781119943372.
6. Schneider, T., J. Teixeira, C. Bretherton, et al. "Climate goals and computing the future of clouds." *Nature Clim Change* 7 (2017): 3–5. https://doi.org/10.1038/nclimate3190.
7. Sellar, A. A., C. G. Jones, J. P. Mulcahy, et al. (2019). "UKESM1: Description and eEvaluation of the U.K. Earth System Model." *Journal of Advances in Modeling Earth Systems* 11 (2019): 4513–4558. https://doi.org/10.1029/2019MS001739

8. Hourdin, Frédéric, Thorsten Mauritsen, Andrew Gettelman, Jean-Christophe Golaz, Venkatramani Balaji, Qingyun Duan, Doris Folini, et al. "The Art and Science of Climate Model Tuning." *Bulletin of the American Meteorological Society* 98 (2017): 589–602. https://journals.ametsoc.org/bams/article/98/3/589/70022/The-Art-and-Science-of-Climate-Model-Tuning.
9. Mauritsen, Thorsten, Jürgen Bader, Tobias Becker, Jörg Behrens, Matthias Bittner, Renate Brokopf, Victor Brovkin, et al. "Developments in the MPI-M Earth System Model Version 1.2 (MPI-ESM1.2) and Its Response to Increasing CO_2." *Journal of Advances in Modeling Earth Systems* 11 (2019): 998–1038. https://agupubs.onlinelibrary.wiley.com/doi/full/10.1029/2018MS001400.
10. Coupled Model Intercomparison Project (CMIP). "A Short Introduction to Climate Models—CMIP & CMIP6." World Climate Research Programme, 2020. https://www.wcrp-climate.org/wgcm-cmip.
11. Voosen, Paul. "Climate scientists open up their black boxes to scrutiny." *Science*, October 2016. https://science.sciencemag.org/content/354/6311/401; Held, Isaac. "73. Tuning to the global mean temperature record." *Isaac Held's Blog*, GFDL, Princeton University, November 28, 2016. https://www.gfdl.noaa.gov/blog_held/73-tuning-to-the-global-mean-temperature-record/.
12. Adapted from IPCC Fifth Assessment Report (AR5), WGI (The Physical Science Basis), Figure 10.1.
13. IPCC. AR5 WGI, 887.
14. Trenberth, Kevin, Rong Zhang, and National Center for Atmospheric Research Staff (eds.). "The Climate Data Guide: Atlantic Multi-decadal Oscillation (AMO)." NCAR—Climate Data Guide, January 10, 2019. https://climatedataguide.ucar.edu/climate-data/atlantic-multi-decadal-oscillation-amo.
15. Data from https://www.psl.noaa.gov/data/correlation/amon.us.long.data.
16. IPCC. AR5 WGI, 801.
17. IPCC. AR5 WGI, 9.5.3.6.
18. Papalexiou, S. M., C. R. Rajulapati, M. P. Clark, and F. Lehner "Robustness of CMIP6 Historical Global Mean Temperature Simulations: Trends, Long-Term Persistence, Autocorrelation, and Distributional Shape." *Earth's Future*, September 2020. https://agupubs.onlinelibrary.wiley.com/doi/epdf/10.1029/2020EF001667.
19. Nijsse, Femke, J. M., Peter M. Cox, and Mark S. Williamson. "Emergent constraints on transient climate response (TCR) and equilibrium climate sensitivity (ECS) from historical warming in CMIP5 and CMIP6 models." *Earth System Dynamics* 11 (2020): 737–750. https://esd.copernicus.org/articles/11/737/2020/.
20. Maher, Nicola, Flavio Lehner, and Jochem Marotzke. "Quantifying the role of internal variability in the temperature we expect to observe in the coming decades." *Environmental Research Letters*, May 12, 2020. https://iopscience.iop.org/article/10.1088/1748-9326/ab7d02.
21. National Research Council. *Carbon Dioxide and Climate: A Scientific Assessment*. Washington, DC: The National Academies Press, 1979. https://www.nap.edu/catalog/12181/carbon-dioxide-and-climate-a-scientific-assessment.

22. Climate Research Board. "Carbon Dioxide and Climate: A Scientific Assessment." National Academy of Sciences, 1979. https://www.bnl.gov/envsci/schwartz/charney_report1979.pdf, page 16.
23. Hausfather, Zeke. "CMIP6: The Next Generation of Climate Models Explained." Carbon Brief, December 2, 2019. https://www.carbonbrief.org/cmip6-the-next-generation-of-climate-models-explained.
24. Tokarska, Katarzyna B., Martin B. Stolpe, Sebastian Sippel, Erich M. Fischer, Christopher J. Smith, Flavio Lehner, and Reto Knutti. "Past Warming Trend Constrains Future Warming in CMIP6 Models." *Science Advances*, March 18, 2020. https://advances.sciencemag.org/content/6/12/eaaz9549.
25. Femke, J. M., et al. "Emergent constraints on transient climate response (TCR)."
26. Zhu, Jiang, Christopher J. Poulsen, and Bette L. Otto-Bliesner. "High Climate Sensitivity in CMIP6 Model Not Supported by Paleoclimate." *Nature News*, April 30, 2020. https://www.nature.com/articles/s41558-020-0764-6.
27. Zelinka, Mark D., Timothy A. Myers, Daniel T. McCoy, Stephen Po-Chedley, Peter M. Caldwell, Paulo Ceppi, Stephen A. Klein, and Karl E. Taylor. "Causes of Higher Climate Sensitivity in CMIP6 Models." *Geophysical Research Letters*, January 16, 2020. https://agupubs.onlinelibrary.wiley.com/doi/full/10.1029/2019GL085782.
28. Meehl, Gerald A., Catherine A. Senior, Veronika Eyring, Gregory Flato, Jean-Francois Lamarque, Ronald J. Stouffer, Karl E. Taylor, and Manuel Schlund. "Context for Interpreting Equilibrium Climate Sensitivity and Transient Climate Response from the CMIP6 Earth System Models." *Science Advances*, June 1, 2020. https://advances.sciencemag.org/content/6/26/eaba1981.
29. National Center for Atmospheric Research. "Increased Warming in Latest Generation of Climate Models Likely Caused by Clouds." Phys.org, June 24, 2020. https://phys.org/news/2020-06-latest-climate-clouds.html.
30. Mauritsen, Thorsten, and Erich Roeckner. "Tuning the MPI-ESM1.2 Global Climate Model to Improve the Match with Instrumental Record Warming by Lowering Its Climate Sensitivity." *Journal of Advances in Modeling Earth Systems* 12 (2020): e2019MS002037. https://agupubs.onlinelibrary.wiley.com/doi/full/10.1029/2019MS002037.
31. Reprinted with permission of AAAS from Tokarska, Katarzyna B., Martin B. Stolpe, Sebastian Sippel, Erich M. Fischer, Christopher J. Smith, Flavio Lehner, and Reto Knutti. "Past Warming Trend Constrains Future Warming in CMIP6 Models." *Science Advances* 6 (2020): eaaz9549. https://advances.sciencemag.org/content/6/12/eaaz9549 © The Authors, some rights reserved; exclusive licensee American Association for the Advancement of Science. Distributed under a Creative Commons Attribution NonCommercial License 4.0 (CC BY-NC) http://creativecommons.org/licenses/by-nc/4.0/.
32. Lewis, Nicholas, and Judith Curry. "The Impact of Recent Forcing and Ocean Heat Uptake Data on Estimates of Climate Sensitivity." *Journal of Climate* 31 (2018): 6051–6071. https://journals.ametsoc.org/jcli/article/31/15/6051/92230/The-Impact-of-Recent-Forcing-and-Ocean-Heat-Uptake.

33. Sherwood, S.C., M.J. Webb, J.D. Annan, K.C. Armour, P.M. Forster, J.C. Hargreaves, et al. "An assessment of Earth's climate sensitivity using multiple lines of evidence." *Reviews of Geophysics* 58 (2020): e2019RG000678. https://agupubs.onlinelibrary.wiley.com/doi/full/10.1029/2019RG000678.
34. National Research Council of the National Academies. *Climate Intervention: Reflecting Sunlight to Cool Earth*: "Comparison of Some Basic Risks Associated with Albedo Modification." The National Academies Press, 2015. https://www.nap.edu/read/18988/chapter/4#39.
35. Kravitz, B., A. Robock, S. Tilmes, O. Boucher, J. M. English, P. J. Irvine, A. Jones, et al. "The Geoengineering Model Intercomparison Project Phase 6 (GeoMIP6): Simulation design and preliminary results." *Geoscientific Model Development.* 8 (2015): 3379–3392. https://gmd.copernicus.org/articles/8/3379/2015/.
36. Lewis, N. "Objectively Combining Climate Sensitivity Evidence." *Climate Dynamics* 60 (2023): 3139–3165. https://doi.org/10.1007/s00382-022-06468-x.
37. Voosen, Paul. "Use of 'too hot' climate models exaggerates impacts of global warming." *Science,* May 4, 2022. https://www.science.org/content/article/use-too-hot-climate-models-exaggerates-impacts-global-warming.
38. Palmer, Tim, and Bjorn Stevens. "The scientific challenge of understanding and estimating climate change." *PNAS: Proceedings of the National Academy of Sciences* 116, no. 49 (December 2, 2019): 24390–24395. https://doi.org/10.1073/pnas.1906691116.
39. Stevens, Bjorn. "2022 Michio Yanai Distinguished Lecture." UCLA, May 5, 2022. https://atmos.ucla.edu/2022-michio-yanai-distinguished-lecture/.
40. Nissan, H., L. Goddard, E.C. de Perez, et al. "On the use and misuse of climate change projections in international development." *WIREs Climate Change* 10, no. 3 (May/June 2019): e579. https://doi.org/10.1002/wcc.579.
41. Jeevanjee, Nadir. "Review of '*Unsettled.*'" C-Change Conversations, June 2021. https://c-changeconversations.org/review-of-unsettled/.
42. See https://leap.columbia.edu/.

CHAPTER 5

1. IPCC. AR5 WGI Section 2.6.2.2.
2. IPCC. AR5 WGI Section 2.6.2.3.
3. IPCC. AR5 WGI Section 2.6.2.4.
4. IPCC. AR5 WGI Section 2.6.4.
5. National Academies of Sciences, Engineering, and Medicine. 2016. *Attribution of Extreme Weather Events in the Context of Climate Change.* The National Academies Press. https://www.nap.edu/catalog/21852/attribution-of-extreme-weather-events-in-the-context-of-climate-change.
6. Fountain, Henry. "Scientists Link Hurricane Harvey's Record Rainfall to Climate Change." *New York Times,* December 13, 2017. https://www.nytimes.com/2017/12/13/climate/hurricane-harvey-climate-change.html.

7. Risser, M. D., and M. F. Wehner. "Attributable human-induced changes in the likelihood and magnitude of the observed extreme precipitation during Hurricane Harvey." *Geophysical Research Letters* 44 (2017): 12, 457–12, 464. https://agupubs.onlinelibrary.wiley.com/doi/full/10.1002/2017GL075888.
8. Intergovernmental Panel on Climate Change (IPCC). 2012. *Managing the Risks of Extreme Events and Disasters to Advance Climate Change Adaptation*. Cambridge and New York: Cambridge University Press. https://www.ipcc.ch/site/assets/uploads/2018/03/SREX_Full_Report-1.pdf.
9. World Meteorological Society. Frequently Asked Questions (FAQ). Accessed November 20, 2020. https://www.wmo.int/pages/prog/wcp/ccl/faq/faq_doc_en.html.
10. IPCC. AR5 WGI Section 2.6.1.
11. USGCRP. *Climate Science Special Report: Fourth National Climate Assessment, Volume I*: Executive Summary, Figure 5.
12. USGCRP. *Climate Science Special Report: Fourth National Climate Assessment, Volume I*: Chapter 6: Temperature Changes in the United States, Figure 6.3.
13. Meehl, G. A., C. Tebaldi, G. Walton, D. Easterling, and L. McDeaniel. "Relative increase of record high maximum temperatures compared to record low minimum temperatures in the U.S." *Geophysical Research Letters* 36, (2009): L23701.
14. Meehl, G. A., C. Tebaldi, and D. Adams-Smith. "US daily temperature records past, present, and future." Proceedings of the National Academy of Sciences, 2016.
15. Meehl, et al. "US daily temperature records past, present, and future."
16. Meehl, et al. "US daily temperature records past, present, and future."
17. USGCRP. *Climate Science Special Report: Fourth National Climate Assessment, Volume I*: "About This Report." https://science2017.globalchange.gov/chapter/front-matter-about/.
18. National Academies of Sciences. "Review of the Draft Climate Science Special Report." The National Academies Press, March 14, 2017. https://www.nap.edu/catalog/24712/review-of-the-draft-climate-science-special-report.
19. Wuebbles, Donald, David Fahey, and Kathleen Hibbard (coordinating lead authors). "U.S. Global Change Research Program Climate Science Special Report (CSSR)." USGCRP, December 2016. https://biotech.law.lsu.edu/blog/Draft-of-the-Climate-Science-Special-Report.pdf.
20. Kuperberg, Michael, and CSSR Writing Team. Letter to Dr. Philip Mote (Chair) and the NAS Committee to Review the Draft Climate Science Special Report. November 2, 2017. https://science2017.globalchange.gov/PDFs/CSSR-NASresponse_110217.pdf.
21. Borenstein, Seth, and Nicky Forster. "Heat Records Falling Twice as Often as Cold Ones, AP Finds." Associated Press, March 19, 2019. https://apnews.com/7d00e38b9ba1470fa526b1da739c5da8.
22. Data from "Global Temperature Report," Earth System Science Center, University of Alabama, November 2023. https://www.nsstc.uah.edu/climate/.
23. Copernicus. "Aerosols: Are SO2 emissions reductions contributing to global warming?" Copernicus.eu, August 1, 2023. https://atmosphere.copernicus.eu/aerosols-are-so2-emissions-reductions-contributing-global-warming.

24. IPCC. AR6 WGI 11.3.2.
25. EPA. "Climate Change Indicators: Heat Waves (Figure 3. Heat Wave Index)." July 2022. https://www.epa.gov/climate-indicators/climate-change-indicators-heat-waves.

CHAPTER 6

1. Mack, Eric. "This Era of Deadly Hurricanes Was Supposed to Be Temporary. Now It's Getting Worse." *Forbes*, October 7, 2020. https://www.forbes.com/sites/ericmack/2020/10/07/this-era-of-deadly-hurricanes-was-supposed-to-be-temporary-now-its-getting-worse/.
2. US Department of Commerce, NOAA. "Saffir-Simpson Hurricane Scale." National Weather Service, June 14, 2019. https://www.weather.gov/mfl/saffirsimpson.
3. Climate Prediction Center Internet Team. "Background Information: North Atlantic Hurricane Season." NOAA Center for Weather and Climate Prediction, Climate Prediction Center, May 22, 2019. https://www.cpc.ncep.noaa.gov/products/outlooks/Background.html.
4. Villarini, Gabriele, and Gabriel A. Vecchi. "North Atlantic Power Dissipation Index (PDI) and Accumulated Cyclone Energy (ACE): Statistical Modeling and Sensitivity to Sea Surface Temperature Changes." *Journal of Climate* 25 (2012): 625–637. https://journals.ametsoc.org/jcli/article/25/2/625/33791/North-Atlantic-Power-Dissipation-Index-PDI-and.
5. Emanuel, K. A. "Environmental factors affecting tropical cyclone power dissipation." *Journal of Climate* 20 (2007): 5497–5509. https://doi.org/10.1175/2007JCLI1571.1.
6. Kossin, J. "Hurricane intensification along United States coast suppressed during active hurricane periods." *Nature* 541 (2017): 390–393. https://doi.org/10.1038/nature20783.
7. Evan, Amato T., Cyrille Flamant, Stephanie Fiedler, and Owen Doherty. "An Analysis of Aeolian Dust in Climate Models." *Geophysical Research Letters* 41 (2014): 5996–6001. https://agupubs.onlinelibrary.wiley.com/doi/full/10.1002/2014GL060545.
8. Vecchi, G. A., T. L. Delworth, H. Murakami, et al. "Tropical cyclone sensitivities to CO_2 doubling: roles of atmospheric resolution, synoptic variability and background climate changes." *Climate Dynamics* 53 (2019): 5999–6033. https://link.springer.com/article/10.1007/s00382-019-04913-y.
9. Department of Atmospheric Science, Tropical Meteorology Project, Colorado State University. "North Atlantic Ocean Historical Tropical Cyclone Statistics." 2020. http://tropical.atmos.colostate.edu/Realtime/index.php?arch&loc=northatlantic.
10. Emanuel, K. A. 2016. Update to data originally published in: Emanuel, K. A. 2007. "Environmental factors affecting tropical cyclone power dissipation." *J. Climate* 20 (22): 5497–5509. https://www.epa.gov/climate-indicators/climate-change-indicators-tropical-cyclone-activity.
11. Melillo, Jerry M., Terese (T. C.) Richmond, and Gary W. Yohe, eds. 2014. *Climate Change Impacts in the United States: The Third National Climate Assessment*. U.S. Global Change Research Program, 841. https://doi.org/10.7930/J0Z31WJ2., 41.

12. Knutson, Thomas R., John L. McBride, Johnny Chan, Kerry Emanuel, Greg Holland, Chris Landsea, Isaac Held, James P. Kossin, A. K. Srivastava, and Masato Sugi. "Tropical Cyclones and Climate Change." *Nature Geoscience* 3 (2010): 157–163. https://www.nature.com/articles/ngeo779.

13. USGCRP. *Climate Change Impacts in the United States: The Third National Climate Assessment.* "Appendix 3: Climate Science Supplement," page 769. http://nca2014.globalchange.gov/report/appendices/climate-science-supplement.

14. Villarini, Gabriele, and Gabriel A. Vecchi. "North Atlantic Power Dissipation Index (PDI) and Accumulated Cyclone Energy (ACE): Statistical Modeling and Sensitivity to Sea Surface Temperature Changes." *Journal of Climate* 25 (2012): 625–637. https://journals.ametsoc.org/jcli/article/25/2/625/33791/North-Atlantic-Power-Dissipation-Index-PDI-and.

15. Data prior to 2009 from Reference 14; data post 2008 as derived by Ryan Maue at www.climatlas.com from HURDAT2 data, https://www.nhc.noaa.gov/data/#hurdat.

16. USGCRP. *Climate Science Special Report: Fourth National Climate Assessment, Volume I: Chapter 9.* https://science2017.globalchange.gov/chapter/9/.

17. National Academies of Sciences. "Review of the Draft Climate Science Special Report." The National Academies Press, March 14, 2017. https://www.nap.edu/catalog/24712/review-of-the-draft-climate-science-special-report.

18. Knutson, Thomas, Suzana J. Camargo, Johnny C. L. Chan, Kerry Emanuel, Chang-Hoi Ho, James Kossin, Mrutyunjay Mohapatra, et al. "Tropical Cyclones and Climate Change Assessment: Part I: Detection and Attribution." *Bulletin of the American Meteorological Society* 100 (2019): 1987–2007. https://journals.ametsoc.org/doi/abs/10.1175/BAMS-D-18-0189.1

19. For another recent analysis showing no long-term trends in Atlantic hurricanes over more than a century, see also Loehle, C., and E. Staehling. "Hurricane trend detection." *Natural Hazards* 104 (2020): 1345–1357. https://link.springer.com/article/10.1007/s11069-020-04219-x#Abs1.

20. Rice, Doyle. "Global warming is making hurricanes stronger, study says." *USA Today*, May 18, 2020. https://www.usatoday.com/story/news/nation/2020/05/18/global-warming-making-hurricanes-stronger-study-suggests/5216028002/.

21. Kossin, James P., Kenneth R. Knapp, Timothy L. Olander, and Christopher S. Velden. "Global Increase in Major Tropical Cyclone Exceedance Probability over the Past Four Decades." *PNAS: Proceedings of the National Academy of Sciences* 117, no. 22 (2020): 11975–11980. https://www.pnas.org/content/early/2020/05/12/1920849117.

22. Weinkle, Jessica, Chris Landsea, Douglas Collins, Rade Musulin, Ryan P. Crompton, Philip J. Klotzbach, and Roger Pielke. "Normalized Hurricane Damage in the Continental United States 1900–2017." *Nature Sustainability* 1 (2018): 808–813. https://www.nature.com/articles/s41893-018-0165-2; Pielke, Roger. "Economic 'normalisation' of disaster losses 1998–2020: a literature review and assessment." *Environmental Hazards*, August 5, 2020. https://www.tandfonline.com/doi/abs/10.1080/17477891.2020.1800440?journalCode=tenh20.

23. Knutson, Thomas, Suzana J. Camargo, Johnny C. L. Chan, Kerry Emanuel, Chang-Hoi Ho, James Kossin, Mrutyunjay Mohapatra, Masaki Satoh, Masato Sugi, Kevin Walsh, and Liguang Wu. "Tropical Cyclones and Climate Change Assessment—Part II: Projected Response to Anthropogenic Warming." *Bulletin of the American Meteorological Society* 101 (2020): E303–E322. https://journals.ametsoc.org/doi/pdf/10.1175/BAMS-D-18-0194.1.

24. The 2020 North Atlantic season was active, with a record number of named storms, but was not unprecedented by almost all other measures. Its ACE was exceeded by 12 earlier years, half of those before 1950.

25. NOAA National Centers for Environmental Information (NCEI). "Tornadoes—Annual 2019." National Climatic Data Center, January 2020. https://www.ncdc.noaa.gov/sotc/tornadoes/201913.
26. NOAA NCEI. "Historical Records and Trends: Ratio of (E)F-0 Tornado Reports to Total Reports." National Climatic Data Center, 2020. https://www.ncdc.noaa.gov/climate-information/extreme-events/us-tornado-climatology/trends.
27. Gensini, V. A., and H. E. Brooks. "Spatial trends in United States tornado frequency." *npj Climate and Atmospheric Science* 1 (2018). https://www.nature.com/articles/s41612-018-0048-2.
28. USGCRP. *Climate Science Special Report: Fourth National Climate Assessment, Volume I*: Chapter 9: Extreme Storms. https://science2017.globalchange.gov/chapter/9/.
29. NOAA NCEI. "Historical Records and Trends: Ratio of (E)F-0 Tornado Reports to Total Reports."
30. Brooks, Harold E., and Charles A. Doswell III. "Deaths in the 3 May 1999 Oklahoma City Tornado from a Historical Perspective." *Weather and Forecasting* 17 (2002): 354–361. https://journals.ametsoc.org/waf/article/17/3/354/40162/Deaths-in-the-3-May-1999-Oklahoma-City-Tornado.
31. Pierre-Louis, Kendra. "As Climate Changes, Scientists Try to Unravel the Effects on Tornadoes." *New York Times*, August 8, 2018. https://www.nytimes.com/2018/08/08/climate/tornadoes-climate-change.html.
32. Adapted from IPCC AR6 WGI Table 12.12. https://www.ipcc.ch/report/ar6/wg1/downloads/report/IPCC_AR6_WGI_Chapter12.pdf.
33. IPCC. AR6 WGI 11.7.1.2. https://www.ipcc.ch/report/ar6/wg1/chapter/chapter-11/.
34. IPCC. AR6 WGI TS.2.3.
35. IPCC. AR6 WGI 11.7.1.4.
36. Kossin, James P., Kenneth R. Knapp, Timothy L. Olander, and Christopher S. Velden. "Global Increase in Major Tropical Cyclone Exceedance Probability over the Past Four Decades." *PNAS: Proceedings of the National Academy of Sciences* 117, no. 22 (2020): 11975–11980. https://www.pnas.org/content/early/2020/05/12/1920849117.
37. PNAS. "Correction for 'Global Increase in Major Tropical Cyclone Exceedance Probability over the Past Four Decades.'" *PNAS: Proceedings of the National Academy of Sciences* 117, no. 47 (2020): 29990. https://www.pnas.org/doi/10.1073/pnas.2021573117.

38. Vecchi, G.A., Landsea, C., Zhang, W., et al. "Changes in Atlantic major hurricane frequency since the late-19th century." *Nature Communications* 12, no. 4054 (2021). https://doi.org/10.1038/s41467-021-24268-5.

CHAPTER 7

1. Li, Ye, Eric J. Johnson, and Lisa Zaval. "Local Warming: Daily Temperature Change Influences Belief in Global Warming." *Psychological Science* 22 (2011). https://journals.sagepub.com/doi/abs/10.1177/0956797611400913.
2. Livingston, Ian, and Jordan Tessler. "Everything You Ever Wanted to Know about Snow in Washington, D.C., Updated." *Washington Post*, February 9, 2018. https://www.washingtonpost.com/news/capital-weather-gang/wp/2018/02/09/everything-you-ever-wanted-to-know-about-snow-in-washington-d-c-updated/.
3. Livingston, Ian. "Snowfall Shows a Sharp Long-Term Decline in the Washington Region, but Some Trends Are Surprising." *Washington Post*, November 29, 2018. https://www.washingtonpost.com/weather/2018/11/29/snowfall-shows-sharp-long-term-decline-washington-region-some-trends-are-surprising/.
4. New, Mark, Martin Todd, Mike Hulme, and Phil Jones. "Precipitation measurements and trends in the twentieth century." *International Journal of Climatology* 21 (2001): 1889–1922. https://rmets.onlinelibrary.wiley.com/doi/pdf/10.1002/joc.680.
5. United States Environmental Protection Agency (EPA). "Climate Change Indicators: U.S. and Global Precipitation." December 17, 2016. https://www.epa.gov/climate-indicators/climate-change-indicators-us-and-global-precipitation.
6. IPCC. AR5 WGI, Section B.1 of Summary for Policymakers.
7. Nguyen, Phu, Andrea Thorstensen, Soroosh Sorooshian, Kuolin Hsu, Amir Aghakouchak, Hamed Ashouri, Hoang Tran, and Dan Braithwaite. "Global Precipitation Trends Across Spatial Scales Using Satellite Observations." *Bulletin of the American Meteorological Society* 99 (2018): 689–697. https://journals.ametsoc.org/bams/article/99/4/689/70305/Global-Precipitation-Trends-across-Spatial-Scales.
8. Huntington, Thomas G. "Climate Warming-Induced Intensification of the Hydrologic Cycle: An Assessment of the Published Record and Potential Impacts on Agriculture." *Advances in Agronomy* 109 (2010): 1–53. https://www.sciencedirect.com/science/article/pii/B9780123850409000013.
9. USGCRP. *Climate Science Special Report: Fourth National Climate Assessment, Volume I*: "Chapter 7: Precipitation Change in the United States." https://science2017.globalchange.gov/chapter/7/.
10. EPA. "Climate Change Indicators: U.S. and Global Precipitation."
11. IPCC AR5 WGI Section 2.6.2.1.
12. EPA. "Climate Change Indicators: U.S. and Global Precipitation."
13. Estilow, T. W., A. H. Young, and D. A. Robinson. "A Long-Term Northern Hemisphere Snow Cover Extent Data Record for Climate Studies and Monitoring." *Earth System Science Data* 7 (2015): 137–142. https://essd.copernicus.org/articles/7/137/2015/essd-7-137-2015.html.

14. National Snow and Ice Data Center. "SOTC: Northern Hemisphere Snow." November 1, 2019. https://nsidc.org/cryosphere/sotc/snow_extent.html.
15. Data from NOAA. "Sea Ice and Snow Cover Extent." NOAA National Centers for Environmental Information, December 8, 2020. https://www.ncdc.noaa.gov/snow-and-ice/extent/snow-cover/nhland/0.
16. EPA. "Climate Change Indicators: River Flooding." December 17, 2016. https://www.epa.gov/climate-indicators/climate-change-indicators-river-flooding.
17. Dai, Aiguo, and National Center for Atmospheric Research Staff. "Palmer Drought Severity Index (PDSI)." NCAR—Climate Data Guide, December 12, 2019. https://climatedataguide.ucar.edu/climate-data/palmer-drought-severity-index-pdsi.
18. EPA. "Climate Change Indicators: Drought." December 17, 2016. https://www.epa.gov/climate-indicators/climate-change-indicators-drought.
19. AR5 WGI Figure 5.13. https://www.ipcc.ch/site/assets/uploads/2018/02/Fig5-13.jpg.
20. Cook, Edward R., Connie A. Woodhouse, C. Mark Eakin, David M. Meko, and David W. Stahle. "Long-Term Aridity Changes in the Western United States." *Science*, November 5, 2004. https://science.sciencemag.org/content/306/5698/1015.
21. IPCC AR5 Section 12.5.5.8.1.
22. Erb, M. P., J. Emile-Geay, G. J. Hakim, N. Steiger, and E. J. Steig. "Atmospheric dynamics drive most interannual U.S. droughts over the last millennium." *Science Advances* 6 (2020). https://advances.sciencemag.org/content/6/32/eaay7268.full.
23. The Climate Assessment for the Southwest (CLIMAS). "Southwest Paleoclimate." Accessed July 8, 2020. https://climas.arizona.edu/sw-climate/southwest-paleoclimate.
24. U.S. Global Change Research Program. "Global Climate Change Impacts in the United States 2009 Report (Legacy Site)." Accessed July 8, 2020. https://nca2009.globalchange.gov/southwest-drought-timeline/index.html.
25. CSSR Section 8.1.1.
26. Associated Press. "California's drought is officially over, Gov. Jerry Brown says." CBS News, April 7, 2017. https://www.cbsnews.com/news/calif-gov-jerry-brown-declares-an-end-to-drought/.
27. NOAA National Centers for Environmental Information, Climate at a Glance: Statewide Mapping, published November 2020, retrieved on December 6, 2020, from https://www.ncdc.noaa.gov/cag/statewide/mapping.
28. Freedman, Andrew. "Western wildfires: An 'unprecedented' climate change fueled event, experts say." *Washington Post*, September 11, 2020. https://www.washingtonpost.com/weather/2020/09/11/western-wildfires-climate-change/.
29. Andela, N., D. C. Morton, L. Giglio, Y. Chen, G. R. van der Werf, P. S. Kasibhatla, R. S. DeFries, et al. "A Human-Driven Decline in Global Burned Area." *Science*, June 30, 2017. https://science.sciencemag.org/content/356/6345/1356.
30. Copernicus Atmosphere Monitoring Service. "How Wildfires in the Americas and Tropical Africa in 2020 Compared to Previous Years." European Centre for Medium-Range Weather Forecasts on behalf of the European Commission. Accessed January 19, 2021. https://atmosphere.copernicus.eu/wildfires-americas-and-tropical-africa-2020-compared-previous-years.

31. Voiland, Adam. "Building a Long-Term Record of Fire." NASA, Earth Observatory. Accessed November 23, 2020. https://earthobservatory.nasa.gov/images/145421/building-a-long-term-record-of-fire.
32. Voiland, Adam. "Building a Long-Term Record of Fire." NASA, Earth Observatory. Accessed November 23, 2020. https://earthobservatory.nasa.gov/images/145421/building-a-long-term-record-of-fire.
33. Abatzoglou, John T., and A. Park Williams. "Impact of Anthropogenic Climate Change on Wildfire across Western US Forests." *PNAS: Proceedings of the National Academy of Sciences* 113 (2016). https://www.pnas.org/content/pnas/113/42/11770.full.pdf.
34. National Parks Service. "Wildfire Causes and Evaluations." US Department of the Interior, November 27, 2018. https://www.nps.gov/articles/wildfire-causes-and-evaluation.htm.
35. Miller, R. K., C. B. Field, and K. J. Mach. "Barriers and enablers for prescribed burns for wildfire management in California." *Nature Sustainability* 3 (2020): 101–109. https://www.nature.com/articles/s41893-019-0451-7; Doerr, Stefan, H. and Cristina Santín. "Global trends in wildfire and its impacts: perceptions versus realities in a changing world." *Philos Trans R Soc Lond B Biol Sci.* 371 (2016): 20150345. https://www.ncbi.nlm.nih.gov/pmc/articles/PMC4874420/.
36. AR5, Box TS.4 of the WGI Technical Summary.
37. AR5 WGI, Box 11.2.
38. Carney, Mark. "Mark Carney: Breaking the Tragedy of the Horizon—Climate Change and Financial Stability." bis.org, September 29, 2015. https://www.bis.org/review/r151009a.pdf.
39. Met Office Hadley Centre (National Meterological Service for the UK). https://www.metoffice.gov.uk/hadobs/hadukp/data/seasonal/HadEWP_ssn.dat.
40. Brown, Simon J. "The drivers of variability in UK extreme rainfall." *International Journal Of Climatology* 38 (2018): e119–e130. https://rmets.onlinelibrary.wiley.com/doi/pdf/10.1002/joc.5356.

CHAPTER 8

1. NOAA. "Relative Sea Level Trend 8518750 The Battery, New York." NOAA Tides & Currents, January 1, 2020. https://tidesandcurrents.noaa.gov/sltrends/sltrends_station.shtml.
2. Bianchi, Carlo Nike, et al. "Mediterranean Sea biodiversity between the legacy from the past and a future of change" in *Life in the Mediterranean Sea: A Look at Habitat Changes*. Ed. Noga Stambler. New York: Nova Science Publishers, 2011. http://dueproject.org/en/wp-content/uploads/2019/01/8.pdf
3. Adapted from Rohde, Robert A. Global Warming Art Project. "Post-Glacial Sea Level Rise." Wikimedia Commons, October 9, 2020. https://commons.wikimedia.org/wiki/File:Post-Glacial_Sea_Level.png.
4. NOAA. "Sea Level Trends." NOAA Tides & Currents, January 1, 2020. https://tidesandcurrents.noaa.gov/sltrends/sltrends.html.

5. CSIRO. "Historical Sea Level Changes—Last Few Hundred Years." CSIRO National Collections and Marine Infrastructure (NCMI) Information and Data Centre, October 12, 2020. https://www.cmar.csiro.au/sealevel/sl_hist_few_hundred.html.
6. AVISO+. "Sea Surface Height Products." CNES AVISO+ Satellite Altimetry Data. Accessed December 1, 2020. https://www.aviso.altimetry.fr/en/home.html.
7. NOAA, NESDIS, STAR. "Laboratory for Satellite Altimetry/Sea Level Rise—Global sea level time series." Accessed December 1, 2020. https://www.star.nesdis.noaa.gov/socd/lsa/SeaLevelRise/LSA_SLR_timeseries_global.php.
8. IPCC. AR5 WGI Figure 3.14.
9. IPCC. AR5 WGI Section 3.7.4.
10. Koonin, Steven E. "A Deceptive New Report on Climate." *Wall Street Journal*, November 2, 2017. https://www.wsj.com/articles/a-deceptive-new-report-on-climate-1509660882.
11. IPCC. "Special Report on the Ocean and Cryosphere in a Changing Climate." https://www.ipcc.ch/srocc/.
12. Hay, Carling C., Eric Morrow, Robert E. Kopp, and Jerry X. Mitrovica. "Probabilistic Reanalysis of Twentieth-Century Sea-Level Rise." *Nature* 517 (2015): 481–484. https://www.nature.com/articles/nature14093.
13. Pierrehumbert, Raymond T. "Climate Science Is Settled *Enough*." *Slate*.
14. Frederikse, Thomas, Felix Landerer, Lambert Caron, Surendra Adhikari, David Parkes, Vincent W. Humphrey, Sönke Dangendorf, et al. "The causes of sea-level rise since 1900." *Nature* 584 (2020): 393–397. https://www.nature.com/articles/s41586-020-2591-3.
15. IPCC. SROCC Section 4.2.2.2.6.
16. IPCC. SROCC Summary for Policymakers, Finding B3.1.
17. Hamlington, B. D., A. S. Gardner, E. Ivins, J. T. M. Lenaerts, J. T. Reager, and D. S. Trossman, et al. "Understanding of contemporary regional sea-level change and the implications for the future." *Reviews of Geophysics* 58 (2020): e2019RG000672. https://agupubs.onlinelibrary.wiley.com/doi/abs/10.1029/2019RG000672.
18. Stammer, Detlef, Robert Nichols, Roderik van de Wal, and the GC Sea Level Steering Team. "WCRP Grand Challenge: Regional Sea Level Change and Coastal Impacts Science and Implementation Plan." CLIVAR (Climate and Ocean: Variability, Predictability and Change), World Climate Research Programme, June 8, 2018. http://www.clivar.org/sites/default/files/documents/GC_SeaLevel_Science_and_Implementation_Plan_V2.1_ds_MS.pdf.
19. Historical data from NOAA. "Relative Sea Level Trend 8518750 The Battery, New York."; projections from IPCC AR5 WGI Figure 13.23.
20. Chambers, D. P., M. A. Merrifield, and R. S. Nerem. "Is there a 60-year oscillation in global mean sea level?" *Geophysical Research Letters* 39 (2012). https://agupubs.onlinelibrary.wiley.com/doi/10.1029/2012GL052885.
21. Tada, Grace Mitchell. "The Rising Tide Underfoot." *Hakai Magazine*, November 17, 2020. https://www.hakaimagazine.com/features/the-rising-tide-underfoot/.

22. "Global Sea Level Rise: What It Means to You and Your Business." Zurich.com, May 21, 2019. https://www.zurich.com/en/knowledge/topics/global-risks/global-sea-level-rise-what-it-means-to-you-and-your-business.
23. IPCC. AR6 WGI, Summary for Policymakers, A1.7. https://www.ipcc.ch/report/ar6/wg1/chapter/summary-for-policymakers/.
24. Sweet, W.V., B.D. Hamlington, R.E. Kopp, et al., "2022: Global and Regional Sea Level Rise Scenarios for the United States: Updated Mean Projections and Extreme Water Level Probabilities Along U.S. Coastlines. NOAA Technical Report NOS 01." National Oceanic and Atmospheric Administration, National Ocean Service. https://oceanservice.noaa.gov/hazards/sealevelrise/noaa-nostechrpt01-global-regional-SLR-scenarios-US.pdf.
25. NOAA. "Relative Sea Level Trend 8518750 The Battery, New York." NOAA Tides and Currents. https://tidesandcurrents.noaa.gov/sltrends/sltrends_station.shtml?id=8518750.
26. Harvey, Fiona. "Greenland's ice sheet melting seven times faster than in 1990s." *Guardian*, December 10, 2019. https://www.theguardian.com/environment/2019/dec/10/greenland-ice-sheet-melting-seven-times-faster-than-in-1990s.
27. NASA Jet Propulsion Laboratory. "Greenland, Antarctica Melting Six Times Faster Than in the 1990s." March 16, 2020. https://www.nasa.gov/centers-and-facilities/jpl/greenland-antarctica-melting-six-times-faster-than-in-the-1990s/.
28. Mankoff, Ken, Xavier Fettweis, Anne Solgaard, et al., "Greenland ice sheet mass balance from 1840 through next week." *Earth System Science Data Copernicus* 13, no. 10 (2021): 5001–5025. https://doi.org/10.22008/FK2/OHI23Z.
29. Greenberg, Jon. "Greenland Lost Ice Faster This Decade Than 80 Years Ago." Politifact, November 3, 2021. https://www.politifact.com/factchecks/2021/nov/03/steven-koonin/greenland-lost-ice-faster-decade-80-years-ago/.
30. Koonin, Steven. "Greenland's Melting Ice Is No Cause for Climate-Change Panic." *Wall Street Journal*, February 17, 2022. https://www.wsj.com/articles/greenland-melting-ice-panic-sheets-global-warming-variance-seal-level-rise-climate-change-carbon-fossil-fuel-11645131739.
31. Ruan, R., X. Chen, J. Zhao, et al. "Decelerated Greenland Ice Sheet melt driven by positive summer North Atlantic Oscillation." *Journal of Geophysical Research: Atmospheres* 124 (2019), 7633–7646. https://doi.org/10.1029/2019JD030689.
32. Rignot, Eric, et al. "Wall Street Journal op-ed by Steven Koonin publishes misleading claims about how climate change influences Greenland ice melt." Climate Feedback, February 24, 2022. https://climatefeedback.org/evaluation/wall-street-journal-steven-koonin-publishes-misleading-claims-climate-change-influences-greenland-ice-melt/.
33. Mankoff K., et al. "Disputing Koonin on Greenland's Melting Ice." *Wall Street Journal*, February 27, 2022. https://www.wsj.com/articles/koonin-greenland-ice-loss-melting-climate-change-11645828198.

34. Danish Meteorological Institute, et al. "Polar Portal Season Report 2022." Polar Portal, March 13, 2023. http://polarportal.dk/en/news/2022-season-report/.

CHAPTER 9

1. McMahon, Jeff. "Rise in Climate-Related Deaths Will Surpass All Infectious Diseases, Economist Testifies." *Forbes*, December 26, 2019. https://www.forbes.com/sites/jeffmcmahon/2019/12/27/climate-related-deaths-in-2100-will-surpass-current-mortality-from-all-infectious-diseases-economist-testifies/; Statement of Michael Greenstone to the United States House Committee on Oversight and Reform, Subcommittee on Environment, hearing on "Economics of Climate Change," December 19, 2019. https://epic.uchicago.edu/wp-content/uploads/2019/12/Greenstone-Testimony-12192019-FINAL.pdf.
2. "World Health Statistics 2018 Monitoring Health for the SDGs." World Health Organization, 2020. https://apps.who.int/iris/bitstream/handle/10665/272596/9789241565585-eng.pdf.
3. EM-DAT. "The International Disasters Database." Centre for Research on the Epidemiology of Disasters—CRED. Accessed December 1, 2020. https://www.emdat.be/.
4. UNDRR and CRED. *Human Cost of Disasters*. Centre for Research on the Epidemiology of Disasters and UN Office for Disaster Risk Reduction, 2019. https://www.undrr.org/media/48008/download.
5. Statement of Michael Greenstone to the United States House Committee on Oversight and Reform, hearing on "The Devastating Health Impacts of Climate Change." August 5, 2020. https://epic.uchicago.edu/wp-content/uploads/2020/08/Greenstone_Testimony_08052020.pdf.
6. Carleton, Tamma A., Amir Jina, Michael T. Delgado, Michael Greenstone, Trevor Houser, Solomon M. Hsiang, Andrew Hultgren, et al. "Valuing the Global Mortality Consequences of Climate Change Accounting for Adaptation Costs and Benefits." National Bureau of Economic Research (NBER), July 2020. https://www.nber.org/papers/w27599.
7. Lomborg, Bjørn. *False Alarm: How Climate Change Panic Costs Us Trillions, Hurts the Poor, and Fails to Fix the Planet*. New York: Basic Books, 2020.
8. Ghebreyesus, Tedros Adhanom. "Climate Change Is Already Killing Us." *Foreign Affairs*, March 12, 2020. https://www.foreignaffairs.com/articles/2019-09-23/climate-change-already-killing-us.
9. "Air pollution." World Health Organization (WHO). 2021. https://www.who.int/health-topics/air-pollution.
10. Flavelle, Christopher. "Climate Change Threatens the World's Food Supply, United Nations Warns." *New York Times*, August 8, 2019. https://www.nytimes.com/2019/08/08/climate/climate-change-food-supply.html.

11. IPCC. "Climate Change and Land." https://www.ipcc.ch/srccl/https://www.ipcc.ch/srccl/.
12. Hasell, Joe, and Max Roser. "Famines." Our World in Data, last modified December 7, 2017. https://ourworldindata.org/famines.
13. Nielsen, R. L. (Bob). "Historical Corn Grain Yields in the U.S." Corny News Network, Purdue University, April 2020. https://www.agry.purdue.edu/ext/corn/news/timeless/YieldTrends.html.
14. Hille, Karl. "Rising Carbon Dioxide Levels Will Help and Hurt Crops." NASA, May 3, 2016. https://www.nasa.gov/feature/goddard/2016/nasa-study-rising-carbon-dioxide-levels-will-help-and-hurt-crops.
15. Dusenge, M. E., A. G. Duarte, and D. A. Way. "Plant carbon metabolism and climate change: elevated CO_2 and temperature impacts on photosynthesis, photorespiration and respiration." *New Phytologist* 221 (2019): 32–49. https://nph.onlinelibrary.wiley.com/doi/full/10.1111/nph.15283.
16. Zhu, Zaichun, Shilong Piao, Ranga B. Myneni, Mengtian Huang, Zhenzhong Zeng, Josep G. Canadell, Philippe Ciais, et al. "Greening of the Earth and Its Drivers." *Nature Climate Change* 6 (2016): 791–795. https://www.nature.com/articles/nclimate3004.
17. IPCC SRCCL, Figure TS.9.
18. IPCC SRCCL Section 5.2.2.1
19. Iizumi, Toshichika, Hideo Shiogama, Yukiko Imada, Naota Hanasaki, Hiroki Takikawa, and Motoki Nishimori. "Crop production losses associated with anthropogenic climate change for 1981–2010 compared with preindustrial levels." *International Journal of Climatology* 38 (2018): 5405–5417. https://rmets.onlinelibrary.wiley.com/doi/10.1002/joc.5818.
20. Food and Agriculture Organization of the United Nations (FAO). "Crops Processed." FAO.org, July 29, 2020. http://www.fao.org/faostat/en/#data/QD/metadata.
21. Food And Agriculture Organization of the United Nations. "FAO Cereal Supply and Demand Brief." December 3, 2020. http://www.fao.org/worldfoodsituation/csdb/en/.
22. USDA Economic Research Service. "Inflation-adjusted price indices for corn, wheat, and soybeans show long-term declines." April 2019. https://www.ers.usda.gov/data-products/chart-gallery/gallery/chart-detail/?chartId=76964
23. Gregorian, Dareh. "Federal Report Says Climate Change Will Wallop U.S. Economy." NBC News, November 24, 2018. https://www.nbcnews.com/news/us-news/federal-report-says-climate-change-will-wallop-u-s-economy-n939521.
24. Shaw, Adam. "Climate Report Warns of Grim Economic Consequences, Worsening Weather Disasters in US." Fox News, November 24, 2018. https://www.foxnews.com/politics/climate-report-warns-of-grim-economic-consequences-more-weather-disasters-in-us.
25. Crooks, Ed. "Climate Change Could Cost US Billions, Report Finds." *Financial Times*, November 23, 2018. https://www.ft.com/content/216b5ed2-ef68-11e8-89c8-d36339d835c0.

26. Davenport, Coral, and Kendra Pierre-Louis. "U.S. Climate Report Warns of Damaged Environment and Shrinking Economy." *New York Times*, November 23, 2018. https://www.nytimes.com/2018/11/23/climate/us-climate-report.html.
27. Diffenbaugh, Noah S., Daniel L. Swain, and Danielle Touma. "Anthropogenic Warming Has Increased Drought Risk in California." Proceedings of the National Academy of Sciences of the United States of America, National Academy of Sciences, March 31, 2015. https://www.ncbi.nlm.nih.gov/pmc/articles/PMC4386330/.
28. Allen, Robert J., and Ray G. Anderson. "21st Century California Drought Risk Linked to Model Fidelity of the El Niño Teleconnection." *npj Climate and Atmospheric Science* 1 (2018). https://www.nature.com/articles/s41612-018-0032-x.
29. Adapted from IPCC WGII AR5, Figure 10.1.
30. Tol, Richard S. J. "The Economic Impacts of Climate Change." *Review of Environmental Economics and Policy* 12 (2018). https://www.journals.uchicago.edu/doi/10.1093/reep/rex027.
31. Hsiang, Solomon, Robert Kopp, Amir Jina, James Rising, Michael Delgado, Shashank Mohan, D. J. Rasmussen, et al. "Estimating Economic Damage from Climate Change in the United States." *Science*, June 30, 2017. https://science.sciencemag.org/content/356/6345/1362.full.
32. Koonin, Steven. "The Climate Won't Crash the Economy." *Wall Street Journal*, November 27, 2018. https://www.wsj.com/articles/the-climate-wont-crash-the-economy-1543276899.
33. Jina, Amir. "Will Global Warming Shrink U.S. GDP 10%? It's Complicated, Says The Person Who Made The Estimate." *Forbes*, December 5, 2018. https://www.forbes.com/sites/ucenergy/2018/12/05/will-global-warming-shrink-u-s-gdp-10-its-complicated-says-the-person-who-made-the-estimate.
34. Zhao, Qi, et al. "Global, regional, and national burden of mortality associated with non-optimal ambient temperatures from 2000 to 2019: A three-stage modelling study." *The Lancet Planetary Health* 5, no. 7 (2021): e415–e425. https://www.thelancet.com/journals/lanplh/article/PIIS2542-5196(21)00081-4/fulltext.
35. IPCC. AR6 WGII Section 16.6.3. https://www.ipcc.ch/report/ar6/wg2/chapter/chapter-16/.
36. Burke, M., S.M. Hsiang, and E. Miguel. "Global non-linear effect of temperature on economic production." *Nature* 527 (2015): 235–239. https://doi.org/10.1038/nature15725.
37. Burke, M., W.M. Davis, and N.S. Diffenbaugh. "Large potential reduction in economic damages under UN mitigation targets." *Nature* 557 (2018): 549–553. https://doi.org/10.1038/s41586-018-0071-9.
38. Lomborg, Bjorn. "Welfare in the 21st century: Increasing development, reducing inequality, the impact of climate change, and the cost of climate policies." *Technological Forecasting and Social Change* 156 (2020): 119981. https://www.sciencedirect.com/science/article/pii/S0040162520304157.

39. Tol, Richard S.J. "A meta-analysis of the total economic impact of climate change." Tinbergen Institute, 2022. https://papers.tinbergen.nl/22056.pdf
40. Dietz, Simon, James Rising, Thomas Stoerk, and Gernot Wagner. "Economic impacts of tipping points in the climate system." *PNAS: Proceedings of the National Academy of Sciences* 118, no. 34 (2021): e2103081118. https://www.pnas.org/doi/full/10.1073/pnas.2103081118.
41. See https://www.whitehouse.gov/cea/written-materials/2023/03/14/methodologies-and-considerations-for-integrating-the-physical-and-transition-risks-of-climate-change-into-macroeconomic-forecasting-for-the-presidents-budget/.
42. Ibid.
43. See https://crudata.uea.ac.uk/~timo/diag/tempdiag.htm.
44. Ritchie, Hannah, Lucas Rodés-Guirao, Edouard Mathieu, Marcel Gerber, Esteban Ortiz-Ospina, Joe Hasell, and Max Roser. "Population Growth." OurWorldInData.org, 2023. https://ourworldindata.org/population-growth.
45. Dattani, Saloni, Lucas Rodés-Guirao, Hannah Ritchie, Esteban Ortiz-Ospina, and Max Roser. "Life Expectancy." OurWorldInData.org, 2023. https://ourworldindata.org/life-expectancy.
46. Roser, Max, and Esteban Ortiz-Ospina. "Literacy." OurWorldInData.org, 2018. https://ourworldindata.org/literacy.
47. Roser, Max, Pablo Arriagada, Joe Hasell, Hannah Ritchie, and Esteban Ortiz-Ospina. "Economic Growth." OurWorldInData.org, 2023. https://ourworldindata.org/economic-growth.
48. Muehlhauser, Luke. "How big a deal was the Industrial Revolution?" Accessed January 18, 2024. https://lukemuehlhauser.com/industrial-revolution/.
49. Golkany, Indur M. "Deaths and Death Rates from Extreme Weather Events: 1900–2008." *Journal of American Physicians and Surgeons* 14, no. 4 (2009): 102–109. https://www.jpands.org/vol14no4/goklany.pdf.
50. Pielke, Roger. "Weather and Climate Disaster Losses So Far in 2022, Still Not Getting Worse." The Honest Broker, July 20, 2022. https://rogerpielkejr.substack.com/p/weather-and-climate-disaster-losses. Updated through 2022 from Aon, "2023 Weather, Climate and Catastrophe Insight." https://www.aon.com/getmedia/f34ec133-3175-406c-9e0b-25cea768c5cf/20230125-weather-climate-catastrophe-insight.pdf.
51. See https://en.wikipedia.org/wiki/Malthusianism.

CHAPTER 10

1. Brady, Dennis and Juliet Eilperin. "In Confronting Climate Change, Biden Won't Have a Day to Waste." *Washington Post*. December 22, 2020. https://www.washingtonpost.com/politics/2020/12/22/biden-climate-change/.
2. Mencken, H. L. *In Defense of Women*. Project Gutenberg. Last updated February 6, 2013. https://www.gutenberg.org/files/1270/1270-h/1270-h.htm.

3. Helm, Burt. "Climate Change's Bottom Line." *New York Times*, January 31, 2015. https://www.nytimes.com/2015/02/01/business/energy-environment/climate-changes-bottom-line.html.
4. The Risky Business Project. "Risky Business: The Economic Risks of Climate Change in the United States." June 2014. http://riskybusiness.org/site/assets/uploads/2015/09/RiskyBusiness_Report_WEB_09_08_14.pdf.
5. Pilke, Roger. "How Billionaires Tom Steyer and Michael Bloomberg Corrupted Climate Science." *Forbes*, January 2, 2020. https://www.forbes.com/sites/rogerpielke/2020/01/02/how-billionaires-tom-steyer-and-michael-bloomberg-corrupted-climate-science.
6. Hausfather, Zeke, and Glen P. Peters. "Emissions—the 'business as usual' story is misleading." *Nature*, January 29, 2020. https://www.nature.com/articles/d41586-020-00177-3.
7. Burgess, Matthew G., et al. *Environmental Research Letters* 16 (2020). https://doi.org/10.1088/1748-9326/abcdd2.
8. "About Us: Who We Are." The National Academies of Sciences, Engineering, and Medicine. Accessed December 1, 2020 https://www.nationalacademies.org/about.
9. The National Academies of Sciences, Engineering, and Medicine. "Climate Change Publications." The National Academies Press. Accessed December 1, 2020. https://www.nap.edu/.
10. McNutt, Marcia, C. D. Mote Jr., Victor J. Dzau. "National Academies Presidents Affirm the Scientific Evidence of Climate Change." The National Academies of Sciences, Engineering, and Medicine, June 18, 2019. http://www8.nationalacademies.org/onpinews/newsitem.aspx?RecordID=06182019.
11. Handler, Philip. "Public Doubts About Science." *Science*, June 6, 1980. https://science.sciencemag.org/content/208/4448/1093.
12. Wunsch, Carl. "Swindled: Carl Wunsch Responds." RealClimate, March 12, 2007. http://www.realclimate.org/index.php/archives/2007/03/swindled-carl-wunsch-responds/comment-page-3/.
13. Revkin, Andrew C. "A Closer Look at Turbulent Oceans and Greenhouse Heating." *New York Times*, August 26, 2014. https://dotearth.blogs.nytimes.com/2014/08/26/a-closer-look-at-turbulent-oceans-and-greenhouse-heating/.
14. Tolstoy, Leo. 1894. *The Kingdom of God Is Within You*. Project Gutenberg, July 26, 2013. https://www.gutenberg.org/files/43302/43302-h/43302-h.htm.
15. "About 350.Org." 350.org. Accessed December 1, 2020. https://350.org/about/.
16. "Climate Change." Union of Concerned Scientists. Accessed December 1, 2020. https://www.ucsusa.org/climate.
17. The Editors of Encyclopaedia Britannica. "Great Drought." *Encyclopædia Britannica*, November 26, 2012. https://www.britannica.com/event/Great-Drought#ref=ref112984.
18. Crichton, Michael. At the International Leadership Forum, La Jolla, CA, April 26, 2002. http://geer.tinho.net/crichton.why.speculate.txt.

CHAPTER 11

1. Smith, Richard. "Peer Review: a Flawed Process at the Heart of Science and Journals." *Journal of the Royal Society of Medicine* 99 (2006): 178–182. https://www.ncbi.nlm.nih.gov/pmc/articles/PMC1420798/.
2. Koonin, Steven. "A 'Red Team' Exercise Would Strengthen Climate Science." *Wall Street Journal*, April 20, 2017. https://www.wsj.com/articles/a-red-team-exercise-would-strengthen-climate-science-1492728579.
3. Holdren, John P. "The Perversity of the Climate Science Kangaroo Court." *Boston Globe*, July 25, 2017. https://www.bostonglobe.com/opinion/2017/07/24/the-perversity-red-teaming-climate-science/VkT05883ajZaTPMbrP3wpJ/story.html.
4. Davidson, Eric, and Marcia K. McNutt. "Red/Blue and Peer Review." Eos, August 2, 2017. https://eos.org/opinions/red-blue-and-peer-review.
5. A Bill to Prohibit the Use of Funds to Federal Agencies to Establish a Panel, Task Force, Advisory Committee, or Other Effort to Challenge the Scientific Consensus on Climate Change, and for Other Purposes, S. 729, 116th Congress (2019). https://www.govtrack.us/congress/bills/116/s729/text.
6. Fourth Session Council of Trent, April 8, 1546. "Canonical Decree Concerning the Canonical Scriptures." https://www.csun.edu/~hcfll004/trent4.html.
7. "Climate Communications Initiative." The National Academies of Sciences, Engineering, and Medicine. Accessed December 1, 2020. https://www.nationalacademies.org/our-work/climate-communications-initiative.
8. *See, for example*, Tol, Richard S. J. "Comment on 'Quantifying the consensus on anthropogenic global warming in the scientific literature.'" *Environmental Research Letters* 11 (2016). https://iopscience.iop.org/article/10.1088/1748-9326/11/4/048001.
9. Jenkins, Holman W., Jr. "Change Would Be Healthy at U.S. Climate Agencies." *Wall Street Journal*, February 4, 2017. https://www.wsj.com/articles/change-would-be-healthy-at-u-s-climate-agencies-1486165226.
10. Kottasová, Ivana. "Oceans Are Warming at the Same Rate as If Five Hiroshima Bombs Were Dropped in Every Second." CNN, January 13, 2020. https://www.cnn.com/2020/01/13/world/climate-change-oceans-heat-intl/index.html.
11. Mills, Mark P. "'Unsettled' Review: The 'Consensus' On Climate." *Wall Street Journal*, April 25, 2021. https://www.wsj.com/articles/unsettled-review-theconsensus-on-climate-11619383653.
12. Jenkins, Holman W., Jr. "How a Physicist Became a Climate Truth Teller." *Wall Street Journal*, April 16, 2021. https://www.wsj.com/articles/how-a-physicist-became-a-climate-truth-teller-11618597216.
13. Thiessen, Mark A. "An Obama scientist debunks the climate doom-mongers." *Wall Street Journal*, May 14, 2021. https://www.washingtonpost.com/opinions/2021/05/14/an-obama-scientist-debunks-climate-doom-mongers/.
14. Yohe, Gary. "A New Book Manages to Get Climate Science Badly Wrong." *Scientific American*, May 13, 2021. https://www.scientificamerican.com/article/a-new-book-manages-to-get-climate-science-badly-wrong/.

15. USGCRP. *Climate Science Special Report: Fourth National Climate Assessment*, Volume I: Chapter 6. https://science2017.globalchange.gov/chapter/6/.
16. EPA. "Climate Change Indicators: Heat Waves." July 2022. https://www.epa.gov/climate-indicators/climate-change-indicators-heat-waves.
17. Royer, Dana, et al. "Wall Street Journal article repeats multiple incorrect and misleading claims made in Steven Koonin's new book 'Unsettled.'" Climate Feedback, April 25, 2021. https://climatefeedback.org/evaluation/wall-street-journal-article-repeats-multiple-incorrect-and-misleading-claims-made-in-steven-koonins-new-book-unsettled-steven-koonin/.
18. Koonin, Steven. "Facebook's 'Fact Checks' Suppress Debate." *Wall Street Journal*, May 16, 2021. https://www.wsj.com/articles/facebooks-fact-checks-suppress-debate-11621194172.
19. Oreskes, Naomi, et al. "That 'Obama Scientist' Climate Skeptic You've Been Hearing About." Scientific American, June 1, 2021. https://www.scientificamerican.com/article/that-obama-scientist-climate-skeptic-youve-been-hearing-about/.
20. Thiessen, Mark A. "An Obama scientist debunks the climate doom-mongers." *Wall Street Journal*, May 14, 2021. https://www.washingtonpost.com/opinions/2021/05/14/an-obama-scientist-debunks-climate-doom-mongers/.
21. See https://steven-koonin.medium.com/.

CHAPTER 12

1. Koonin, Steven E. "The Tough Realities of the Paris Climate Talks." *New York Times*, November 4, 2015. https://www.nytimes.com/2015/11/04/opinion/the-tough-realities-of-the-paris-climate-talks.html.
2. Office of the Press Secretary, The White House. "U.S. Leadership and the Historic Paris Agreement to Combat Climate Change." National Archives and Records Administration, December 12, 2015. https://obamawhitehouse.archives.gov/the-press-office/2015/12/12/us-leadership-and-historic-paris-agreement-combat-climate-change.
3. Titley, David. "Why Is Climate Change's 2 Degrees Celsius of Warming Limit So Important?" The Conversation, March 20, 2020. https://theconversation.com/why-is-climate-changes-2-degrees-celsius-of-warming-limit-so-important-82058.
4. Tol. "The Economic Impacts of Climate Change."
5. Carbon Brief Staff. "Two Degrees: The History of Climate Change's Speed Limit." Carbon Brief, December 8, 2014. https://www.carbonbrief.org/two-degrees-the-history-of-climate-changes-speed-limit.
6. "'The Father of the 2 Degrees Limit': Schellnhuber Receives Blue Planet Prize." Potsdam Institute for Climate Impact Research, October 19, 2017. https://www.pik-potsdam.de/news/press-releases/201cthe-father-of-the-2-degrees-limit201d-schellnhuber-receives-blue-planet-prize.

7. Van Vuuren, Detlef P., Jae Edmonds, Mikiko Kainuma, Keywan Riahi, Allison Thomson, Kathy Hibbard, George C. Hurtt, et al. "The Representative Concentration Pathways: An Overview." *Climatic Change* 109 (2011). https://link.springer.com/article/10.1007/s10584-011-0148-z.
8. "World Population Prospects 2019." United Nations, Department of Economic and Social Affairs Population Dynamics, 2020. http://esa.un.org/unpd/wpp/DataQuery/.
9. GDP data from the IMF (https://www.imf.org/en/Publications/WEO/weo-database/2020/October/download-entire-database), energy data from the EIA (https://www.eia.gov/international/overview/world), and population data from the World Bank (https://data.worldbank.org/indicator/SP.POP.TOTL). Data for Germany and the World are only for 1990 onward.
10. Kahan, Ari. "EIA projects nearly 50% increase in world energy usage by 2050, led by growth in Asia." US Energy Information Administration (EIA): Today in Energy, September 24, 2019. https://www.eia.gov/todayinenergy/detail.php?id=41433.
11. Ritchie, Hannah, and Max Roser. "CO_2 and Greenhouse Gas Emissions." Our World in Data, May 11, 2017. https://ourworldindata.org/co2-and-other-greenhouse-gas-emissions.
12. "Synthesis report on the aggregate effect of the intended nationally determined contributions." UN, FCCC, Conference of the parties twenty-first session, October 30, 2015. http://unfccc.int/resource/docs/2015/cop21/eng/07.pdf.
13. "2030 Emissions Gaps." Climate Action Tracker, CAT Emissions Gaps, September 23, 2020. https://climateactiontracker.org/global/cat-emissions-gaps/.
14. Emissions data from 2019 Emissions Gap Report and projections from climateactiontracker.org.
15. UNEP (2019). Emissions Gap Report 2019. *Executive summary.* United Nations Environment Programme, Nairobi. https://www.unenvironment.org/resources/emissions-gap-report-2019.
16. Allan, Bentley B. "Analysis | The E.U.'s Looking at a 'Carbon Border Tax.' What's a Carbon Border Tax?" *Washington Post*, October 23, 2019. https://www.washingtonpost.com/politics/2019/10/23/eus-looking-carbon-border-tax-whats-carbon-border-tax/.
17. IEA. *Net Zero by 2050*. IEA, 2021. https://www.iea.org/reports/net-zero-by-2050, License: CC BY 4.0.
18. Odgerel, Batt, et al. "A Critical Assessment of the IEA's Net Zero Scenario, ESG, and the Cessation of Investment in New Oil and Gas Fields." Energy Policy Research Foundation, June 2023. https://eprinc.org/a-critical-assessment-of-the-ieas-net-zero-scenario-esg-and-the-cessation-of-investment-in-new-oil-and-gas-fields/.
19. Tinker, Scott W. "The road to Glasgow is paved with bad assumptions." The Hill, October 28, 2021. https://thehill.com/opinion/energy-environment/578925-the-road-to-glasgow-is-paved-with-bad-assumptions/.
20. Enerdata. "World Energy & Climate Statistics—Yearbook 2023." Enerdata, accessed January 18, 2024. https://yearbook.enerdata.net/natural-gas/gas-consumption-data.html.

21. Tinker, "The road to Glasgow is paved with bad assumptions."
22. Khrishnan, Mekala, et al. "The economic transformation: What would change in the net-zero transition." McKinsey, January 25, 2022. https://www.mckinsey.com/capabilities/sustainability/our-insights/the-economic-transformation-what-would-change-in-the-net-zero-transition.
23. Kelly, Michael. "The Feasibility of a Net-zero Economy for the USA by 2050." The Global Warming Policy Foundation, note 39, 2023. https://www.thegwpf.org/content/uploads/2023/03/Kelly-Net-Zero-Progress-Report-USA-.pdf.

CHAPTER 13

1. "Inventory of U.S. Greenhouse Gas Emissions and Sinks." Environmental Protection Agency, April 13, 2020. https://www.epa.gov/ghgemissions/inventory-us-greenhouse-gas-emissions-and-sinks.
2. Liu, Z., P. Ciais, Z. Deng, et al. "Near-real-time monitoring of global CO_2 emissions reveals the effects of the COVID-19 pandemic." *Nature Communications* 11 (2020). https://www.nature.com/articles/s41467-020-18922-7.
3. US Energy Information Administration (EIA). "Table 1.1. Primary Energy Overview." EIA, Monthly Energy Review. Accessed December 1, 2020. https://www.eia.gov/totalenergy/data/browser/.
4. Koonin, Steven E., and Avi M. Gopstein. "Accelerating the Pace of Energy Change." *Issues in Science and Technology* 27 (2011). https://issues.org/koonin/.
5. Stout, David. "Gore Calls for Carbon-Free Electric Power." *New York Times*, July 18, 2008. https://www.nytimes.com/2008/07/18/washington/18gorecnd.html.
6. Hamilton, Alexander or James Madison (as credited on site). "Federalist No. 62." Library of Congress, Research Guides, Federalist Papers: Primary Documents in American History, Federalist Nos. 61–70. Accessed December 1, 2020. https://www.congress.gov/resources/display/content/The+Federalist+Papers#TheFederalistPapers-62.
7. US Energy Information Agency (EIA). "Frequently Asked Questions (FAQS)." EIA, November 2, 2020. https://www.eia.gov/tools/faqs/faq.php?id=427&t=3; US Energy Information Administration (EIA). "Coal explained—Coal prices and outlook." EIA, October 9, 2020. https://www.eia.gov/energyexplained/coal/prices-and-outlook.php.
8. Green, Miranda. "Analysis: Trump Solar Tariffs Cost 62K US Jobs." The Hill, December 3, 2019. https://thehill.com/policy/energy-environment/472691-analysis-trump-solar-tariffs-cost-62k-us-jobs.
9. Agence France-Presse. "EU Approves Anti-Dumping Penalty on Chinese Light Bulb." *Industry Week*, October 15, 2007. https://www.industryweek.com/the-economy/regulations/article/21956186/eu-approves-antidumping-penalty-on-chinese-light-bulb.
10. Cart, Julie. "California's 'Hydrogen Highway' Never Happened. Could 2020 Change That?" CalMatters, January 9, 2020. https://calmatters.org/environment/2020/01/why-california-hydrogen-cars-2020/.

11. International Energy Agency (IEA). "Clean Energy Innovation." IEA, July 2020. https://www.iea.org/reports/clean-energy-innovation.
12. Jansen, Malte, et al. "Q1 2021: When the Wind Goes, Gas Fills in the Gap." Electric Insights Quarterly Reports. https://reports.electricinsights.co.uk/q1-2021/when-the-wind-goes-gas-fills-in-the-gap/. Figure 13.4 was created from data obtained from https://www.gridwatch.templar.co.uk/download.php.
13. Dowling, Jacqueline A., Katherine Z. Rinaldi, Tyler H. Ruggles, Steven J. Davis, Mengyao Yuan, Fan Tong, Nathan S. Lewis, and Ken Caldeira. "Role of Long-Duration Energy Storage in Variable Renewable Electricity Systems." *Joule* 4, no. 9 (2020): 1907–1928. https://www.sciencedirect.com/science/article/pii/S2542435120303251.
14. Moch, Jonathan M., and Henry Lee. "The Challenges of Decarbonizing the U.S. Electric Grid by 2035." Policy Brief, February 2022. https://www.belfercenter.org/publication/challenges-decarbonizing-us-electric-grid-2035.
15. Bipartisan Policy Center. "BPC Roundtable Series Exploring Energy Permitting Reform." November 2, 2023. https://bipartisanpolicy.org/blog/bpc-roundtable-series-exploring-energy-permitting-reform/.
16. See https://www.llnl.gov/archive/news/lawrence-livermore-national-laboratory-achieves-fusion-ignition.
17. National Research Council, *Review of the Department of Energy's Inertial Confinement Fusion Program: The National Ignition Facility* (Washington, DC: The National Academies Press, 1997), p. 37. https://doi.org/10.17226/5730.
18. IEA. "The Role of Critical Minerals in Clean Energy Transitions." IEA, 2021. https://www.iea.org/reports/the-role-of-critical-minerals-in-clean-energy-transitions. Figure 13.5 is a work derived by the author, Steven Koonin, from IEA material, and the author is solely liable and responsible for this derived work. The derived work is not endorsed by the IEA in any manner.
19. Ibid.
20. US Treasury. "Risks and Considerations for Businesses and Individuals with Exposure to Entities Engaged in Forced Labor and Other Human Rights Abuses Linked to Xinjiang, China." Updated July 13, 2021. https://ofac.treasury.gov/media/911311/download?inline.
21. Kaplan, Thomas, et al. "U.S. Bans Imports of Some Chinese Solar Materials Tied to Forced Labor." *New York Times*, June 24, 2021. https://www.nytimes.com/2021/06/24/business/economy/china-forced-labor-solar.html.
22. Timbie, James, et al. "Progress on Critical Materials Resilience." Hoover Institution, July 5, 2023. https://www.hoover.org/research/progress-critical-materials-resilience.
23. S&P Global. "The Future of Copper: Will the looming supply gap short-circuit the energy transition?" https://www.spglobal.com/marketintelligence/en/mi/Info/0722/futureofcopper.html.

CHAPTER 14

1. McCormick, Ty. "Geoengineering: A Short History." Foreign Policy, September 3, 2013. https://foreignpolicy.com/2013/09/03/geoengineering-a-short-history/.
2. Garber, Megan. "The Scientist Who Told Congress He Could (Literally) Make It Rain." *The Atlantic*, May 4, 2015. https://www.theatlantic.com/technology/archive/2015/05/the-scientist-who-told-congress-he-could-literally-make-it-rain/392219/.
3. Battisti, D., et al. 2009 *IOP Conf. Ser.: Earth Environ. Sci.* 6 452015. https://iopscience.iop.org/article/10.1088/1755-1307/6/45/452015.
4. Revkin, Andrew C. "Science Adviser Lays Out Climate and Energy Plans." *New York Times*, April 9, 2009. https://dotearth.blogs.nytimes.com/2009/04/09/science-adviser-lists-goals-on-climate-energy/.
5. "Geoengineering the Climate: Science, Governance and Uncertainty." The Royal Society, September 1, 2019. https://royalsociety.org/topics-policy/publications/2009/geoengineering-climate/.
6. National Research Council of the National Academies. *Climate Intervention: Reflecting Sunlight to Cool Earth.* Washington, DC: The National Academies Press, 2015. https://www.nap.edu/catalog/18988/climate-intervention-reflecting-sunlight-to-cool-earth.
7. National Research Council of the National Academies. *Climate Intervention: Carbon Dioxide Removal and Reliable Sequestration.* Washington, DC: The National Academies Press, 2015. https://www.nap.edu/catalog/18805/climate-intervention-carbon-dioxide-removal-and-reliable-sequestration.
8. Fialka, John. "NOAA Gets Go-Ahead to Study Climate Plan B: Geoengineering." E&E News: Climatewire, January 23, 2020. https://www.eenews.net/climatewire/2020/01/23/stories/1062156429.
9. American Physical Society, APS Panel on Public Affairs. 2011. *Direct Air Capture of CO_2 with Chemicals.* American Physical Society. https://www.aps.org/policy/reports/assessments/upload/dac2011.pdf.
10. Keith, David W., Geoffrey Holmes, David St. Angelo, and Kenton Heidel. "A Process for Capturing CO_2 from the Atmosphere." *Joule* 2 (2018): 1573–1594. https://www.cell.com/joule/fulltext/S2542-4351(18)30225-3.
11. Buis, Alan. "Examining the Viability of Planting Trees to Help Mitigate Climate Change." NASA, Global Climate Change: Vital Signs of the Planet, November 11, 2019. https://climate.nasa.gov/news/2927/examining-the-viability-of-planting-trees-to-help-mitigate-climate-change/.
12. Schwarber, Adria. "Moniz Making Case for $11 Billion Carbon Removal Initiative." American Institute of Physics, November 19, 2019. https://www.aip.org/fyi/2019/moniz-making-case-11-billion-carbon-removal-initiative.
13. Busch, Wolfgang, Joanne Chory, Joseph Ecker, Julie Law, Todd Michael, and Joseph Noel. "Harnessing Plants Initiative." Salk Institute for Biological Studies, 2020. https://www.salk.edu/harnessing-plants-initiative/.

14. US Geological Survey. "Cool Earthquake Facts." USGS. Accessed November 27, 2020. https://www.usgs.gov/natural-hazards/earthquake-hazards/science/cool-earthquake-facts.
15. Pacala, S., and R. Socolow. "Stabilization Wedges: Solving the Climate Problem for the Next 50 Years with Current Technologies." *Science*, August 13, 2004. https://science.sciencemag.org/content/305/5686/968.full.
16. Morris, Stan. "Doing more with less CO_2." AHEAD Energy Corporation, 2020. http://www.aheadenergy.org/.

CHAPTER 15

1. See https://reason.com/video/2022/08/31/do-we-need-to-rapidly-convert-to-renewables-to-save-the-planet-a-soho-forum-debate/.
2. See https://steamboatinstitute.org/past-debates/.
3. Nordhaus, William D. "Climate Change: The Ultimate Challenge for Economics." Nobel Prize Lecture, December 8, 2018. https://www.nobelprize.org/uploads/2018/10/nordhaus-lecture.pdf.
4. Moss, Todd. "Why 'the Fridge' Continues to Resonate." Eat More Electrons, February 23, 2023. https://toddmoss.substack.com/p/why-the-fridge-continues-to-resonate.
5. Downs, Anthony. "Up and down with ecology—The 'issue-attention cycle.'" *The Public Interest* 28 (1972): 38–50. http://sciencepolicy.colorado.edu/students/envs_5720/downs_1972.pdf.
6. Prabhu, Sunil. "'Colonial Mindset': PM Slams Pressures on India over Climate Pledges." *India News*, November 27, 2021. https://www.ndtv.com/india-news/pm-modi-in-the-name-of-environment-various-pressures-created-on-india-all-this-is-result-of-colonial-mentality-2626202.
7. Al Jazeera. "Africa being 'punished' by fossil fuel investment ban – Niger." *Al Jazeera*, June 15, 2022. https://www.aljazeera.com/news/2022/6/15/africa-punished-by-investment-clamp-on-fossils-says-niger.
8. Barber, Harriet. "'Eco-anxiety': The fear of environmental doom and how to overcome it." *Telegraph*, October 31, 2021. https://www.telegraph.co.uk/global-health/climate-and-people/eco-anxiety-fear-environmental-doom-overcome/.
9. Dillarstone, Hope, Laura J. Brown, and Elaine C. Flores. "Climate change, mental health, and reproductive decision-making: A systematic review." *PLOS Climate* 2, no. 11 (2023): e0000236. https://doi.org/10.1371/journal.pclm.0000236.
10. Milman, Oliver. "Suicides indicate wave of 'doomerism' over escalating climate crisis." *Guardian*, May 19, 2022. https://www.theguardian.com/environment/2022/may/19/climate-suicides-despair-global-heating.
11. Hindustan Times. "Sri Lanka turmoil: Timeline of worst economic crisis since independence." *Hindustan Times*, July 9, 2022. https://www.hindustantimes.com/world-news/sri-lanka-turmoil-timeline-of-worst-economic-crisis-since-independence-101657374767747.html.

12. Nordhaus, Ted, and Saloni Shah. "In Sri Lanka, Organic Farming Went Catastrophically Wrong." *Foreign Policy*, March 5, 2022. https://foreignpolicy.com/2022/03/05/sri-lanka-organic-farming-crisis/.
13. Guterres, António. "Secretary-General's remarks to High-Level opening of COP27." United Nations, November 7, 2022. https://un.org/sg/en/content/sg/speeches/2022-11-07/secretary-generals-remarks-high-level-opening-of-cop27.
14. "Don't overstate 1.5 degrees C threat, new IPCC head says." DW.com, July 30, 2023. https://www.dw.com/en/climate-change-do-not-overstate-15-degrees-threat/a-66386523.

CLOSING THOUGHTS

1. As has been pointed out recently by one leader of the field; Emanuel, K. "The Relevance of Theory for Contemporary Research in Atmospheres, Oceans, and Climate." *AGU Advances* 1 (2020): e2019AV000129. https://agupubs.onlinelibrary.wiley.com/doi/epdf/10.1029/2019AV000129.

INDEX

T

U

ABOUT THE AUTHOR

PHOTO BY KELLY KOLLAR

Dr. Steven E. Koonin is one of America's most distinguished scientists, a member of the National Academy of Sciences, and a leader in United States science policy.

Currently a professor at New York University, Dr. Koonin holds appointments in the Stern School of Business, the Tandon School of Engineering, and the Department of Physics. He founded NYU's Center for Urban Science and Progress, which focuses research and education on the acquisition, integration, and analysis of big data for big cities.

Dr. Koonin served as Undersecretary for Science in the US Department of Energy under President Obama, where his portfolio included the climate research program and energy technology strategy. He was the lead author of the US Department of Energy's Strategic Plan (2011) and the inaugural Department of Energy Quadrennial Technology Review (2011). Before joining the government, Dr. Koonin spent five years as chief scientist for BP, researching renewable energy options to move the company "beyond petroleum."

For almost thirty years, Dr. Koonin was a professor of theoretical physics at Caltech. He also served for nine years as Caltech's vice president and provost, facilitating the research of more than three hundred science and engineering faculty and catalyzing the development of the world's largest optical telescope, as well as research initiatives in computational science, bioengineering, and the biological sciences.

In addition to the National Academy of Sciences, Dr. Koonin's memberships include the American Academy of Arts and Sciences, the Council on Foreign Relations, and JASON, the group of scientists who solve technical problems for the US government; he served as JASON's chair for six years. He chaired the National Academies' Divisional Committee for Engineering and Physical Sciences from 2014 to 2019, and since 2014 he has been a trustee of the Institute for Defense Analyses. He is currently an independent governor of the Lawrence Livermore National Laboratory and has served in similar roles for the Los Alamos, Sandia, Brookhaven, and Argonne National Laboratories. He is a Senior Fellow at Stanford's Hoover Institution.

Dr. Koonin has a BS in physics from Caltech and a PhD in theoretical physics from MIT. He is an award-winning classroom teacher and his public lectures are noted for their clarity in conveying complex subjects. He is the author of the classic 1985 textbook *Computational Physics*, which introduced methodology for building computer models of complex physical systems. He has published some two hundred peer-reviewed papers in the fields of physics and astrophysics, scientific computation, energy technology and policy, and climate science, and has been the lead author on multiple book-length reports, including two National Academies studies.

Through a series of articles and lectures that began in 2014, Dr. Koonin has advocated for a more accurate, complete, and transparent public representation of climate and energy matters.